Péter Érdi
Complexity Explained

Péter Érdi

Complexity Explained

With 129 Figures

Péter Érdi
Kalamazoo College
Center for Complex Systems Studies
Department of Physics, Department of Psychology
Academy Street 1200
49006 Kalamazoo
USA
perdi@kzoo.edu

and

KFKI Research Institute for Particle and Nuclear Physics
Hungarian Academy of Sciences
Department of Biophysics
Konkoly-Thege út 29–33
1121 Budapest
Hungary
erdi@rmki.kfki.hu

ISBN-13 978-3-540-35777-3   e-ISBN-13 978-3-540-35778-0

DOI 10.1007/978-3-540-35778-0

Library of Congress Control Number: 2007934800

© 2008 Springer-Verlag Berlin Heidelberg

This work is subject to copyright. All rights are reserved, whether the whole or part of the material is concerned, specifically the rights of translation, reprinting, reuse of illustrations, recitation, broadcasting, reproduction on microfilm or in any other way, and storage in data banks. Duplication of this publication or parts thereof is permitted only under the provisions of the German Copyright Law of September 9, 1965, in its current version, and permission for use must always be obtained from Springer. Violations are liable to prosecution under the German Copyright Law.

The use of general descriptive names, registered names, trademarks, etc. in this publication does not imply, even in the absence of a specific statement, that such names are exempt from the relevant protective laws and regulations and therefore free for general use.

Typesetting and Production: LE-TEX Jelonek, Schmidt & Vöckler GbR, Leipzig
Cover design: Erich Kirchner, Heidelberg

Printed on acid-free paper   SPIN 11545286   89/3180/YL

9 8 7 6 5 4 3 2 1

springer.com

To the memory of my parents

Preface

This book is, of course about complexity. The title of the book, as you may recognize was motivated (excuse me for using this very mild expression) by Daniel Dennett's Consciousness Explained [130]. Dennett's intention was to explain consciousness as the emergent product of the interaction among constituents having physical and neural character. The goal of this book is to explain how various types of complexity emerge due to the interaction among constituents. There are many questions to be answered, how to understand, control, decompose, manage, predict the many-faced complexity. After teaching this subject for several years I feel that the time has come to put the whole story together.

The term "complex system" is a buzzword, but we certainly don't have a single definition for it. There are several predominant features of complexity. Complex processes may show unpredictable behavior (which we still try to predict somehow), may lead to uncontrolled explosion (such in case of epilepsy, earthquake eruptions or stock market crashes). One of the characteristic feature of simple systems is, that there is a single cause which implies a single effect. For large class of complex systems it is true that effects are fed back to modify causes. Biological cells belong to this class. Furthermore they are open to material, energetic and information flow by interaction with their environment, still they are organizationally closed units. Another aspect of complexity is the question how collective phenomena emerge by some self-organized mechanisms. Thomas Schelling's model, which suggests that strong racial prejudice is not needed to generate urban segregation, is paradigmatic.

There is a remarkable and unique statistical feature of certain complex systems. Generally we expect that there is an average, (say, the average height of people), and the deviation from this average is symmetric. Biologists have found that the Gaussian distribution can be applied in many, many cases. In a large number of social systems (but not only there) we see another type of pattern, occasionally called as the 80/20 rule. About 80 percent of the income is made by 20% of people, 80% (well, 70 or 85) of flights are landing on twenty percent of the airports, while there are many small airports with a few flights per day. A large number of scientific papers are written by a small number of scientists, and so on. Such kinds of phenomena, which don't really have a characteristic size, are described by an asymmetric (skew) so-called power law distribution. The brilliant best-seller of Douglas Hofstadter on Gödel, Escher and Bach published in 1979 emphasized self-reference and loops, actually he calls a strange loop. Loops, specifically feedback loops were studied by cybernetics, an abandoned scientific discipline, which emphasized that effects may feed back to influence causes. Such kinds of systems, which are characterized by "circular causality" certainly could be qualified to be called as complex.

I would like to mention three books, which influenced the way of my writing. Heinz R. Pagels' (1939–1988) posthumous book "The Dreams of Reason: The Computer and the Rise of the Sciences of Complexity" [395], a very exciting book about chaos, complexity, neural networks, cognitive science, and philosophy of science. I think I remember the excitement I felt when I read at least half of the book in the same breath in a trans-Atlantic flight, actually the first time from Michigan to Europe, in the incredible fall of Eastern Europe in 1989.

John Casti's "Paradigms lost" [94] showed me that it is possible to mention different fields in the same book from philosophy of science via molecular biology and origin of life, theory of evolution, sociobiology, linguistics, cognitive science, foundation of mathematics, to quantum physics and cosmology. I have a somewhat overlapping list.

I learned from Michael Arbib's "The Metaphorical Brain" [18] how to use and not use mathematical formalism. Some pages of his book are filled with equations, and then you may find fifty pages without any formulas. So, I extracted the implicit message to be "Don't be afraid to use math when it helps to explain your ideas, and don't be afraid to avoid mathematics when you can convey your ideas without it."

I have heard about the notions of complex systems, simulation methods, and thinking in models in the late sixties from my undergraduate mentor, Pál Benedek, and later had numerous conversation about the complexity of the brain with János Szentágothai.

It happened that this has been my sixth year to teach complex systems and related fields at Kalamazoo College for undergraduates. Kalamazoo College was awarded by a Henry R. Luce Professorship, and I have had the privilege to serve here to build a program about complex systems. I learned (hopefully) a lot during these years, and the book grew up from my class notes. I benefited very much from the interaction with my colleague Jan Tobochnik. We have a mutual interest in understanding and making others understand problems, many of them are related to complex systems.

Previously I taught a History of Complex Systems Research class at the Department of History and Philosophy of Science at Eötvös University, Budapest (Hungary), when I served there as a Széchenyi professor, and the plan of writing a book about complexity emerged in that period.

I am particularly indebted to two of my Hungarian graduate students, Gábor Csárdi, Balázs Ujfalussy, who helped a lot in preparing the text and figures. Zsófi Huhn, Tamás Nepusz helped to complete specific sections.

I deliberately adopted/adapted texts, figures, ideas from works published earlier with a large set of coworkers, such as Ildikó Aradi, Michael Arbib, George Barna, Fülöp Bazsó, Tamás F. Farkas, Csaba Földy, Mihaly Hajos, Tamás Kiss, Máté Lengyel, Gergő Orbán, Zoltán Somogyvári, Katherine Strandburg, Krisztina Szalisznyó, János Tóth, Ichiro Tsuda and László Zalányi.

I acknowledge the discussion of many details with George Kampis and József Lázár. The latter also helped to tell some Hungarian anecdotes, etc. in English. I am grateful to Kati Pető and Péter Bruck for their explicit and implicit contributions. Comments from Michael Arbib, Jancsi (Chaim) Forgacs, Viktor Jirsa, Robert Kozma, Ichiro Tsuda, and Günter Wagner are acknowledged. Francesco Ventriglia kindly read the whole manuscript, and made a lot of comments.

I got a lot of motivations from the ELMOHA circle.

I got a lot of help from my students at Kalamazoo. Trevor Jones and Andrew Schneider accepted the painstaking task to correct the grammar. I enjoyed very much numerous conversations among others with Dan Catlin, Griffin Drutchas, Brad Flaugher, Richard Gejji, Elizabeth Gillstrom, Justin Horwitz, Hannah Masuga, Elliot Paquette, Bobby Rohrkemper, Clara Scholl and Jen Watkins.

I would like to thank to Thomas Ditzinger, the editor of the Springer to encourage me to write the book in the style I chose. Comments from Christian Caron also acknowledged.

While the book is dedicated to the memory of my parents, I think they would have suggested me to do it to my children. Zsuzsi and Gábor, the book is dedicated also to you.

We have a long experience with my wife, Csuti, to cope with the complexity of life. I benefited very much from her support, love and wisdom. I am not sure I could have completed this book without her deep understanding. It is difficult to express my gratitude.

Kalamazoo, Michigan and Budapest-Csillebérc *Péter Érdi*
June 2007

Contents

1 **Complex Systems: The Intellectual Landscape** 1
   1.1 The Century of Complexity? 1
   1.2 Characteristics of Simple and Complex Systems 5
       1.2.1 System and Its Environment 5
       1.2.2 Simple Systems 6
       1.2.3 Complex Systems 7
   1.3 Connecting the Dots 20

2 **History of Complex Systems Research** 25
   2.1 Reductionist Success Stories Versus the Importance
       of Organization Principles 25
       2.1.1 Reductionism and Holism in Quantum Physics 25
       2.1.2 Reductionism and Complexity in Molecular Biology ... 29
   2.2 Ancestors of present day complex system research 35
       2.2.1 Systems Theory 35
       2.2.2 Cybernetics 37
       2.2.3 Nonlinear Science in Action: Theory of Dissipative
             Structures, Synergetics and Catastrophe Theory 45

3 **From the Clockwork World View to Irreversibility
   (and Back?)** 57
   3.1 Cyclic Universe Versus Linear Time Concept:
       the Metaphysical Perspective 57
       3.1.1 Cyclic Universe 57
       3.1.2 Linear Time Concepts 59
   3.2 The Newtonian Clockwork Universe 61
       3.2.1 The Mechanical Clock 61
       3.2.2 Kepler's Integral Laws 66
       3.2.3 Newton's Differential Laws, Hamilton Equations,
             Conservative Oscillation, Dissipation 68
   3.3 Mechanics Versus Thermodynamics 75

	3.3.1	Heat Conduction and Irreversibility 75

- 3.3.1 Heat Conduction and Irreversibility 75
- 3.3.2 Steam Engine, Feedback control, Irreversibility 77
- 3.3.3 The First and Second Laws of Thermodynamics 77
- 3.4 The Birth of the Modern Theory of Dynamical Systems 79
- 3.5 Oscillations 81
  - 3.5.1 The Lotka–Volterra Model 81
  - 3.5.2 Stable Oscillation: Limit Cycles 83
  - 3.5.3 Quasiperiodic Motions: A Few Words About the Modern Theory of Dynamical Systems 86
- 3.6 The Chaos Paradigm: Then and Now 87
  - 3.6.1 Defining and Detecting Chaos 87
  - 3.6.2 Structural and Geometrical Conditions of Chaos: what Is Important and What Is Not? 91
  - 3.6.3 The Necessity of Being Chaotic 94
  - 3.6.4 Controlling Chaos: Why and How? 96
  - 3.6.5 Traveling to High-Dimension Land: Chaotic Itinerary 98
- 3.7 Direction of Evolution 100
  - 3.7.1 Dollo's Law in Retrospective 100
  - 3.7.2 Is Something Never-Decreasing During Evolution? 102
- 3.8 Cyclic Universe: Revisited... and Criticized 105

4 The Dynamic World View in Action 109
- 4.1 Causality, Teleology and About the Scope and Limits of the Dynamical Paradigm 109
  - 4.1.1 Causal Versus Teleological Description 110
  - 4.1.2 Causality, Networks, Emergent Novelty 112
- 4.2 Chemical Kinetics: A Prototype of Nonlinear Science 113
  - 4.2.1 On the Structure – Dynamics Relationship for Chemical Reactions 118
  - 4.2.2 Chemical Kinetics as a Metalanguage 119
  - 4.2.3 Spatiotemporal Patterns in Chemistry and Biology 120
- 4.3 Systems Biology: The Half Admitted Renaissance of Cybernetics and Systems Theory 130
  - 4.3.1 Life itself 130
  - 4.3.2 Cells As Self-Referential Systems 131
  - 4.3.3 The Old–New Systems Biology 133
  - 4.3.4 Random Boolean Networks: Model Framework and Applications for Genetic Networks 135
- 4.4 Population Dynamic and Epidemic Models: Biological and Social 140
  - 4.4.1 Connectivity, Stability, Diversity 140
  - 4.4.2 The Epidemic Propagation of Infections and Ideas 144
  - 4.4.3 Modeling Social Epidemics 146
- 4.5 Evolutionary Dynamics 147
- 4.6 Dynamic Models of War and Love 148

		4.6.1 Lanchaster's Combat Model and Its Variations 148

      4.6.1 Lanchaster's Combat Model and Its Variations 148
      4.6.2 Is Love Different from War? 151
  4.7 Social Dynamics: Some Examples 154
      4.7.1 Segregation dynamics 154
      4.7.2 Opinion Dynamics 157
  4.8 Nonlinear Dynamics in Economics: Some Examples 159
      4.8.1 Business Cycles 159
      4.8.2 Controlling Chaos in Economic Models 161
  4.9 Drug Market: Controlling Chaos 162

5 The Search for Laws: Deductive Versus Inductive 165
  5.1 Deductive Versus Inductive Arguments 165
  5.2 Principia Mathematica and the Deductive Approach:
      From Newton to Russell and Whitehead 167
  5.3 Karl Popper and the Problem of Induction 169
  5.4 Cybernetics: Bridge Between Natural and Artificial 169
  5.5 John von Neumann: The Real Pioneer
      of Complex Systems Studies 170
  5.6 Artificial Intelligence, Herbert Simon
      and the Bounded Rationality 175
  5.7 Inductive Reasoning and Bounded Rationality:
      from Herbert Simon to Brian Arthur 178
  5.8 Minority Game 180
  5.9 Summary and "What Next?" 182

6 Statistical Laws: From Symmetric to Asymmetric 185
  6.1 Normal Distribution 185
      6.1.1 General Remarks 185
      6.1.2 Generation of Normal Distribution: Brownian Motion .. 187
      6.1.3 Liouville Process, Wiener and Special Wiener Process,
           Ornstein–Uhlenbeck Process 188
  6.2 Bimodal and Multimodal Distributions 190
  6.3 Long Tail Distributions 191
      6.3.1 Lognormal and Power Law Distributions:
           Phenomenology 191
      6.3.2 Generation of Lognormal and Power Law Distributions 194

7 Simple and Complex Structures: Between Order
and Randomness 201
  7.1 Complexity and Randomness 201
  7.2 Structural Complexity 203
      7.2.1 Structures and Graphs 204
      7.2.2 Complexity of Graphs 208
      7.2.3 Fractal Structures 212
  7.3 Noise-Induced Ordering: An Elementary Mathematical Model 217

	7.4 Networks Everywhere: Between Order and Randomness 219
		7.4.1 Statistical Approach to Large Networks 219
		7.4.2 Networks in Cell Biology 221
		7.4.3 Epidemics on Networks 223
		7.4.4 Citation and Collaboration Networks in Science
		and Technology 225

8 **Complexity of the Brain: Structure, Function
and Dynamics** 237
	8.1 Introductory Remarks..................................... 237
	8.2 Windows on the Brain 238
		8.2.1 A Few Words About the Brain–Mind Problem 238
		8.2.2 Experimental Methods: A Brief Review............. 239
	8.3 Approaches and Organizational Principles 241
		8.3.1 Levels.. 241
		8.3.2 Bottom Up and top Down 242
		8.3.3 Organizational Principles.......................... 243
	8.4 Single Cells 247
		8.4.1 Single Cells: General Remarks 247
		8.4.2 Single Cell Modeling: Deterministic
		and Stochastic Framework.......................... 250
	8.5 Structure, Dynamics, Function 255
		8.5.1 Structural Aspects................................ 255
		8.5.2 Neural Rhythms 261
		8.5.3 Variations on the Hebbian Learning Rule:
		Different Roots 283
	8.6 Complexity and Cybernetics: Towards a Unified Theory
	of Brain–Mind and Computer 289
		8.6.1 Cybernetics Strikes Back 289
		8.6.2 From Cognitive Science to Embodied Cognition 291
		8.6.3 The Brain as a Hermeneutic Device 296
		8.6.4 From Neurons to Soul and Back..................... 299

9 **From Models to Decision Making** 305
	9.1 Equation-Based Versus Agent-Based Model 305
		9.1.1 Motivations....................................... 305
		9.1.2 Artificial Life 306
		9.1.3 Artificial Societies 311
		9.1.4 Agent-Based Computational Economics 316
	9.2 Game Theory: Where We Are Now? 318
		9.2.1 Classical game theory 318
		9.2.2 Evolutionary Game Theory 322
	9.3 Widening the Limits to Predictions: Earthquake, Eruptions
	Epileptics Seizures, and Stock Market Crashes 328
		9.3.1 Scope and Limits of Predictability 328

	9.3.2	Phenomenology 329
	9.3.3	Statistical Analysis of Extreme Events 338
	9.3.4	Towards Predicting Seizures 341
	9.3.5	Towards Predicting Market Crashes: Analysis of Price Peaks 344
	9.3.6	Dynamical Models of Extreme Events 345

10 How Many Cultures We Have? 353
   10.1 Complexity as a Unifying Concept 353
      10.1.1 Systems and Simulations 353
      10.1.2 The Topics of the Book in Retrospective: Natural and Human Socioeconomic Systems 354
   10.2 The Ingredients of Complex Systems 357
   10.3 Complexity Explained: In Defense of (Bounded) Rationality .. 359

References 365

Index 393

1

Complex Systems: The Intellectual Landscape

1.1 The Century of Complexity?

"I think the next century will be the century of complexity."
Stephen Hawking (Complexity Digest 2001/10, 5 March 2001.)

The term "complexity" became a buzzword in the last decade, or so. We see that the science of complexity has roots both in natural and social sciences. The term "complex" appears as an adjective in very different contexts. "Complex structures", "complex networks", "complex processes", "complex information processing", "complex management", etc.

One aspect of complexity is related to the *structure* of a system. In elementary chemistry the most fundamental organization is the structure of molecules, where the elements are atoms, and the relations are chemical bindings between them. Biologists study structures at very different levels of hierarchical organization, from molecular biology via the brain to population dynamics. The neuronal network of the brain consists of neurons connected by synapses (though extrasynaptic communications cannot be excluded). Food webs describe the relationship among species: living creatures should eat other living creatures to survive. An example of structures studied by psychologists is the so-called semantic memory. Semantic memory describes abstract relationship among concepts; e.g., Cairo is the capital of Egypt. In our mind there is a network of words connected by associations. Nodes of the networks are concepts, they are connected by associations, which are the edges of a network. Computer scientists adopt measures to characterize the static structural complexity of software. Programs, containing more cycles, are supposed to be more complex. The selective, functional and evolutionary advantage of the hierarchically organized structures (composed of subsystems, again composed of their own subsystems, etc.) was emphasized by one of the pioneers of complexity, Herbert Simon (1916–2001) [467].[1]

Structuralism was one of the most popular approaches (with a peak in the nineteen sixties) in a bunch of social scientific disciplines, such as linguistics, anthropology, ethnography, literary theory, etc. Its fundamental aim was to describe the relationship among the elements in a system. A predominant contribution to the initiation of the idea of structuralism was Ferdinand de Saussure's (1857-1913) approach, [128] who considered language, as a system of its elements. Claude Levy Strauss, and his structural anthropology [313] was motivated by structural linguistics. Interestingly, he was also influenced by celebrated mathematical methods applied in cybernetics, and information theory, which were flourished in the nineteen-forties and fifties. To analyze the structure of social groups (first the native Bororo tribes in Brazil), he used the terms "elementary and complex structures and even semi-complex structures".

Another emblematic scientist, Jean Piaget (1896-1980), a Swiss psychologist, applied structuralism to mental development. The human mind (and brain) is certainly an extremely complex structure. Piaget classified the child development to four classes, the last one around 11 years is characterized with the ability of abstract thinking. The transitions between the stages is driven by errors. The accumulated errors require the reorganization of the cognitive structure. Please note, that Piaget was concerned with development, so he was interested not in static, but dynamic structures. Also, Piaget assumed (and he was right), that knowledge is not only acquired from outside but constructed from inside. We shall return to this topic a few hundred pages later, after we have discussed the structural, functional and dynamic complexity of brain and mind.

---

[1] There once was two watchmakers, named Hora and Tempus, who manufactured very fine watches. Both of them were highly regarded, and the phones in their workshops rang frequently. New customers were constantly calling them. However, Hora prospered while Tempus became poorer and poorer and finally lost his shop. What was the reason?

The watches the men made consisted of about 1000 parts each. Tempus had so constructed his that if he had one partially assembled and had to put it down – to answer the phone, say – it immediately fell to pieces and had to be reassembled from the elements. The better the customers liked his watches the more they phoned him and the more difficult it became for him to find enough uninterrupted time to finish a watch.

The watches Hora handled were no less complex than those of Tempus, but he had designed them so that he could put together sub-assemblies of about ten elements each. Ten of these subassemblies, again, could be put together into a larger sub-assembly and a system of ten of the latter constituted the whole watch. Hence, when Hora had to put down a partly assembled watch in order to answer the phone, he lost only a small part of his work, and he assembled his watches in only a fraction of the man-hours it took Tempus.

The late mathematician of Yale University, Charles E. Rickart (1914–2002) [430] carefully reviewed structuralism from the perspective of mathematics, which "*does not mean* a formal mathematical treatment of the subject". While structuralism was an intellectually challenging, optimistic movement, it "has often been criticized for being ahistorical and for favoring deterministic structural forces over the ability of individual people to act". From Wikipedia.

The most important mathematical technique to represent the structural relationship among the elements of natural and social systems is *graph theory*. Graph theory offers a natural way to *represent* systems with individual nodes and connection between nodes. Chemists use graphs to represent their molecules, having the atoms as nodes, and chemical bonds as edges. To represent three-dimensional objects by two-dimensional graphs (in order to visualize molecules on paper, blackboard and/or computer screen) some geometric information (as distances, and angles) are sacrificed, while the topological information (i.e., the existence and non-existence of bonds between atoms) is preserved. Graph theory proved to be very successful in analyzing structures in many disciplines. There is a single mathematician, Frank Harary (1921–2005) who, in addition to his contribution to the development of graph theory itself, had papers on the application of graph theory to anthropology, biology, chemistry, computer science, geography, linguistics, music, physics, political science, psychology, social science. Harary investigated many problems which are nowadays reanalyzed in terms of network theory.

Dynamical complexity is about *temporal processes*. Here is an arbitrary list of related concepts: clockwork Universe, heat death, arrow of time, time reversal, eternal recurrence, biological clock, heart beat, neural rhythmicity, weather prediction, epilepsy, ... Irreversibility and periodicity are recurring themes. There is no strict correlation between structural and dynamical complexity. Robert May published a paper in 1976 in Nature with the title "Simple mathematical models with very complicated dynamics" [334], which clearly explained a mechanism of the emergence of chaos in simple mathematical equations: models even with simple structure may lead to complicated dynamics.

The *algorithmic information* complexity (introduced by the legendary mathematician Andrey Nikolaevich Kolmogorov (1903–1987) and extended by Gregory Chaitin) of some computable object is the length (measured in number of bits) of the shortest algorithm that can be used to compute it. So, the shorter the algorithm, the simpler the object.

The notion of *cognitive* complexity has been related to personality theory [272]. It has been used as a basis for discussion on the complexity of personal constructions of the real world (and particularly of other people) in psychology. People have mental models about their social environment. A subject should rate a number of people around her on a number of attributes, generally in a bipolar axis. The complexity of the world view of a subject can be measured by this test. A subject, who assigns to all their friends positive attributes and to their enemies negative attributes would have a less complex, basically one-dimensional mental model of their acquaintances. (S)he has only friends and enemies. A subject with the ability to see people as a mixture of "good" and "bad" characteristic properties has a higher "cognitive complexity".

There are many more disciplines related to complexity issues: (such as, say computational complexity, ecological complexity, economic complexity, organizational complexity, political complexity, social complexity), just to mention some of them. Browsing the web you may see that almost all of them have their journals, conferences, etc. So, it would be difficult to deny that complexity theory has become very popular.

Roughly speaking we know that the big success stories of 20th century science are related to the reductionist research strategy. Particle physics and molecular biology were the most fortuitous disciplines, both apply predominantly the reductionist research strategy. Reductionism is a coherent view, which suggests that chemistry is based on physics, biology is based on chemistry, psychology on biology, and sociology on (bio)psychology. A more extreme view claims that finally everything is "physical", any aspect of life and mind is basically a physical thing. Molecular biology emerged in search for the structure of genes, and the application of the reductionist strategy implied big progress in reducing genetics to molecular biology.

While the differences between the 20th century sciences and complexity are significant, neither one has a hegemony. The former, including biology, were dominated by mechanistic reductionism.

Mechanistic reductionism suggested, that the universe, including life, were considered as "mechanisms". Consequently, understanding any system required the application of the mental strategy of engineering: the whole system should be reduced to its parts. Knowing the parts was thought to imply the complete understanding of the whole.

The science of complexity suggests that while life is in accordance with the laws of physics, physics cannot predict life. Therefore, in addition to reductionism, a more complete understanding of complex dynamical systems requires some holism. The (w)holistic approach is interested in *organization principles*. One of the most important key concept of this approach is the notion of "emergent properties": system's properties emerge from the interaction of its parts. Holists like to tell, that the whole is somehow (?) greater than the sum of its parts. In extreme form, holism not only denied that life can be understood by physical-chemical laws, but suggested the existence of some "non-material agent".

1.2 Characteristics of Simple and Complex Systems

1.2.1 System and Its Environment

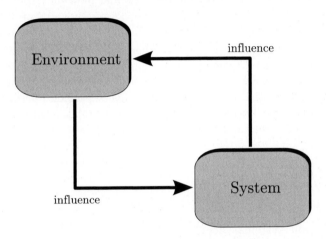

Fig. 1.1. System and its environment.

A system is a delineated part of the universe which is distinguished from the rest by a real or imaginary boundary (Fig. 1.1). The system approach integrates the reductionist and the holistic approaches. There are *closed systems* which are maintained by internal forces, and not influenced by external forces. A piece of stone is an isolated system, and its shape is preserved by increasing a force to threshold, when it might be subject to disintegration by rupture. The properties of the stone can be studied by neglecting its interaction with the environment.

The founder of the "General systems theory", Ludwig von Bertalanffy (1901–1972) emphasized that, as opposed to stones, and to other isolated systems, the majority of biological and social systems are *open systems*. The behavior of an isolated system is completely explainable from within. Living structures, and other open systems, should be considered, as systems being in permanent interaction with their environment to ensure normal performance.

Systems theory is always concerned about the boundary between a system and its environment. It is clear where the boundary of a stone is. Also, we might intuitively expect that the chemical composition of a stone will not change for tomorrow. But its temperature is influenced by the sunshine. Roughly speaking a piece of stone is closed for material flow, but it is subject to energy flow. The stone is not really an isolated system, since it has the same temperature as its environment. They are in thermal equilibrium. More precisely, the change in the temperature of the environment implies an energy flow to equilibrate the temperature difference. Our temperature is not the same as that of our environment, and our internal processes ensure to maintain the temperature difference. We are "open systems", dynamic structures maintained by permanent material, energetic and information flow with our environment.

To give an appropriate description there are three concepts we should know: a system, its environment and the interaction between them. So from the perspective of system theory, we should know how to characterize the state of the system, the properties of the universe, which affect the system excluding the system itself , and the interactions/relationships between the system and its environment.

1.2.2 Simple Systems

What are simple systems? Here are some characteristic properties of simple systems:

- Single cause and single effect

- A small change in the cause implies a small change in the effects

- Predictability

Common sense thinking and problem solving often adopts the concept of "single cause and a single effect". It is probably not a very big exaggeration to say that both the classical engineer's and the medical doctor's fundamental approach was based on this concept. Common sense also suggests that small changes in the cause imply small changes in the effect. It does not literally mean (as sometimes is mentioned) that there is a linear relationship between the cause and the effect, but it means that the system's behavior will not be surprising, its behavior is *predictable*. With a somewhat more technical terminology, small changes in the parameters (or in the structure of the system) do not qualitatively alter the system's behavior, i.e., the system is "structurally stable".

Intuitively it seems to be obvious that there are simpler and less simpler patterns of binary strings. A very simple pattern is 01010101, which shows biperiodicity. Assuming some "continuity hypothesis", the behavior of the continuation of the pattern is predictable. It is not predictable, however, how to continue a randomly generated string. There are patterns, which are not simple and not random.

Simple Versus Complex Systems and the Deborah Number

Marcus Reiner, the founding father of rheology defined a non-dimensional number, the Deborah number D, as

$$D := \frac{\text{time of relaxation}}{\text{time of observation}}. \qquad (1.1)$$

The difference between non-changing "solids" and flowing materials "fluids" is then defined by the magnitude of D. If the time of observation is very large, or, conversely, if the time of relaxation of the material under observation is very small, you see the material flowing. On the other hand, if the time of relaxation of the material is larger than your time of observation, the material, for practical purpose is solid. The name resembles us to Prophetess Deborah [425], who sang (even before Heraclitus's "panta rei") "The mountains flowed before the Lord". Deborah knew not only that everything flows. She also knew that with the infinite observation time God can see the flowing of those objects what the man in her short lifetime cannot see.

We may conclude that the complexity of a stone should increase with the length of its observation.

1.2.3 Complex Systems

Here are some characteristic properties of complex systems:

- Circular causality, feedback loops, logical paradoxes and strange loops
- Small change in the cause implies dramatic effects
- Emergence and unpredictability

Circular Causality

Circular causality in essence is a sequence of causes and effects whereby the explanation of a pattern leads back to the first cause and either confirms or changes that first cause; Example: A causes B causes C that causes or modifies A. The concept itself had a bad reputation in legitimate scientific circles, since it was somehow related to use "vicious circles" in reasoning. It was reintroduced to science by Cybernetics (see Sect. 2.2.2), emphasizing feedback. In a feedback system there is no clear discrimination between "causes" and "effects", since the output influences the input.

Feedback

Feedback is a process whereby some proportion of the output signal of a system is passed (fed back) to the input. So, the system itself contains a loop. Feedback mechanisms fundamentally influence the dynamic behavior of a system. Roughly speaking negative feedback reduces the error or deviation from a goal state, therefore has stabilizing effects. Positive feedback which increases the deviation from an initial state, has destabilizing effects. Natural, technological and social systems are full with feedback mechanisms (Fig. 1.2).

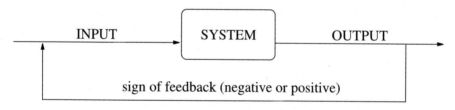

Fig. 1.2. Systems with feedback.

Systems with feedback loops are used by engineers (to serve them more justice) to stabilize the operation of plants. They use sensors and actuators to measure and control their system. The Greek Ktesibios, who lived in Alexandria, revolutionized the measurement of time by building a water clock. To achieve his goal he had to invent a regulator valve to maintain the level of the water in a tank at a constant level. Maybe he was the first who consciously used feedback control. (see later, such as Fig. 3.4). The toilet uses negative-feedback to fill itself up with water when flushed. In this case an impairment of the control system may show positive feedback, which implies non-required overfill. A simple example of feedback inhibition is a connected thermostat-heater system. A sensor detects the temperature, and when the temperature reaches a predetermined value, the thermostat signals the furnace to switch off. When the temperature drops below an other predetermined value, the furnace is turned back on.

For chemical reactions, positive feedback is related to the concept of *autocatalysis*. Autocatalytic reactions have a specific property: n components generate $n + m$ components, $m > 0$. For such reactions there is a self-amplifying mechanism between the concentration of a component and its velocity of formation: the velocity is increasing function of its own concentration.

In the "simplest case" the velocity is proportional with the concentration. "Simplest case" means that $n = 1$ and $m = 1$, so one component (molecules) produces two molecules by using some other component A:

Linear autocatalysis:
$$A + X \longrightarrow 2\,X, \qquad \text{velocity} = k[A][X].$$

There are higher-order autocatalytic reactions, too. Quadratic autocatalysis:
$$A + 2\,X \longrightarrow 3\,X, \qquad \text{velocity} = k[A][X]^2.$$

Chemists often use a different terminology. They (we) call "linear" autocatalysis as "quadratic" and "quadratic" autocatalysis as "cubic", respectively to reflect the *the molecularity* of the reaction. Two molecules collide in the first case, three collide in the second.

There is a big qualitative difference in the behavior of the linear and the quadratic autocatalytic system. The first reaction is just equivalent to the model of population growth posited by Malthus more than two hundred years ago. The *assumption* of the model is that the rate of the increase of the population is proportional to the actual size of the population. The model leads to exponential growth (Fig. 1.3). It means that in the limit of infinite time the population size will be infinite. By modifying the assumption a little,[2] and assuming that the encounter of two individuals is necessary to have a third individual leads to super-exponential behavior, specifically to "explosion". The term "explosion" here means that during **finite** time period the population size will be infinite (Fig. 1.3).

Autocatalytic reactions have an important role in getting so-called exotic chemical behavior (as periodicity or deterministic aperiodicity (chaos) in the

---

[2] Any models are based on assumptions. Model builders like to play the game to study the effects of changes in the assumptions.

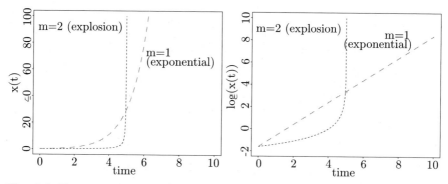

Fig. 1.3. Exponential growth (linear autocatalysis) and explosion (quadratic autocatalysis). *Left*: linear scale. *Right*: logarithmic scale.

concentration, as it will be discussed later in several contexts. We shall return to the difference between exponential and super-exponential dynamics in the Sect. 9.3.6.

Biological networks use feedback in all hierarchical levels of the organization. Jacob and Monod outlined a network theory of genetic control in prokaryotes (prokaryotes are simple cells, which don't contain nucleus, while eukaryotes do) in 1961 [357, 255]. The Operon model is the classical model for the cellular metabolism, growth and differentiation (for the legacy and historical analysis of this seminal work see [361]). Now there are detailed mathematical models [576] which by taking into account the network structure of the lactose operon regulatory system (Fig. 1.4) are able to reflect the fundamental bistable property of the system.

Bistability occurs in many natural, social and technological systems, when a system is able to exist in either of two, so-called steady states, and when there is an abrupt jump from one state to the other. Bistability is a property of certain nonlinear systems, and such kinds of phenomena were demonstrated and analyzed in different fields, such as from phase transition in physics to jumps in the oil price and to interpretation of ambiguous figures, etc. We shall return to this topic in Sect. 2.2.3. Abrupt jumps from one state to the other were subjects of interdisciplinary studies, what we shall discuss later under names of "phase transition", "synergetics" and "catastrophe theory".

A simple example for positive feedback in economics is the relationship between *income* and *consumption*. By increasing income per capita there is an increase in consumption. Increased consumption has of course a positive effect in income (if nothing else changes, or "ceteris paribus" using a favorite term of economists).

Interestingly, Brian Arthur, who took the lion's share in popularizing complex systems thinking in economics, and specifically to use positive feedback

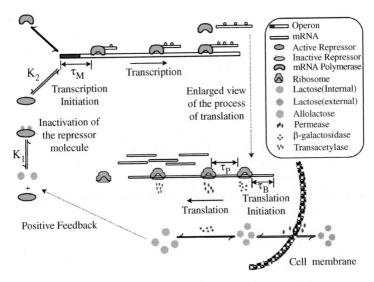

Fig. 1.4. Schematic representation of the lactose operon regulatory system. From [576].

(or increasing return, using again the terminology of economists) [25] was strongly influenced by the Monod-Jacob feedback model in molecular biology, and the theoretical studies on autocatalytic reactions, which proved to be important ingredients of producing oscillatory patterns in chemical and biological systems.[3]

The sociologist Merton coined the term "Matthew effect" [351] as a mechanism for amplifying small social differences. He paraphrased the Gospel of St. Matthew 25:29[4] to explain why and how already established scientists are able to dramatically increase their resources compared to others.

The stability and occasionally abrupt change in governmental policies and institutions can be explained by negative and positive feedback processes (see Baumgartner and Jones 2002), [51]. Negative feedback processes ensure the stability of institutions, while positive feedback, i.e., self-amplifying processes

---

[3] "I started to read about enzyme reactions and the writings of Jacques Monod, a French molecular biologist and Nobel Prize winner. He had written a book called Chance and Necessity where small events could get magnified by positive feedbacks and lead to different enzyme reaction paths. I began to realize that the counterpart in economics to positive feedback was increasing returns. I started reading the physics of positive feedback, and particularly the work of the German Hermann Haken at Stuttgart and the Belgian Ilya Prigogine, a man I am very fond of ..." http://www.dialogonleadership.org/Arthur-1999.html.

[4] "To him who hath shall be given and from him who hath not, shall be take away even what he hath", from the parable of the three servants.

may help to diffuse new ideas. The more politicians talk about an issue, such as health-care system or prescription drug coverage, the more the public will be concerned with it. Examples of actions implying stability: counter-inflatory effects of the Federal Reserve, unemployment compensation, price supports for farmers, etc. The political scientists Baumgartner and Jones cite the mathematician-economist Brian Arthur to explain, that the "increasing return" mechanism also operates in politics. (It is interesting to see the propagation of ideas through disciplines: Arthur's economic theory was motivated by physical-chemical and biochemical models, while his economic models fertilized thinking in political science.)

One mechanism of the decision-makers is based on mimicking others behavior, a strategy labeled as "go with the winner". Nobel prize winner (2005) Thomas Schelling [458] (go with him) set a celebrated model of segregation. In his model there are two types (say, red and blue, or white and black) of agents, who live in a two-dimensional grid. The agents may choose to remain or to move to a free space. With the model it is possible to investigate the effect of different rules. Schelling concluded that even without having a racist attitude, a slight preference to live "among your owns" may lead global segregation. Schelling models will be discussed in Sect. 4.7.1.

Schelling studied a variety of social phenomena, where the individual decision was determined by the behavior of the others in the group. We are ready to cross the street when the light is red, if others around us do the same, we buy similar goods as our neighbors, and so on. Schelling intuitively has foreseen that there are situations, which basically have two outcomes, i.e., they show bistability; adopting Schelling's terminology, two macrobehaviors (which emerged from the micromotivations of the individuals). Either all people cross the street, (since the positive feedback among the people amplified the action), or everybody decides to wait. Either everybody starts to applause after a performance, or a weak applause decays very rapidly. Which of the two possibilities will be realized, it is determined by the number of people involved. If it is larger than a *critical mass* than (well, almost) everybody will do the same.[5]

Logical Paradoxes

Logical paradoxes are also intimate ingredients of complexity. While this subsection will speak about paradoxes, first let's discuss a case which turns out to be non-paradoxical. The so-called liar paradox goes back to Epimenides, who

---

[5] Phil Ball [40] enthusiastically popularizes how the concepts of physics can be used to predict the collective behavior of a large group of people.

lived in the sixth century BC. He wrote: A person from the island of Crete asserts:

> All Cretans are liars.

We can conclude that if he is telling the truth, then he is lying. One might believe that if he is lying, then he is telling the truth. But the statement is not paradoxical at all: if there has ever been a Cretan who made a true statement then Epimenides's sentence is simply false without implying its own truth. This is a paradox only if Epimenides is the only Cretan.

A paradox may emerge if the sentence is *self-referential*: A man says that he is lying. Is what he says true or false? The core of this paradox is:

> This sentence is false.

A self-referential sentence talks about itself. Another form of the paradox is when a single sentence is not self-referential, but it refers to the next sentence, which refers back to the first one, in a controversial way:

> The next sentence is false.
> The preceding sentence is true.

> Russel's Barber Paradox:
>
> In a small town, a barber cuts the hair of all people who do not cut their own hair, and does not cut the hair of people who cut their own hair. Does the barber cut his own hair? Suppose he does cut his own hair. But by the second half of the first sentence, he does not cut the hair of people who cut their own hair, including the barber himself. Contradiction. Suppose he does not cut his own hair. Then by the first half of the first sentence he does cut the hair of people who don't cut their own, including the barber himself. Contradiction.

Strange Loops

Probably the (sorry, not too young) Reader will agree with me, that the most popular (which certainly does not mean that the most extensively read) book on complexity issues (though this terminology is hardly used in the book) is Hofstadter's Gödel, Escher, Bach: an Eternal Golden Braid [238]. The book is about *"Strange loops"*. Strange loop is a sophisticated version of self-reference. As Hofstadter says: "whenever, by moving upwards (or downwards) through the levels of some hierarchical system, we unexpectedly find ourselves right back where we started". Escher's famous Waterfall (Fig. 1.5) is an example for strange loop: the water flows continuously down, still returns again to the top of the waterfall. Our intuitive notion that water flows downhill is challenged.

Fig. 1.5. Escher's Waterfall.

Hofstadter uses the strange loop as a paradigm to interpret paradoxes in logic, and founds it as a main organizational principle (not his terminology) not only in mathematics and computer science, but also in molecular biology, cognitive science, and in political science.

Feedback cycles, strange loops and iterative algorithms (recursions), are elements leading to complexity. When I checked what Wikipedia writes about it, and found a subsection about recursive humor, I laughed:

Recursion
See "Recursion".

Recursive algorithms might lead to certain type of complex structures which have the property of *self-similarity* (Fig. 1.6).

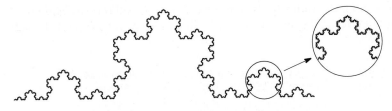

Fig. 1.6. Self-similarity: a property of (many) fractals in which the pattern of the whole occurs in each part. Actually this is a Koch curve.

Small Changes Imply Dramatic Effects

The expression "butterfly effect" became well-known, even in the popular culture. As we may read in http://www.anecdotage.com/index.php?aid=1358 (checked in June 16th, 2007):

> Butterfly Effect
>
> While working on a weather prediction problem one day in 1961, Edward Lorenz, using a computer to model the weather with a system of 12 basic equations, ran a program from the middle rather than the beginning – and entered parameters to three decimal places, rather than the normal six. Lorenz was surprised to see how differently the weather patterns evolved. Such minute changes, he supposed, might be caused by something as trivial as the beating of a butterfly's wing. Lorenz had discovered the so-called "butterfly effect", and was soon embroiled in chaos theory ...

We know its geographic variations:

- The flap of a Butterfly's Wings in Brazil sets off a Tornado in Texas.

- A butterfly flapping its wings in Kansas could trigger a typhoon in Singapore or a downpour on your summer party.

- A man sneezing in China may set people to shoveling snow in New York.

- A butterfly flaps its wings in Asia, the action may eventually alter the course of a tornado in Kansas.

- A butterfly flapping its wings in Tokyo could cause a cyclone in the Caribbean.

- A butterfly flapping its wings in South America can affect the weather in Central Park.

- The possibility of large storm in New England may be caused by a butterfly wing flap in China.

Edward Lorenz himself gave a lecture in a session on Global Atmospheric Research Program of the meeting of the American Association for the Advancement of Science in 1972 with the title "Predictability: Does the Flap of a Butterfly's Wings in Brazil set off a Tornado in Texas?" Somewhat more technically speaking he suggested that certain dynamical systems show "sensitive dependence on initial conditions", and small errors are subjects of dramatic amplification. Implicitly, the effect of a flap in the reality is similar to a round-off error in his model.

The "butterfly effect" is at most a hypothesis, and it was certainly not Lorenz's intention to change it to a metaphor for the importance of small event. It is used at least in metaphoric sense to explain stock market crashes, but also political events (how the tiny change in the mind of the designer of the ballot paper in Palm Beach for the US Presidential election in 2000 led to the result, which finally was settled by the Supreme Court).

Dynamical systems that exhibit sensitive dependence on initial conditions produce remarkably different solutions for two initial values that are close to each other. Sensitive dependence on initial conditions is one of the properties to exhibit *chaotic* behavior. In addition, at least one further implicit assumption is that the system is bounded in some finite region, i.e., the system cannot blow up. When one uses expanding dynamics, a way of pull-back of too much expanded phase volume to some finite domain is necessary to get chaos. This property is also true for linear systems: $dx/dt = ax$, $a > 0$, the solutions are written down by $x(t) = x(0)\exp(at)$, which means exponential separation of the difference of two close initial points. This example is just an unbounded expanding system. The complicated geometry behind the generation of chaos, and the stretching and folding of trajectories leading to divergence will be discussed in Sect. 3.6.2.

A large class of chaotic motions, which seem phenomenologically irregular, lead to *strange attractors* as opposed to simpler motions, e.g., damped or sustained oscillations which tends to simpler attractors, such as points, or closed curves. Strange attractors have *fractal* structure. Fractals often (but not always) are self-similar. Strange attractors have complex structure, but sometimes even simple rules, simple algorithms (such as the logistic difference equation, a very simple example of a first-order difference equation) may lead to such kinds of patterns:

$$x_{n+1} = r x_n (1 - x_n), \quad x_0 = \xi \tag{1.2}$$

where r is the only parameter, and ξ is the starting point of the iteration.

The equation is recursive or iterative mapping: the result of the mapping is fed back to the equation, producing a new result, i.e., a new x_n, etc. What is interesting, that depending on the value of the control parameter r, the mapping leads to qualitatively very different results (Fig. 1.7). The general form of the first-order iterative mappings is $x_{n+1}(r) = f(x_n(r), r)$, $x(0) = x_0$.

How sensitive are the solutions with respect to the control parameter r and the initial value x_0?

Bifurcations analysis seeks answer to the first question. More precisely, it is used for studying the changes in the long term qualitative behavior of the system. Chaos is related both in changes in the parameter value, and sensitivity to the initial conditions as well. There are parameter windows where chaos may occur, and a basic property of chaotic systems is the sensitive dependence of the solution on initial conditions.

Chaos is a fundamentally important example what we might call *dynamic complexity*. It became extremely popular in the 1980s. This was the period when personal computers started to be used and simple (or not so simple) equations were solved numerically and the results were visualized. In many labs students and their teachers played with their computers and found that visualization of scientific results is important and possible. For a black-and-white version see Fig. 7.11 in Chap. 7.

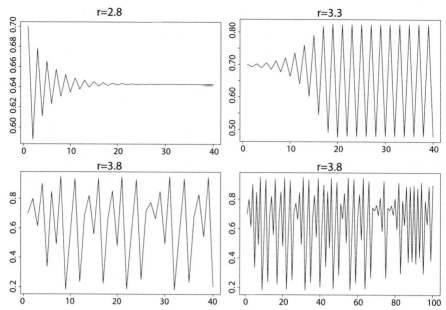

Fig. 1.7. For $r = 2.8$ the system shows a damping oscillatory approach to a fixed point. For $r = 3.3$ there is a non-damping, self-sustained oscillation between two points. For $r = 3.8$ there is a chaotic, irregular behavior. The *right bottom* part of the figure shows the results of a somewhat longer simulation.

Emergence and Unpredictability

István Örkény: The meaning of life

If we string up a bunch of red hot peppers, we end up with a wreath of red hot peppers. On the other hand, if we don't string up a bunch of red peppers, we will not end with a wreath of red hot peppers. The peppers will be just as many, just as red as hot, but they will not be a wreath of hot peppers.

Is it the string? No, it is not a string. The string, as we know is a third-rate factor of little importance. What then?

Whoever falls to thinking about this and makes sure not to let his thoughts stary from true path may come to discover many great truths.

István Örkény[391] , Hungarian writer, wrote among others grotesque with the title "One Minute Stories"

1.2 Characteristics of Simple and Complex Systems

The problem of "emergence of complexity" has at least two different levels. First, what parts of a system do together that they would not do by themselves. One molecule of H_2O is not liquid, one neuron is not conscious, one amino acid is certainly not alive, one sheep is not a herd, one soccer player is not a team (well, eleven players are also not necessarily a team, as desperate Hungarian soccer fans know these days, more than fifty years after the great period of Ferenc Puskás and his teammates). How do system properties arise from the properties of the parts connected? Sometimes system properties emerge due to the *local* interactions among the elements, without any external command, so the mechanism is called *self-organization*. In other case (say, soccer), some external stimulation might help (a little bit). But even the best coach is unable to build a world champion team from a set of ungifted players.

Second, there exists the question about the "evolution of complexity" occurring in phylogenetic time scale. Theory of evolution suggests that there is a mechanism called "natural selection" which explains the development of life forms. While it does not predict precisely the future forms, as the theory is sometimes criticized, Peter Medawar, a Nobel-prize winner biologist answered the critics in 1973:

"...finds Darwinian theory still at fault from a strictly methodological point of view. Darwinians have yet to produce a theory which makes specific predictions possible. I think the justice of this criticism really depends on how specific the predictions have to be. Let us imagine, as is not improbable, that a metabolic product which we shall call gorgonzolin of the mold Penicillium gorgonzoli is a powerful antibiotic. From what one knows of the genetic system of bacteria it is already quite possible to predict that if strains of streptococci or staphylococci are cultivated in sublethal concentrations of gorgonzolin a new variant will eventually appear which is entirely resistant to the action of gorgonzolin. It is true that gorgonzolin has not yet been discovered, but a great many fungal antibiotics have been, and I predict that what has turned out to be true of all of them would turn out to be true of gorgonzolin also." [345]

Interestingly, the lack of ability to predict does not imply qualms regarding the relevance of a scientific discipline. While there is some progress in predicting the onset of epileptic seizures, eruption of earthquakes and crash of stock market, or at least it is possible to analyze the limits of predictability, (as we shall discuss in Sect. 9.3.1) nobody claims that neuroscience, geophysics and economics are not legitimate scientific disciplines.

1.3 Connecting the Dots

The intention of our book is to explain complexity, by connecting the scattered dots of concepts. The spirit of complex systems research can be understood by analyzing the ideas, concepts and methods related to different disciplines. While these disciplines sometimes cooperated and competed with others, in other times they neglected each other.

Complex systems research certainly was not main stream during the incredible success of quantum physics, relativity theory and molecular biology (and the 20th century physics and biology dramatically changed the world), there were however several scientific disciplines which searched for "organization principles".

In Chap. 2 the history of complex systems research is reviewed. Bertalanffy's systems theory was interested in finding common principles in different disciplines, so he tried to revive the concept of the "unity of science". He suggested that to explain life phenomena theoretical biology should be based on the notion of open systems. The founding fathers of the general systems did not want to establish a "general theory of everything", which would not have any content. They hoped to integrate the different perspectives, and also to analyze and synthesize different forms of complexity.

The birth, rise, and fall, or better saying, dissolution of cybernetics, the science of "communication and control in animals and machines", was about the theory of the computers and brains. The mathematician Norbert Wiener, and the neurophysiologist Warren McCulloch are the founding fathers of cybernetics. It was an optimistic movement, with some overambitious goals. It used many concepts, which became popular again related to both the science of "complex adaptive systems" around the Santa Fe Institute, and to the discipline of cognitive science.

Cybernetics is striking back now, and many of its ideas appear from brain theory and cognitive science to complex systems theory, see Sect. 8.6.1.

In the nineteen seventies three disciplines, each of them with the goal of being interdisciplinary and emphasizing "nonlinearity" became fashionable in Europe. The connotation of the word "nonlinearity" was that "linear" is uninteresting, and nonlinearity leads to different and nice patterns in time and space. I think, three European schools, each of them had leaders with strong personality, dominated the field. The theory of "dissipative structures" grew out from non-equilibrium thermodynamics and was labeled with the name of Ilya Prigogine, and his "Brussels school". Hermann Haken (Stuttgart) used the term "synergetics" to deal with systems, where order is an emergent property of macroscopic systems due to the interactions of elementary constituents. "Catastrophe theory" is a mathematical theory for classifying abrupt qualitative changes in the behavior of a system due to small changes in the circumstances. It emerged from the works of the French mathematician René Thom. Thom's approach was deeply deterministic, while the other two schools took into account random effects and fluctuations to get ordered structures.

Chapter 3 is about the complexity of temporal patterns, i.e., about notions, such as periodicity, reversibility, irreversibility and chaos. The concept of time is strongly culture-dependent, and there is an evolution of the concept from the ancient Mayan cyclic world view via the Newtonian clockwork worldview and pessimistic view of the eventual heat death of the Universe to the modern concept of irreversibility. Mechanical clocks not only measured the passing time but served as the model of the eternal, periodically moving Universe. The collision between the "reversible" mechanics and the "irreversible" thermodynamics led to the birth of statistical physics, which now seems to be a very efficient conceptual and mathematical tool to understand collective phenomena beyond the kingdom of strict physics. Nature and society is full with periodic phenomena. A very popular model of oscillatory changes is the Lotka-Volterra model. It explained, more precisely gave an insight into the mechanism of the prey-predator dynamics. The model was able to describe the periodic change of the quantity of "big" and "small" fishes in the Adriatic sea, under the qualitative assumptions that the small fish has infinite resource (say, plankton), the big fishes' only food are small fishes, and there is a natural decay of the big fishes. This toy model, and its generalizations, are popular tools in modeling competitive and cooperative interactions, in biological, economical and sociological context. One of the most important ingredients of any theory of complex processes, as we mentioned earlier, is *chaos theory*. Chaos theory was a big fashion, although as it is happens with any fashion, its popularity is decaying. Chaotic phenomena are, in any case very important, they give a unique insight to understand the mechanism of generating a large class of unpredictable events. There are big universal questions: does biological and even cosmical evolution has an irreversible direction, or are there any arguments that they are cyclic phenomena?

Chapter 4 illustrates the scope of the dynamic world view, which grew out from mechanics. As in the mechanics of mass points, the state of a point is determined by coordinates, the composition vector of chemical components determines the "chemical state", a vector, which defines the quantities of the different species. The temporal change of the state due to the interaction among chemical components or among species is described by the laws of chemical kinetics, and the laws of population dynamics, respectively. Such kind of causal modeling approach helps to understand biological and social processes as well, as biological pattern formation, propagation of epidemics and ideas, evolutionary dynamics, opinion formation, attitude changes, business cycle, etc.

Chapter 5 is about the role of deductive and inductive reasoning and strategies in decomposing and constructing complex systems. Cybernetics and artificial intelligence is mentioned in terms of combining different reasoning strategies through the contributions of two intellectual giants, John von Neumann and Herbert Simon. Chapter 6 is about randomness. Biological variation was dominated by the Gaussian distribution. Deviation from the average behavior (which is in this case equivalent with the "most probable" state) is symmetric. Deviation from average height is symmetric: roughly the same number of people are shorter and taller than the average. Asymmetric (skew) distributions have strongly different properties. The distribution of wealth of people is not symmetric. There are much less people with large income than with a low one. The popular 80–20 thumb rule says that 20% of the population owns 80% of the wealth. There are a set of models, which lead to such kinds of skew distributions, such as lognormal distribution and power law distribution.

In many respects, complex structures are neither purely regular, deterministic nor completely random structures. Chapter 7 is opened by discussing structural complexity, and the ways of measuring it. Then self-organizing mechanisms and algorithms are discussed, the interaction between deterministic and random effects are specifically studied. Randomness and complexity in artworks (of Jackson Pollock) are also discussed. Networks have many excellent properties to represent natural, technological, social systems, so it is easy to accept why network theory is so important in analyzing complex systems. In addition to a brief review of its main concepts, a specific class of networks, i.e., citation networks in technology, will be explained.

In colloquial sense we often say that the brain is the most complex structure. Chapter 8 discusses the brain-mind-computer trichotomy. Both the philosophical and experimental backgrounds are briefly reviewed. Then the basic principles of neural organization, and the frameworks of the fundamental computational models are summarized. We argue that the tradition of cybernetics combined with the perspective of complex systems theory offers a framework of a brain-mind-computer theory.

Chapter 9 gives some more ideas on how to use complex systems in practice. The comparative analysis of equation-based and agent-based modeling strategies is given. What we can hope now from game theory? Once it was applied for analyzing political conflicts in terms of predicting the outcome of possible strategies depending of the strategies of the other players, and later was successfully applied to evolution. Evolutionary game theory offers causal mechanisms for the emergence of cooperation and social norms. Complex systems theory, in accordance with the ambition of the elder days general systems theory is interested in finding similarities among the (un?)-predictability of such kinds of phenomena, as epileptic seizure, eruption of earthquakes, and stock market crashes.

In the final, tenth chapter it will be summarized how the ingredients of complex systems contribute to explain complexity. We argue that complex systems research offers a perspective to (well, not to bridge, but at least) narrow the gap between science and the humanities. While the "Age of Reason" is over, the author agrees with those who work on a new, "third culture", and believe that "bounded rationality" and the acceptance of our fallibility is the only way for mankind to survive ourselves.

2

History of Complex Systems Research

2.1 Reductionist Success Stories Versus the Importance of Organization Principles

There is no doubt that the superstars of 20th century science are in atomic physics (including nuclear and particle physics) and molecular (if you wish "particle") biology. Both disciplines were driven by searching for the constituents of the organized whole. The "take home message" of the lessons from the history of science is that methodological reductionism, the analytical decomposition of structures to parts, should be completed by searching for organizational principles, too.

2.1.1 Reductionism and Holism in Quantum Physics

Capsule history of early atomic physics

Early atomism assumed that matter is composed of indivisible and unchangeable atoms. Later it turned out that atoms were made of even smaller building blocks. First, the existence of electrons were demonstrated, and the Millikan experiment showed that its mass is very small. Since atoms have neutral charge, positive particles should also exist. When it turned out that atoms are composed of parts, models were constructed to describe the relationship among these parts. A series of atom models were created for describing the distribution of negative and positive charges. The interactions of newer and newer data led to more and more refined models. In 1904 Thomson suggested the "Plum Pudding Model". Positively charged fluid was assumed to be the pudding, and electrons were scattered in this fluid. Ernest Rutherford's (1871–1937) experiments led to the idea of nucleus: mass and positive charge are concentrated to very small place. Niels Bohr (1885–1962), a passionate soccer player, adopted the quantum assumptions (1900) of Max Planck (1858–1947) and Albert Einstein (1879–1955) and postulated in 1913 that electrons circulate around the nucleus without energy loss (radiation), and there are jumps from one state to another with energy changes prescribed by the Planck-Einstein formula. The quantum model of the atom takes into account the particle-wave duality of the electrons suggested by de Broglie (1892–1987). Based on this assumption the location of the electrons had a probability character. Erwin Schrödinger (1887–1961), while first tried to save the classical world view, set wave equations for the motion of electrons (1925). Instead of giving a precise, deterministic description of the motion of electrons around the nucleus, a cloud of points were derived. Max Born(1882-1970) suggested that the cloud should be interpreted as the probability of the electrons being in a certain place.

There is a direct connection between atom physics and the science of complexity, since Murray Gell-Mann has been working on both fields. Quarks (and leptons) are supposed to be the most fundamental types of particle. Quarks can not occur in isolation, the must be bound to another quark or antiquark. This phenomenon is called quark confinement. Murray Gell-Mann got the Nobel prize in 1969 for explaining the interaction of quarks by the theory called *quantum chronodynamics*. Gell-Mann's interest turned later to complex systems. He served as a founder of the Santa Fe Institute, and wrote a popular book with the title "The Quark and the Jaguar: Adventures in the Simple and the Complex" [197].

A Few Words About Quantum Mechanics

Inference phenomena measured by electron diffraction confirmed that electrons may have a wave character, so the atoms are no longer seen as *discrete* entities only, but they have also *continuous* wave nature. Quantum mechanics solved this paradox: Werner Heisenberg (1901–1976) adopted a new formalism and developed his famous uncertainty principle and quantum mechanics proved to be an extremely successful discipline.

The uncertainty principle says, that one cannot assign full precision values for certain pairs of observable variables, such as the position and momentum (i.e., mass multiplied by velocity) of a single particle at the same time even in theory, and gives a quantitative relationship for the measure of uncertainty:

$$\Delta x \Delta p \geq \frac{\hbar}{2}. \qquad (2.1)$$

Here Δx and Δp are the uncertainty of the measurement of position and momentum, respectively, $\hbar$ is the Planck constant divided by 2π.

The general implication of the relationship was that quantum mechanics is inherently indeterministic.

Broadly speaking, quantum mechanics incorporates four classes of phenomena that classical physics cannot account for: (i) the quantization (discretization) of certain physical quantities, (ii) wave-particle duality, (iii) the uncertainty principle, and (iv) quantum entanglement.

Atomism Versus Holism in Physics

The concept of wave-particle duality challenged our naive view. The naive view suggested that electrons, as other particles, are discrete, localized entities. Since things we sense directly seem to be localized and discrete, one might believe the elementary particles we do not sense directly are also localized and discrete. The *principle of local action* prescribes that if A and B are spatially distant things, then an external influence on A has no immediate effect on B.

Entanglement is one of the core concepts of current quantum physics, it challenged the universal validity of the atomism, and basically implies the separability of localized particles. In certain composite systems the state of the individual components can not be separated, so it should be considered as a holistic system.

The story goes back to the Einstein, Rosen and Podolsky (EPR) paradox, and is related to the concept of locality. EPR showed that under certain circumstances quantum mechanics violates locality. Since they did not believe that this effect, which Einstein later called "spooky action at a distance", could really happen, they implied that quantum mechanics was incomplete.

David Bohm (1917–1994) suggested the "local hidden variable" theory. He disproved von Neumann's analysis about the impossibility of completing quantum mechanics by introducing hidden variables. However, classical quantum physics worked well and proved to be extremely useful for calculating the behavior of the physical systems, so the whole non-locality problem was left for philosophers.

John Bell (1928–1990) following Bohm's spirit, established an inequality, which is valid under local realism but not under quantum mechanics. Basically he suggested an experiment to decide whether or not hidden variables may exist. The intrinsic non-locality of quantum mechanics has been demonstrated later by experiments, but there are still ongoing debates about the interpretation of these results. In any case, quantum mechanics put an end of atomism. The material world is a whole, *a whole, which is not made out of parts* [414]. To put it in another way: there are objects which are not wholly decomposed into more elementary parts.

As I learned from Péter Hraskó [242] in our first informal gathering of ELMOHA, (ELmélet-MOdell-HAgyomány in Hungarian, Theory-Model-Tradition in English): realism, locality, induction hypothesis cannot be true together. More about to laymen see Chap. 7 "How real is the real world" in John Casti's Paradigm Lost [94] explains beautifully the story of the paradox of quantum reality.

Emergence and organizational principles in quantum mechanics

> In some theories of particle physics, even such basic structures as mass, space, and time are viewed as emergent phenomena, arising from more fundamental concepts such as the Higgs boson or strings. In some interpretations of quantum mechanics, the perception of a deterministic reality, in which all objects have a definite position, momentum, and so forth, is actually an emergent phenomenon, with the true state of matter being described instead by a wave-function which need not have a single position or momentum.
>
> http://en.wikipedia.org/wiki/Emergence (checked on 16 June 2007)

Hardcore physicists, [13, 207, 305, 120] stated that the wonderful elementary laws of physics are not sufficient to explain emerging complexity. As Anderson formulated: "the ability to reduce everything to simple fundamental laws does not imply the ability to start from those laws and reconstruct the universe". Laughlin and Pines state that "emergent physical phenomena regulated by higher organizing principles have a property, namely their insensitivity to microscopics, that is directly relevant to the broad question of what is knowable in the deepest sense of the term".

The debate about the indispensable role of organization principles to explain emerging complexity is not over. The reductionist method proved to be very successful. Wolfenstein feels [566] that the fundamental emerging macroscopic patterns should be understood by the fundamental physical equations.

Actually he might be right: "the solution may require a collaboration of reductionists and emergentists, if they can be persuaded to talk with one another".

2.1.2 Reductionism and Complexity in Molecular Biology

Capsule History of Early Molecular Biology

From Mendel to the Double Helix

It is known that modern genetics started with Gregor Mendel's (1823–1884) experiments around 1865 which led him to the discovery of heritability. The laws of heredity say that physical traits are determined by factors (what we now call genes) passed on by both parents, and that these factors are passed in a predictable pattern from one generation to the next. Mendel's laws were rediscovered around 1900 (at the same time when Planck assumed the quantum hypothesis).

Max Delbrück (1906–1981) was a German physicist who moved to the United States in 1937, where he started to study the basic rules of inheritance in a simple organism (bacterial viruses, also called as bacteriophages, or more shortly, phages). Since there were no direct methods for studying the chemical nature of the genes, Delbrück's speculated about the atomic structure of a gene, and explained mutation as a quantum jump, and also introduced the standard experimental techniques. The question to be answered was how heritable information is stored in cells. Proteins, composed of 20 different amino acids seemed to be much more likely candidates, than desoxyribonucleic acid (DNA), a heteropolymer built of four types of monomers. Though DNA was isolated even in the middle of nineteen century, it was only in 1944, when Oswald Avery found that chromosomes and genes are made of DNA. Delbrück motivated one of the fathers of quantum mechanics, Schrödinger, to think on the basis of life and inheritance [460]. He assumed that the gene is like an aperiodic one-dimensional crystal. Linus Pauling (1901-1994) probably the most influential chemist of the 20th century (who applied quantum mechanical theory to explain chemical bonds) has already seen that the DNA had a helical structure. There is a (not so) controversial story that Rosalind Franklin's (1920-1958) data obtained by X-ray crystallography (and given out without Franklin's knowledge) played a critical role in the discovery what Watson and Crick made. As everybody knows they suggested that DNA has a double helix structure. They also adopted data which showed that among the four nucleotides there are two pairs, adenine–thymine and guanine–cytosine, which occur in equal proportions. This is called Chargaff's rule. These data led them to the concept of base-pairing, which was the supportive pillar of their whole argument.

Genetic Code

The problem of the genetic code was to find a relationship between DNA structure and protein structure. It turned out that tri-nucleotide units (codons) code individual amino acids. There are $4^3 = 64$ different codon combinations and it was a surprise that many codons are redundant, and an amino acid maybe coded by two or more codons. Though the genetic code shows some variations, all the genetic codes used in living creatures on the Earth show a remarkable similarity: the genetic code should have evolved in very early times.

Central Dogma, Genetic Reductionism and Their Critique

The research program of "molecular biology" suggested that the replication, transcription and translation of the genetic material should and could be explained by chemical mechanisms. Crick's central dogma of molecular biology stated that there was a unidirectional information flow from DNA via RNA (ribonucleic acid) to proteins. First, in the process of *replication* the information in the DNA is copied. Second, during *transcription* DNA codes for the production of messenger RNA. In the third phase (*processing*) RNA migrates from the cell nucleus to the cytoplasm. Fourth, messenger RNA carries coded information via ribosomes for protein synthesis (*translation*). The schema of the central dogma is:

$$DNA \longrightarrow RNA \longrightarrow protein$$

While the central dogma was enormously successful in discovering many detailed chemical processes of life phenomena, philosophically it suggested, as Crick himself wrote [114], that "the ultimate aim of the modern movement in biology is to explain all biology in terms of physics and chemistry". The central dogma led to *genetic determinism*. While certain phenotypes can be mapped to a single gene, the extreme form of genetic determinism, which probably nobody believes, would state that all phenotypes are purely genetically determined. In "Not in Our Genes" [437], Richard Lewontin, a controversial combatant hero of genetics and evolutionary biology with Steve Rose and Leon Kamin attacked genetic determinism. Another hero, Richard Dawkins criticized the authors by accusing them of fighting with strawman [125]. The general validity of the central dogma was challenged and falsified by Howard Temin (1934-1994) who found that RNA can be copied to DNA by an enzyme, called reverse transcriptase [509]. The central dogma was modified:

$$\text{DNA} \longleftrightarrow \text{RNA} \longrightarrow \text{protein}$$

Temin's and (a few others') finding about reverse transcription might have more dramatic consequences if the second step, the RNA $\longrightarrow$ protein information transfer, would be also reversible. The existence of such kind of reversibility would make the inheritance of *acquired traits* possible, i.e., the *Lamarckian* mechanism. Since the second step is not reversible, Temin's discovery did not shake molecular biology. After about an eight year fight Temins discovery was accepted, and it contributed to the success of genetic engineering.

Genetic determinism has lost its attraction as a unique explanation for the appearance of specific phenotypic traits. After 50+ years of extensive research in molecular biology, there is a very good understanding of the intricate mechanisms that allow genes to be translated into proteins. However, this knowledge has given us very little insight about the *causal chains* that link genes to the morphological and other phenotypic traits of organisms [360]. Also, human diseases due to genetic disorders are the results of the interaction of many gene products. One generally used method to understand the performance of a complex genetic networks is the *knockout* technique. It is often applied in mice, when a single gene is deleted. Occasionally there are unexpected results: a gene that is assumed to be essential to a functions was inactivated or removed, but the knockout might have no effect, or even a surprising one. Knockout experiments implied disappointing results, partially due to pleiotropy (i.e., when a single gene influences multiple phenotypic traits), or gene redundancy (when multiple copies of the same gene can be found in the genome).

> Genetic reductionism, in particular, has been abandoned as a useful explanatory scheme for understanding the phenotypic traits of complex biological systems. Genes are increasingly studied today because they are involved in the genetic program that unfolds during development and embryogenesis rather than as agents responsible for the inheritance of traits from parents to offspring.
>
> M.H.V. Van Regenmortel:
> Biological complexity emerges from the ashes
> of genetic reductionism. See [532].

From Reductionism to Systems Biology

As a reaction to something that some people might have seen as the "tyranny of molecular biology", the systems thinking has been revitalized in the last several years. Systems thinking correctly states that while reductionist research strategy was very successful, it underestimates the complexity of life. It is clear, that decomposing, dissecting and analyzing constituents of a complex system is indispensable and extremely important. Molecular biology achieved a lot to uncover the structures of many chemical molecules and chemical reactions among the molecules behind life processes. The typical molecular biologist's approach suggests that there is an "upward causation" from molecular states to behavior. The complex systems perspective does not deny the fundamental results of molecular biology, but emphasizes other principles of biological organization. Several of these principles will now be discussed briefly.

Downward Causation and Network Causality

"Downward causation" is a notion which suggests that higher level systems influence lower level configurations. Classical molecular biology deals exclusively with upward mechanisms of causation (from simple events to more complicated ones) and neglects completely the explanatory role of downward causation. Since we know that both molecules and genes form complicated networks or feedback loops, it is difficult to defend the concept that there is nothing else in science than a linear chain of elementary steps leading from cause to effects [533]. The methodologically successful reductionism is never complete, as Popper suggested: there is always some "residue" to be explained.

The concept of downward causation was offered as a philosophical perspective to the brain-mind problem. Specifically, Roger Sperry (1913–1994) suggested that mental agents can influence the neural functioning [476, 477]. Sperry was criticized by stating that the postulate that physiological mechanisms of the brain are directly influenced by conscious processes is unclear [142]. Alternatively, it was suggested by the Hungarian neuroscientist John Szentágothai (1912–1994), that the nervous system can be considered as being open to various kinds of information, and that there would be no valid scientific reason to deny the existence of downward causation, or more precisely, a two-way causal relationship between brain and mind [499].

Robustness

Biological systems show remarkable robustness, i.e., they maintain functional performance and phenotypic stability both for external perturbation and internal fluctuations [486]. Robustness in biological systems at the cellular level is related to the celebrated concept of "homeostasis"[1], what a biological system should show in order to survive. The interplay between negative and positive feedback is the mechanism of maintaining homeostatic robustness.

"There is no new thing under the sun". The old – and many times well-operating – concept of homeostasis [91] suggests that a certain state of the internal medium [56] is totally maintained. The notion of homeokinesis [249, 575] was suggested to serve better than homeostasis as it captured the **dynamics** of control mechanisms for the self-maintenance of organisms. As a compromise between homeostasis and chaos, Tsuda et al. (1991) [525] assumed that biological organisms keep a "homeochaotic" state to adapt dynamically to a variable non-stationary environment. Homeochaos may play a role in evolutionary processes: it was identified as the mechanism of the evolution of symbiosis ([250]); the strong instability in low-dimensional chaos is smoothed out, and dynamic stability is sustained in high-dimensional chaos.

David Krakauer from the Santa Fe Institute, and his close colleagues have investigated the tradeoff between robustness and evolvability (see e.g., [298]) in a series of papers. Robustness is certainly a more vague concept than the mathematically precisely defined notions of stability (stability of states, orbits etc). Krakauer [297] gave a classification of different mechanisms for preserving function. One of them is modularity.

Modularity

Cells as *structural units* form functionally separable *modules* [232]. Modules have strong internal connections, and looser external connections to their environment. Cellular function should emerge from the molecular interactions taking place in cells. These functions cannot be predicted by studying the isolated components alone.[2]

Biochemical modules are composed of many types of molecules, such as DNA, RNA, proteins, small molecules etc. Are modules real functional elements? They probably are. One way of verifying the existence of functional modules in vivo is to reconstitute the structure/function in vitro. Certain modules, such as the ones responsible for DNA replication, protein synthesis and

---

[1] Pubmed search showed 155498 results on 10 April 2006, 158002 on 10 June 2006; and 172623 on 16 June 2007.

[2] I have heard the old biochemist joke first in a lecture by the then leading Hungarian biochemist, F. Bruno Straub. "Let's imagine you have a simple radio set, you disassemble it, you put in a mortar and pulverize it, than you take to chromatography to see what components you find and even you may guess how much of them, and then now, you are supposed to find out how a radio in fact works."(thanks to Jóska Lázár).

glycolysis were successfully reconstituted. There are modules for which the reconstruction from purified components is difficult, for these one possible strategy is to transplant the module into a different type of cell. The fundamental module, which makes a cell excitable, i.e., ion channels and pumps necessary to action potential generation, have been transplanted into non-excitable cells, and made the cell excitable. So, this technique opens the possibility toward synthetic biology. A third way is reconstitution *in silico*. A celebrated example of this theoretical reconstruction is the mechanism of the signal (i.e., action potential) generation and propagation in nerve cells. Hodgkin and Huxley in 1952 assumed that some phenomenological relationship for the voltage-dependent conductance of the K^+ and Na^+ ions and described the dynamics by semi-empirical equations. That time there was no information about the structure and dynamics of ion channels which mediate the ion transport through the cell membrane. Still, a phenomenological module was sufficient to predict the signal generation.

Cellular modules are certainly not rigid, and there might be overlap between modules containing common components. A complete understanding of a module requires the synthesis of phenomenological and molecular analysis. We learned from the experience of Herbert Simon's watchmakers that modularization has an evolutionary advantage.

Modules are key intermediate structures in the organizational hierarchy of cells. It is known that some cellular components are conserved across species while others evolve rapidly. Functional modules, i.e., integrated activity of highly interacting cellular components carry out many biological functions, and they may be conserved during evolution.

It seems to be clear that in spite of the enormous success of the reductionist research strategy, biological function can very rarely be attributed to an individual molecule. Biological functions should be understood as the emergent product of interactions among different types of molecules. Also, molecular biology neglects the temporal aspects, the dynamic character of organization.

Systems biology emphasizes (i) the interactions among cell constituents and (ii) the dynamic character of these interactions. Systems biology emerged in the last several years and, partially unwittingly, returned to its predecessors, systems theory and cybernetics. The history of these early disciplines will briefly be reviewed soon, while for systems biology see Sect. 4.3.

Table 2.1. Here is a somewhat arbitrary list of disciplines and their pioneers, who contributed very much what we call now complex systems research. Game theory will be discussed in Sects. 5.5 and 9.2.2.

Discipline	Pioneers
Systems Theory	von Bertalanffy
Cybernetics	McCulloch, Wiener
Game Theory	von Neumann
Theory of Dissipative Structures	Prigogine
Synergetics	Haken
Catastrophe Theory	Thom

2.2 Ancestors of present day complex system research

2.2.1 Systems Theory

Systems theory was proposed by Ludwig von Bertalanffy (1901–1972), a biologist who worked on the basic principles of life and searched universal laws of organization.

Basic Concepts of the Systems Approach

1. The System. A system is a whole that functions by virtue of the interaction of its parts. A system is defined by its elements and the relationship among these elements.

2. Analytic and Synthetic Methods. Systems approach integrates the analytic and synthetic methods by taking into account the interaction of the system with its environment.

3. Closed versus Open Systems.

   a) Closed systems do not interact with other systems.

   b) Open systems interact with other systems outside of themselves.

> "Living forms are not in being, they are happening, they are the expression of a perpetual stream of matter and energy which passes through the organism and at the same time constitutes it."
>
> Bertalanffy's conceptual model of the living organism as an open system has had revolutionary implications for behavioral and social sciences.

Systems theory is interested in *similarities* and *isomorphism*, not in the differences of various systems. The basic assumption is that physical, chemical, biological and psychological systems are governed by the same fundamental principles. The theory partially grew up from Bertalanffy's own studies on biological growth. According to his law of growth the temporal change of the body mass of an animal can be described by the equation:

$$L(t) = L_{\max} - (L_{\max} - L(0))\exp(-kt), \qquad (2.2)$$

where $L(t)$ is the actual mass, $L(0)$ is the initial mass, and $L_{\max}$ is an upper limit to the growth.

Exponential growth can be detected, as he mentioned, in single bacterial cells, in populations of bacteria of animals or humans, and in the progress of scientific research measured by the number of publications, etc.

I think, the most important element in von Bertalanffy's concept is that he emphasized the necessity of **organization principles** to understand the behavior of living organisms and social groups.

Bertalanffy worked first on theoretical biology in Vienna. While he opposed the logical positivism of the Viennese Circle, he attended their meetings. After his immigration to North America in 1950, he co-founded the Society for General Systems Research (SGSR) in 1956 among others with Kenneth Boulding.[3] and Anatol Rapoport.[4]

---

[3] Kenneth Boulding (1910–1993) suggested that economics should be investigated within the framework of general systems, and in evolutionary context.

[4] Anatol Rapoport applied mathematical models for biological and social phenomena. He worked also in game theory, and won two competitions by his Tit-for-Tat strategy (cooperate first, then respond with the opponent's previous answer.) See later in Sect. 9.2.2.

> Bertalanffy's General System Theory:
>
> "(1) There is a general tendency towards integration in the various sciences, both natural and social.
>
> (2) Such integration seems to be centered in a general theory of systems.
>
> (3) Such theory may be an important means of aiming at exact theory in the nonphysical fields of science.
>
> (4) Developing unifying principles running 'vertically' through the universe of the individual sciences, this theory brings us nearer to the goal of the unity of science.
>
> (5) This can lead to a much-needed integration in scientific education."

2.2.2 Cybernetics

Warren McCulloch: The Real Pioneer of Interdisciplinarity

Warren McCulloch (1898–1969) was one of the Founding Fathers of the movement and scientific discipline of cybernetics, who had a particular personality, a very original individual, a polymath. He learned philosophy, became an MD, and got education in mathematical physics and physiological psychology, as well. McCulloch was an experimentalist, a theoretician, a premodern scientist, a philosopher, and maybe a magician. The interest that shaped his work and life was a question, as the title of one of his papers reflects: "What is the number that a man may know it and a man that he may know a number?" [340].

Between 1941 and 1952 (i.e., in the initial period and during the golden age of cybernetics) he was at the Neuropsychiatric Institute of the University of Illinois in Chicago. Than he moved to the Department of Electrical Engineering at MIT, to work on brain circuits. The abbreviation EE of the department, however, had a different meaning to him. McCulloch founded a new field of study based on this intersection of the physical and the philosophical. This field of study he called "experimental epistemology", the study of knowledge through neurophysiology. The goal was to explain how a neural network produces ideas. (See Fig. 2.1.)

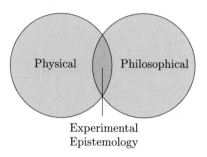

Fig. 2.1. McCulloch's view.

His entire scientific activity was a big experiment to give a logic-based physiological theory of knowledge. Assuming that (1) the brain performs logical thinking (2) which is described by logic, the implication is that the operation of the brain could and should be described by logic.[5]

Style

The editors of the scientific journals of our age would have strong difficulties and most likely repugnance for his essayistic writings. In these articles he mixed physiology, logic, literature and psychiatry, and his personality was also involved. Demokritos, Charles Pierce, Josiah Willard Gibbs, Rudolph Magnus, Immanuel Kant, Sir Charles Sherrington, Clerk Maxwell: these names can be found on a single page of a paper on analyzing the physics and metaphysics of knowledge.

The McCulloch–Pitts (MCP) Model

In 1943 McCulloch and the prodigy Walter Pitts (1926–1969) published a paper with the title "A Logical Calculus of the Ideas Immanent in Nervous System", which was probably the first experiment to describe the operation

---

[5] McCulloch's papers are collected with the title "Embodiments of Mind" [341].

of the brain in terms of interacting neurons [342], for historical analysis see [19, 2, 407].

The MCP model was basically established to capture the logical structure of the nervous system. Therefore cellular physiological facts known even that time were intentionally neglected.

The MCP networks are composed by multi-input $(x_i, i = 1, \ldots, n)$ single output (y) threshold elements. The state of one element (neuron) of a network is determined by the following rule: $y = 1$, if the weighted sum of the inputs is larger than a threshold, and $y = 0$, in any other case:

$$y = \begin{cases} 1, \text{ if } \sum_i w_i x_i > \Theta \\ 0, \text{ otherwise.} \end{cases} \qquad (2.3)$$

Such a rule describes the operation of all neurons of the network. The state of the network is characterized at a fixed time point by a series of zeros and ones, i.e., by a binary vector, where the dimension of the vector is equal with the number of neurons of the network. The updating rule contains an arbitrary factor: during one time step either the state of *one* single neuron or of the *all* neurons can be modified. The former materialize asynchronous or serial, the latter synchronous or parallel processing.

Obviously, the model contains neurobiological simplifications. The state is binary, the time is discrete, the threshold and the wiring are fixed. Chemical and electrical interactions are neglected, glia cells are also not taken into consideration. McCulloch and Pitts showed that a large enough number of synchronously updated neurons connected by appropriate weights could perform many possible computations.

Since all Boolean functions can be calculated by loop-free (or feed-forward) neuronal networks, and all finite automata can be simulated by neuronal networks (loops are permitted, i.e., recurrent networks), von Neumann adapted the MCP model to the logical design of the computers. The problem of the brain-computer analogy/disanalogy was a central issue of early cybernetics, in a sense revived by the neurocomputing boom from the mid-eighties. More precisely, the metaphor has two sides ("computational brain" versus "neural computer"). There are several different roots of the early optimism related to the power of the brain-computer analogy. We will review two of them. First, both elementary computing units and neurons were characterized as digital input-output devices, suggesting an analogy at even the elementary hardware level. Second, the equivalence (more or less) had been demonstrated between the mathematical model of the "control box" of a computer as represented by the state-transition rules for a Turing machine, and of the nervous system as represented by the McCulloch-Pitts model. Binary vectors of "0"s and "1"s represented the state of the computer and of the brain, and their temporal behavior was described by the updating rule of these vectors. In his posthumously published book The Computer and the Brain, John von Neumann [543] emphasized the particular character of "neural mathematics": "... The logics and mathematics in the central nervous system, when viewed as languages, must structurally be essentially different from those languages to which our common experience refers..."

The MCP model (i) introduced a formalism whose refinement and generalization led to the notion of finite automata (an important concept in computability theory); (ii) is a technique that inspired the notion of logic design of computers; (iii) was used to build neural network models to connect neural structures and functions by dynamic models; (iv) offered the first modern computational theory of brain and mind.

McCulloch served as the chairman of a series of conferences (1946-1953) (sponsored by and named after the Macy Foundation), where at the beginning the mathematician Norbert Wiener (1894–1964) also played an important role. Cybernetics was very American. It was labeled (together with genetics) as bourgeois pseudoscience in the Soviet Union of Stalin. (I find remarkable the coincidence that there was only several days difference between Churchill's Iron Curtain speech in Fulton and the first Macy conference on cybernetics (5 March, 8–9 March 1946). The last conference was held several weeks after Stalin's death. Interestingly, but not very surprisingly, after the decline of cybernetics in the U.S it became popular in the Soviet scientific community. Maybe it is not literally true, that cybernetics became a dirty word in the US, but some people say, "well, it is nothing else but computer science", others somehow identify it with robotics.

Wiener: "A Misunderstood Hero of the Information Age"?

The same year the MCP model was published, another supporting pillar of the emerging cybernetics appeared. The paper entitled "Behavior, Purpose and Teleology" by Arturo Rosenblueth, Norbert Wiener, and Julian Bigelow [443] gave the conceptual framework of **goal-directed** behavior both in technological and biological context. Looking back from now the paper is strange in several respects. It was published in Philosophy of Science, did not contain a single formula, figure or reference. In any case, the paper emphasized that purposeful behavior can exist both in engineered and biological systems without assuming the Aristotelean "final cause". Purposeful behavior can be explained by present causes, but the causation acts in a circular manner.

The general principles of *feedback control* were understood by engineers, and autonomous control systems were used to replace human operators.

Rosenblueth worked with Walter Cannon (who popularized the concept of "homeostasis"), and he considered living processes as self-regulated ones. Wiener and Bigelow were involved in developing antiaircraft guns by using negative feedback control during the second world war.

Cybernetics, as a scientific discipline has been named by Wiener, in his book "Cybernetics", with the subtitle "Control and Communication in the Animal and the Machine" [558]. While the physiologists already knew that the involuntary (autonomous) nervous systems control Bernard's "internal milieu", Wiener extended the concept suggesting that the voluntary nervous system may control the environment by some feedback mechanisms. The theory of goal-oriented behavior promised a new framework to understand the behavior of animals, humans, and computers just under design and construction that time.

Conway and Siegelman in their book ("Dark Hero of the Information Age. In Search of Norbert Wiener, the Father of Cybernetics") [110] analyzed how Wiener's dark personal history led to a break among the founding fathers of cybernetics, followed by the dissolution of cybernetics into other disciplines.

Michael B. Marcus, a former student of Wiener put his supervisor's whole activity in a different context [328]. Wiener was a well-accepted mathematician, who worked on functional analysis and on the stochastic processes before Kolmogorov gave its systematic formulation. Wiener studied a model of the Brownian motion, a classical model if the theory of stochastic processes, which is called now as the Wiener process. We also know the Wiener-Khintchine relationship, which helps to analyze stationary stochastic processes. It connects the temporal domain with the frequency domain, i.e., shows how to transform the autocorrelation function of a stationary time series to power spectrum by means of a Fourier transform. No doubt that Wiener was interested in philosophy, mathematics, mathematical physics, biology and literature. Marcus says: "There was nothing 'dark' about Norbert Wiener's mathematics or morals".

The Cybernetics Movement

The Macy conference series was organized to understand the feedback mechanisms in biological, technological and social systems, by the aid of concepts like circular causality and self-regulations. The conferences had interdisciplinary character, and Wiener and von Neumann in particular made claims that their theories and models would be of utility in economics and political science. It is interesting to note that no economist or political scientist attended any of the ten conferences. While cyberneticians spoke on behalf of physics, (well, a strange physics, not a physics of matter and energy, but a physics of information, program, code, communication and control) there was no professional physicist among them. Max Delbrück (who was trained, as we already know, a hard core physicist, but was already working on applying physics to biology) was invited, since von Neumann felt that the arising molecular genetics will be interesting from mathematical point of view. Delbrück did not like the conference, and never returned. As Jean-Pierre Dupuy [137] analyzes, it is one of the most striking ironies in the history of science, that the big attempt of molecular biology to reduce biology to physics happened by using the vocabulary of the cyberneticians. "Cybernetics, it seems, has been condemned to enjoy only posthumous revenge" ([137], pp. 78).

The main topics of the conferences were [137]:

- Applicability of a Logic Machine Model to both Brain and Computer

- Analogies between Organisms and Machines

- Information Theory

- Neuroses and Pathology of Mental Life

- Human and Social Communication

No doubt that cybernetics was an intellectually appealing, ambitious discipline, partially victim of its own ambition. But many of its tenets survived under the names of other disciplines, and I think, cybernetics now strikes back.

Cybernetics: 50 Years After

0,1 Versus Symbol Manipulation

The members of the next generation following cyberneticians, mostly just their students, shifted the emphasis from the structural approach to the functional one. If you wish, they formulated the antithesis: "To put the scientific question, we may paraphrase the title of a paper by Warren McCulloch [340]. As Newell and Simon wrote [375]: What is a symbol, that intelligence may use it, and intelligence, that it may use a symbol?"

Consequently, the pioneers of the artificial intelligence (AI) research substituted McCulloch and Pitts' binary strings of zeros and ones by more general symbols. Procedures on physical symbol systems were viewed the necessary and sufficient means for general (i.e natural and artificial) intelligent action. While the symbolistic paradigm became predominant, the perspectives of the cyberneticians and AI researchers did not separate immediately, but the debate became very sharply related to the Perceptron battle.[6] We shall tell more about the story in Sect. 5.6.

Second-Order cybernetics: Autonomous System, Role of Observer, Self-Referential Systems

It is easy to conceive that the movement of cybernetics was driven, at least implicitly, by the grand utopia that Metaphysics, Logic, Psychology and Technology can be synthesized into a unified framework. While the keywords of the early cybernetics (identified, say, with the first five meetings), were communication and control, the "second order cybernetics" (initiated by Heinz von Foerster and Roger Ashby), considered that the observer and the observed are the parts of the same system, and the result of the observation depends on the nature of their interaction.

Heinz von Foerster (1911–2002), born and raised in Vienna, who was the secretary of the last five Macy conferences. (He served between 1958-1975 as a director of the very influential Biological Computer Laboratory at the University of Illinois at Urbana-Champaign). he constructed and defended the concept of second-order cybernetics. As opposed to the new computer science and control engineering, which became independent fields, the second order cybernetics emphasized the concepts of autonomy, self-organization, cognition, and the role of the observer in modeling a system. Cybernetic systems, such as organisms and social systems are studied by an other cybernetic system, namely by the observer. Von Foerster was a radical *constructivist*. According to this view, knowledge about the external world is obtained by preparing models on it. The observer constructs a model of the observed system, therefore their interactions should be understood "by cybernetics of cybernetics", or "second-order" cybernetics. It is difficult to reconstruct the story, but it might be true that a set of cyberneticians, who felt the irreducible complexity of the system-observer interactions, abandoned to build and test formal models, and used a verbal language using metaphors. They were the subjects

---

[6] The Perceptron is a mathematical construction of an adaptive neural network being able to learn and classify inputs. It was defined by Rosenblatt [442] by adding to the MCP rule a learning rule by modifying synaptic weights. Minsky and Papert proved in 1969 [353] that a single layer Perceptron cannot solve the "exclusive OR" problem. Perceptrons were assumed to be able to classify only linearly separable patterns. The implication of the critique was the serious restriction on funding neural network research. However, the critique is not valid for multilayer neural networks.

of well-founded critics for not studying specific phenomena [236]. Constructivism is an important element of new cognitive systems, as we shall discuss in Sect. 8.6.2.

Ross Ashby (1903-1972) [28, 29] (the latter has a freely downloadable electronic version) was probably first to use the term "self organization", and contributed very much to cybernetics and system science. One of his main conceptual achievements was to make a difference between an object, and a system defined on an object ([29], p. 39):

> Object Versus System:
>
> "At this point we must be clear about how a 'system' is to be defined. Our first impulse is to point at the pendulum and to say, the system is that thing there. This method, however, has a fundamental disadvantage: every material object contains no less than an infinity of variables and therefore of possible systems. The real pendulum, for instance, has not only length and position; it has also mass, temperature, electric conductivity, crystalline structure, chemical impurities, some radioactivity, velocity, reflecting power, tensile strength, a surface film of moisture, bacterial contamination, an optical absorption, elasticity, shape, specific gravity, and so on and on. Any suggestion that we should study 'all' the facts is unrealistic, and actually the attempt is never made. What is necessary is that we should pick out and study the facts that are relevant to some main interest that is already given. The system now means not a thing, but a list of variables."

As Dupuy explains [137], cybernetics was built on the beliefs that

> "1. Thinking is a form of computation. The computation involved is not the mental operation of a human being who manipulates symbols in applying rules, such as those of addition or multiplication; instead it is what a particular class of machines do – machines technically referred to as 'algorithms'. By virtue of this, thinking comes within the domain of the mechanical.
>
> 2. Physical laws can explain why and how nature – in certain of its manifestations, not restricted exclusively to the human world – appears to us to contain meaning, finality, directionality, and intentionality."
>
> ([137], pp. 3–4)

2.2 Ancestors of present day complex system research

The mistakes of the cyberneticians led the next generation of thinkers to ignore their work. The development of a scientific theory of brain and mind was thus significantly delayed. The perspective of cybernetics now slowly returns. We discuss this question after learning more about the arguably more complex system, i.e., about the brain, in Sect. 8.6.1.

2.2.3 Nonlinear Science in Action: Theory of Dissipative Structures, Synergetics and Catastrophe Theory

From the late 1960s nonlinear science propagated from math to applied sciences. It culminated in the mid 1980s, when PCs appeared on the desk of each young researcher. Nonlinear differential equations, iterative maps, stochastic models, cellular automata, as models of many natural and social phenomena started to be investigated. New visualization tools, color coded representations of the properties of the equations were used, and people adored to play with it. Several schools competed with each other.

The *theory of dissipative structures* labeled with the name of Ilya Prigogine and his "Brussels school" grew out from the thermodynamic theory of open systems, and intended to describe the formation of (temporal, spatial and spatiotemporal) patterns first in physico-chemical, later, more ambitiously as well in biological and social systems. *Synergetics* was founded by Hermann Haken, in Stuttgart, Germany. The goal has been to find general principles governing self-organization of elements independently of their nature. A variety of disciplines such as physics (lasers, fluids, plasmas), meteorology, chemistry (pattern formation by chemical reactions), biology (morphogenesis, brain, evolution theory, motor coordination), computer sciences (synergetic computer), sociology (e.g., regional migration), psychology and psychiatry were approached. Haken's synergetics grew up from his research in laser physics. Synergetics extended the concept of phase transition (which is a jump-like change in some variables) between so-called nonequilibrium structures. Somewhat earlier, in Bur sur Yvette, (a suburb of Paris) René Thom established catastrophe theory. One of his big goals was to explain the mathematical basis of morphogenesis of biological organisms. Though the schools did not often refer to each others' works, there is a big overlap in the phenomena they studied. The transitions among different dynamical states are the common themes. While the theory of dissipative structures and of synergetics used both deterministic and stochastic models and emphasized the role of fluctuations in switching systems from one state to another, catastrophe theory was purely deterministic.[7]

---

[7] A stochastic version of catastrophe theory was elaborated by Cobb [101]. Multistationarity in deterministic models might be associated (at least approximately), to the multimodality of stationary (being continued)

From Thermodynamics to the Theory of Dissipative Structures

Classical thermodynamics (better saying thermostatics)[8] is interested in isolated systems, i.e., systems without being influenced by flow of matter and/or energy. The two basic laws of thermodynamics state that (1) energy is conserved; (2) physical and chemical processes degrade energy. Following Sadi Carnot[9] the second law of thermodynamics was formulated by Clausius. He defined a measure of irreversibility, called entropy. The second law is formulated as

$$\frac{dS}{dt} \geq 0, \tag{2.4}$$

where S is the entropy and t is time.

As Boltzmann pointed out in a series of discussions[10] the second law has probabilistic character. Boltzmann derived a relationship between entropy, i.e a macroscopic quantity, and the micro states of matter. Entropy is the measure of different configurations of micro states materializing the same macro state. Macro states which could be related to more configurations are more probable, so they occur in a closed system with a higher probability. This relationship is given in his famous formula:

$$S = k \, \log W, \tag{2.5}$$

where k is the Boltzmann constant, and W is the thermodynamic probability (i.e., number of possible configurations) of a macro state. The extension of the theory for open systems required to define an internal entropy production dS_i/dt within the system, and dS_e/dt, which characterizes the entropy flux between the system and its environment. While $dS_i/dt \geq 0$ is postulated, the entropy flux across the border remains unspecified. There is no reason to exclude the possibility when it is negative and large, so

$$\frac{dS}{dt} = \frac{dS_i}{dt} + \frac{dS_e}{dt} \leq 0. \tag{2.6}$$

(continued from Page 45) distributions or probability density functions. It is generally assumed that (i) the number of equilibrium points in the deterministic model coincides with the number of extreme points of the density functions, (ii) equilibrium points can be associated with the location of maxima of the density functions; (iii) stable equilibrium points coincide with maxima, unstable equilibrium points coincided with minima of density functions. A change in the number of equilibrium points corresponds to the change of the extreme points in the density functions. See also Sect. 6.2.

[8] Classical thermodynamics does NOT use the concept of time, it is a truly static theory. Its history characterized by Clifford Truesdell as tragicomical [520].

[9] Many members of the Carnot family (an old Burgundy bourgeoisie family) are known from history of science and politics: http://www.carnot.org/ENGLISH/carnot%20family.htm.

[10] About the debates with Zermelo see Sect. 3.3.3.

Disorder may be reducing in non-isolated systems. (Of course the total entropy, that of the open system and of the environment, would not decrease.) Energy flowing through the system makes it possible to produce "dissipative structures" in an open system, which is not possible in isolated systems. Temporal structures (such as multiple steady states, limit cycle oscillation in chemical systems), and spatial structures, such as spatial periodicity, waves and fronts were studied first in physical-chemical systems, and occasionally in social sciences as well. A specific model, i.e., the so-called Brusselator model of an oscillatory chemical system showing limit cycle behavior will be presented in Sect. 3.5.2. Here the internal process is described by nonlinear differential equations, but for the emergence of self-sustained oscillation continuous interaction with the environment is also needed.

Synergetics

Synergetics has been interested in the extension of the theory of phase transition of equilibrium states (such as between e.g., liquid and gas phases) for transitions among nonequilibrium stationary states. The characteristic variable of the transition is called the *order parameter* [228].

The basic principles of synergetics are easily illustrated in light of the example of Bénard convection (Fig. 2.2). In this case a liquid is heated from below. Since there is a temperature difference between the bottom and top surface, a macroscopic movement of the liquid begins in accordance with a specially ordered pattern. The molecules move in such a way that a rolling movement within the liquid becomes identifiable. Because of the increase in temperature, the liquid expands and the specific weight of the single molecules decreases, which implies an upward movement of the liquid elements. Up until a certain temperature, the upward movement can not overcome the internal friction. The liquid remains, therefore, in a macroscopic resting condition.

The Slaving Principle

Probably the most important concept of synergetics is the "slaving principle". This principle connects the few numbers of macroscopic variables to the large number of microscopic ones, and ensures that dynamics can be described by a low-dimensional system. Of course, there is a bidirectional relationship between the macroscopic and microscopic variables.

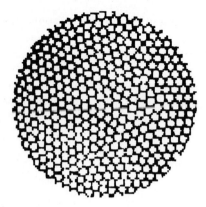

Fig. 2.2. Bénard cell: an example for a beautiful self-organized spatial patterns.

"Phase Transition" in Chemical Systems

The Schlögl model of first-order phase transition is given by the reaction

$$A + 2X \underset{k_1^-}{\overset{k_1^+}{\rightleftharpoons}} 3X, \qquad B + X \underset{k_2^-}{\overset{k_2^+}{\rightleftharpoons}} C, \qquad (2.7)$$

where A, B and C are external components, i.e., a component that is held at constant concentration. (This can experimentally be realized by a constant supply from a reservoir). X is the only internal component. With the notation $a \equiv (k_1^+/k_1^-)[A]$, $k \equiv (k_2^+/k_1^-)[B]$, $b \equiv (k_2^-/k_1^-)[C]$ the deterministic model is

$$-\frac{dx(t)}{dt} = x^3 - ax^2 + kx - b \equiv R(x). \qquad (2.8)$$

Without the loss of generality, (2.8) could be rewritten as

$$-\frac{dx(t)}{dt} = x^3 - \lambda x - \mu, \qquad (2.9)$$

since the quadratic term can always be eliminated. For the fixed points we have the equation

$$-x_{eq}^3 + \lambda x_{eq} + \mu = 0. \qquad (2.10)$$

The two-dimensional parameter space can separated into two regions by the equation defining the only triple root:

$$-4\lambda^3 + 27\mu^2 = 0 \qquad (2.11)$$

An analogy with the theory of phase transitions can be seen. The phases are represented by the fixed points. The triple root may be associated with the "critical point". Since the constitutive equation of the van der Waals gases is also a third order polynomial, $R(x)$ can be associated with the equation

$$P = \frac{RT}{V} - \frac{a_1}{V^2} + \frac{a_2}{V^3} \qquad (2.12)$$

by making the following correspondences:

$$x \leftrightarrow V^{-1}, \quad k \leftrightarrow RT, \quad a \leftrightarrow a_1, \quad b \leftrightarrow p,$$

where V is the volume, p is the pressure, R is the Raoult constant and T is the temperature.

The curve delimiting the two "phases" (i.e., the regimes, where there are one and three solutions respectively) is shown in Fig. 2.3. Furthermore, Fig. 2.4 shows the dependence of the possible fixed points on one parameter, actually on μ, while the other parameter, λ, is fixed. The curve has a characteristic S-shape, which indicates the existence of multistationarity. More precisely, for a value μ, $\mu_1 \leq \mu \leq \mu_2$ there are three fixed points, two of them are stable, and one unstable.

It is often mentioned that there is direction-dependent phenomenon, i.e hysteresis. This is intended to mean that the jump from the regime of the "low" fixed points to the regime of the "high" fixed points and the jump back from the "high" regime to the "low" regime does not happen at the same parameter values. The phenomenon should not be overemphasized, since the parameters

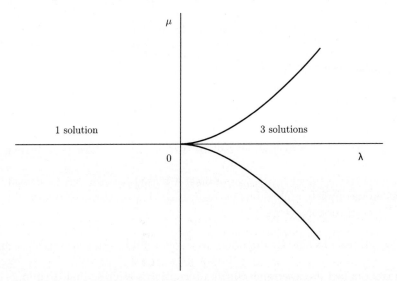

Fig. 2.3. The two-dimensional parameter space is classified into two regions (one solution and three solutions).

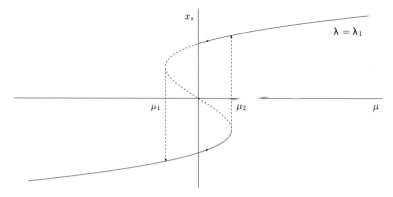

Fig. 2.4. Jumps from the regime of "low" stationary states to the regime of "high" stationary states show a hysteresis.

don't depend on time. It is more informative to say that a bistable system can be used to classify the set (actually the interval) of the initial values. We shall go back to this question soon with catastrophe theory.

Multistable Perception

Bistability is strongly related to multistable perception of ambiguous figures. These figures have two interpretations, and the observer flip back and forth between interpretations. The Necker cube is an old example of ambiguous figures. Ambiguous patterns have common properties:

- A pattern can be perceived in two different ways.

- The time, while a perceived alternative, remains stable and is characteristic for the pattern, but may vary from person to person.

- There is no reason to assume that the two alternatives have equal strengths.

- The patterns might be subject of bias. A biased pattern may be considered as an incomplete ambiguous pattern. If the bias is stronger than a threshold, no reversion occurs.

- This threshold may be direction-dependent, and hysteresis might occur.

- Random factors determine which alternative is realized first. Priming (i.e., the showing first a strongly biased alternative) influences the result of the first perception.

- There is a transient period (1–3 min) for reaching the stationary value of the frequency of switching.

- Reversion can be influenced by conscious effort, but cannot be suppressed.

Discontinuous phase transition proved to be an appropriate model of switching between alternative percepts. Hysteresis effect (which now should be understood by looking at the figures subsequently, so real time, and history really matters - which is not (!) the case in strict bifurcation problems) can be modeled by changes in the potential landscape. [301]. Ditzinger and Haken introduced both a deterministic and a stochastic model [132, 133] for describing the oscillation in the perception of ambiguous patterns. The basic model assumes that there are two prototype patterns encoded by two linearly independent vectors, where the components of a vector encode the different features of a pattern. The state of the system is characterized by the perception amplitudes d_1 and d_2, and the dynamics of pattern recognition for the case the two unbiased patterns are given as

$$\dot{d}_1 = d_1(\lambda_1 - Ad_1{}^2 - Bd_2{}^2), \tag{2.13}$$
$$\dot{d}_2 = d_2(\lambda_2 - Bd_1{}^2 - Ad_1{}^2), \tag{2.14}$$

where λ_1 and λ_2 are the "attention parameters". If the attention parameters are time-independent, the recall process (governed by an appropriate potential function) leads to some fixed point attractor. However, we can assume time-dependent attention parameters with the dynamics:

$$\dot{\lambda}_1 = a - b\lambda_1 - cd_1{}^2, \tag{2.15}$$
$$\dot{\lambda}_2 = a - b\lambda_2 - cd_2{}^2. \tag{2.16}$$

Then *linear stability analysis* shows that in a certain region of the parameters there are periodic solutions, so oscillation of the perception occur. The model was extended for showing how oscillation of perception happens in the presence of a bias. Change in the bias implies different potential functions (see Fig. 2.5), which determine the recognition dynamics.

Catastrophe Theory

Catastrophe theory (CT) was fashionable in the 1970s and 1980s. It belongs to dynamical systems theory, originated from the qualitative theory of differential equations, and it is not related to apocalyptic events. The French mathematician René Thom classified the sudden jumps (called "catastrophes") in

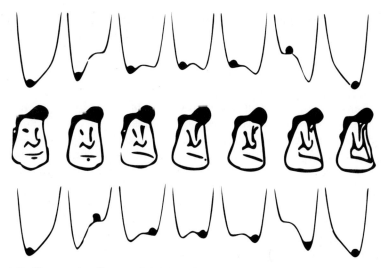

Fig. 2.5. Hysteresis effect modeled in the potential landscape. Based on Kruse et al. [301] and Ditzinger and Haken [132].

the state of certain systems due to changes in the circumstances (parameters). Actually when the number of variables is not larger than three, and the number of control parameters are smaller than or equal to five, then with one more restriction, i.e., when the dynamics is governed by a potential gradient,

$$\dot{x}(t,p) = f(x(t,p),p) = -\frac{\partial V(x(t,p),p)}{\partial p}, \qquad x(0) = x_0, \qquad (2.17)$$

there were only seven families of functions

$$p \mapsto \text{stationary solution}$$

(x and p denote the state vector and the parameter vector, respectively). The negative sign in the equation reflects the physical convention: a particle is assumed to move downhill in a potential well. The seven types of catastrophes were given strange names (fold, cusp, swallowtail, butterfly, hyperbolic umbilic, elliptic umbilic and parabolic umbilic). Catastrophe landscapes demonstrate that gradual and sudden changes in behavior occur in the same system under different circumstances (i.e., changes in p).

There were two types of applications of CT. First, there were low-dimensional equations, belonging to a class of gradient systems. The cubic Schlögl equation is a simple example for cusp catastrophe. Defining the potential function as

$$V = (x^4)/4 - \lambda/2x^2 - \mu x \qquad (2.18)$$

and substituting to (2.17) leads to (2.9).

Second, experimental data (or more often hypothetical data) were interpreted by CT. Applied catastrophe theory's way of thinking is well represented by the next example to model oil price.

A Catastrophe Theory-Based Oil Price Model

An example on how hypothetical data was interpreted by catastrophe theory is illustrated on the example of oil-price modeling [569]. The tacit assumption is that oil prices have either low or high values, there are two separated regimes. Occasionally small changes in the circumstances imply jumps from one regime to the other. Two control parameters were defined, and the general cusp catastrophe with two control parameters were visualized in Fig. 2.6. Then there is a story which tells us the possible scenarios of the jumps. (See the caption of the figure.) The whole modeling procedure is intuitive rather than technical.

Catastrophe Theory: Was the Baby Thrown out with the Bath Water?

Catastrophe theory became the victim of its large success and maybe of the ambition of its pioneers (in addition to Thom, Christopher Zeeman, a British mathematician popularized both the theory and applications [580]). Zahler and Susmann [577] sharply criticized catastrophe theory, and most of its applications. They claimed that such kinds of modeling efforts should be restricted to science and engineering, and has almost nothing to do with biology and social sciences. While catastrophe theory disappeared from the field of applications, the celebrated mathematician Vladimir Arnold contributed to the deepening mathematical foundations of the theory [24]. The emotional attitude behind the heated debate was certainly related to the methodological discrepancies between natural and social science. However, the attack was somewhat misdirected. First, Thom and Zeeman trained and worked as mathematicians. Second, while it was true that some applications were over-dimensionalized or not justified, the attack weakened the general position of those who tried to use mathematical models in social sciences.

The Triumphant Nonlinear Dynamics: Books for Teaching

With the all the successes and misinterpretations, nonlinear dynamics, a special branch of mathematics became an extensively used framework to understand, predict and control phenomena from condensed matter physics to

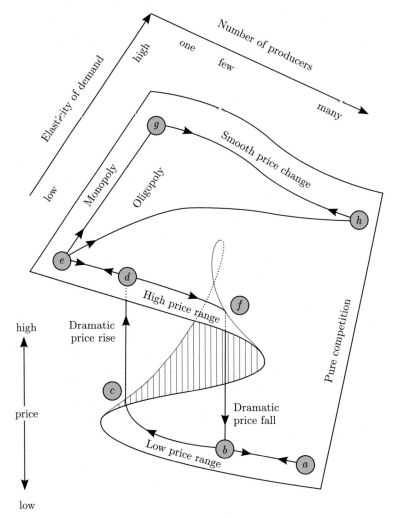

Fig. 2.6. The latitude and longitude in this case represent the elasticity of demand and level of competition in the crude oil market. The height of the landscape represents the price of oil. The model illustrates situations involving monopoly, oligopoly, and pure competition. The folded nature of the landscape surface suggests the existence of conditions supporting high and low price ranges. Paths such as $(a \to b \to c \to d \to e)$ on the landscape surface illustrate how decreasing competition can lead to sudden increases in price. Paths such as $(e \to d \to f \to b \to a)$ reflect sudden price declines due to increasing competition as new suppliers enter the market place. Increasing elasticity of demand can also lead to gradual changes in price (paths $e \to h$ and $e \to g$) under appropriate conditions. Adapted from *www.kkrva.se/Artiklar/003/woodcock.html* .

chemical reactions, from enzyme kinetics to population dynamics, from ecology to evolution, from brain dynamics via personal psychology to sociodynamics, and from economics back to astrophysics etc. There is no doubt, that the heros of the last four decades contributed to these successes very much by affecting people with theories and providing them with a forum on conferences, in book series etc. One of the best textbooks was written by Strogatz [491] and it is used in many courses on nonlinear dynamics. Another excellent textbook that is also suitable for undergraduate teaching is Atlee Jackson's book [253]. Concerning the applications of nonlinear dynamics, two books published in the late eighties dominated mathematical biology, Edelstein-Keshet's and Murray's monographs [143, 368]. During the writing of this book another textbook was published by Ellner and Guckenheimer [150] and I am sure it will be popular too, since it helps to teach applied mathematics to motivated biologists.

(Non)linear models of chemical reactions (both deterministic and stochastic ones) were reviewed in our book written with János Tóth [164], while theories and experiments grown up from the observation of oscillating concentration patterns were reviewed in [152]. My experience is that Joshua Epstein's thin book [153] on transferring basic mathematical biological models (such as of population dynamics and of epidemiology) to social problems (arms race, combat, drug propagation, propagation of ideas, etc.) is first-rate. A set of papers on dynamical systems approach to social psychology was edited by [529]. The conceptual and mathematical framework of synergetics was applied to sociodynamics [552], in particular e.g., for group dynamics, opinion formation, urban evolution, etc.

The most general phenomena in nonlinear dynamical systems are the self-sustained periodic behavior. Oscillations occur in all types of systems. Clocks, pacemakers, rhythms, cycles are everywhere. But we also have to fact to relentless irreversibility. The dichotomy of irreversibility and periodicity is discussed in the next chapter.

3

From the Clockwork World View to Irreversibility (and Back?)

3.1 Cyclic Universe Versus Linear Time Concept: the Metaphysical Perspective

3.1.1 Cyclic Universe

The concept of an "eternal recurrence" may have always existed, and was formulated in ancient Egypt. Recurrence was taught by the Pythagoreans: "If one were to believe the Pythagoreans, with the result that the same individual things will recur, then I shall be talking to you again sitting as you are now, with this pointer in my hand, and everything else will be just as it is now." (Eudemus, Frag. 272 in [279]). The main idea is that the universe does not have a final state and exhibits a periodic motion. Classical Greek and Roman art and literature does not have much awareness of the past and future. They emphasize the existence of eternal, time-independent values.

Buddhism has the notion called "sa.msaara", the famed spinning of the Wheel of Life or Birth-Death Cycle of Being, illustrated by Fig. 3.1.

The ancient Hindu religion also had a cyclic time concept, and a sophisticated cosmology containing steps of endlessly repeating creation and reabsorption of the world. One of the attractive properties of the cyclic time concept, or the notion of "eternal recurrence", is that there is no need to explain the beginning.

This idea seems to occur in the Old Testament. The writer of Ecclesiastes 1:9 says: "The thing that hath been, it is that which shall be; and that which is done is that which shall be done: and there is no new thing under the sun."

Fig. 3.1. Wheel of life expresses the cyclic view of time and life of Buddhism. Based on http://downloads.wisdompubs.org/website_downloads/WheelofLife.jpg..

The eternal recurrence claims that history repeats itself, and it has been one of Nietzsche's most important thoughts. He argued that the universe does not have a final state (goal); if there were such a final state it would have reached it already. Interestingly, while Nietzsche analyzed the eternal recurrence in metaphysical context, the French mathematician Henri Poincaré proved a theorem on dynamical systems, called the "recurrence theorem". The theorem states that under certain conditions the state of a dynamical system goes arbitrarily close to its initial state. (Stephen Brush, an excellent historian of science addressed the hidden relationship between Nietzsche and Poincaré. [80, 81].) The recurrence paradox (it is a paradox) seemed to be in contradiction with the principle of irreversibility, see Sect. 3.3.3.

Nietzsche about Pythagoreans:

At bottom, indeed, that which was once possible could present itself as a possibility for a second time only if the Pythagoreans were right in believing that when the constellation of the heavenly bodies is repeated the same things, down to the smallest event, must also be repeated on earth: so that whenever the stars stand in a certain relation to one another a Stoic again joins with an Epicurean to murder Caesar, and when they stand in another relation Columbus will again discover America.

Nietzsche learned physics to support his metaphysical statement by physical arguments. If the number of energy centers are finite, even in the infinite time and space, then the same configuration should repeat again and again, he stated. While Nietzsche believed that his arguments were against the mechanistic world view, I think Brush is right and Nietzsche's effort should be interpreted as the qualitative anticipation of Poincaré's recurrence theorem.

The preeminent Argentine writer Jorge Luis Borges (1899–1986) was fascinated by the notion of cyclic time: "I eternally return to the theory of Eternal Recurrence" [67]. He was, however, well informed about the revolution in mathematics, and explains how George Cantor's set theory destroys the basis of Nietzsche theorem. Cantor, Borges's Cantor, states that the number of points in the Universe are entirely infinite, even as the number of points in a single meter, or in a fragment of meter. The eternal recurrence has only a small probability and this probability tends to zero.[1]

As it was mentioned earlier, the most attractive feature of cyclic universe models is that they don't have to explain the beginning. There is a recent excitement and debate among the leading cosmologists about the cyclic universe concept suggested several years ago, as it will be discussed in Sect. 3.8.

3.1.2 Linear Time Concepts

Linear time concepts are based on the view that there is a beginning followed by some other events and there is an end. Past, present and future.

Zarathustra from ancient Persia may have initiated the appearance of the linear time concept in the Western thinking. Judaism declared linear time concepts and created *historical thinking*: events can be ordered in sequence, from the beginning[2] via the middle to the end. Christianity and Islam inherited this view.

Our everyday perspective of time, i.e., remembering the past end expecting the future is expressed in Fig. 3.2.

Linear time concepts have been reflected in the notion of an *arrow of time*. Macroscopic physical processes are irreversible, and this irreversibility is manifested by the *thermodynamic arrow of time*. Later in this chapter we shall discuss how the founders of the laws of thermodynamics, mostly Ludwig

---

[1] In a much more mythological way, Mircea Eliade analyzed the concept of eternal recurrence in terms of history of religion [149].
[2] "In the beginning gods created the heaven and the earth" Genesis 1:1.

Fig. 3.2. Knowing the past and expecting the future. Adapted from [127].

Boltzmann, explained the macroscopical irreversibility of physical processes. Two big Victorian narratives (Charles Darwin (1809–1882) and Karl Marx (1818–1883) [123, 330]) stated the *biological* and *historical arrow of time*.³ The *cosmological arrow of time* describes the direction of the expansion of the Universe, and the *psychological arrow of time* expresses that we feel ourselves as travelers from the past to the future.

State space and attractors: a few remarks

It is comfortable to consider a state of a dynamical system as a point in the state space. The dimension of the system is the number of independent characteristic variables of the system. In certain situations the state of the system converges to some fixed point. The convergence can be monotonous, but a damped pendulum (showing less and less displacement from the resting state) also tends to a fixed point. The physical reason of this convergence is dissipation. Without dissipation a pendulum would not damp. Fixed points are the simplest form of attractors. Closed curves also may serve as attractors, called *limit cycles*. Nonlinear systems can possess these periodic attractors. Trajectories that do not tend to point or periodic attractors are called *strange attractors* and are associated with chaotic processes. Chaotic processes are often labeled as causal but unpredictable, and they are considered as complexity generators. These concepts will be explained in a more detailed way later in this chapter.

---

³ Another famous "fixed-point theorem" (please, don't take it literally. Such kind of mathematical theorems give conditions for the existence of a fixed point) in history was Fukuyama's much debated book "The End of History and the Last Man" [190].

3.2 The Newtonian Clockwork Universe

3.2.1 The Mechanical Clock

Feedback Control in Ancient Water Clocks

Measuring the passage of time was first related to the apparent motion of celestial bodies. A *calendar* is a system that gives names to hierarchical units of time. While years, months and days have natural (physical, biological) basis, the week cycle seems to have a cultural origin. *Clocks* typically measure time intervals within a day. Sundials were calibrated based on the direction of shadows, gradually melting wax also was able to measure the continuously passing time. Water clocks showed the time by measuring the regular outflow of water from a container to some scaled vessel. These outflow clocks were called *clepsydra* ("water-thief"). Since the velocity of the outflow of the water depends on the level of the water, some regulatory mechanism was necessary to keep the level at constant value. Ktesibios revolutionized the measurement of time when he invented a new water clock, where the flow of water was held steady by a feedback-controlled water supply valve. The appearance of feedback control in the ancient technologies was propagated by Otto Mayr [335], and a mathematical analysis of the Ktesibios' water clock (and also for some mechanical clocks discussed later) was given in [311]. A reconstruction of Ktesibios' clock is shown in Fig. 3.3, while Fig. 3.4 explains the operation of the feedback loops. The goal was to ensure that the flow of water into a measuring vessel should happen with constant velocity, independently of the volume of the water in the "upper" vessel.

The control system ensures that the (small) deviation form the steady flow of the water (which actually implies the increase of the h_m at constant speed) decays exponentially to the steady state.

Mechanical Clock and Feedback Control: The "Verge-and-Foliot" Escapement

In the second half of the 13th century a new technology appeared in England, France and Italy: the mechanical clock. It showed time independently of the season, the lengths of days etc. However, initially mechanical clocks had disadvantages as well, since they were heavy and weight-driven. Also, they were large, expensive, and in the beginning not very accurate. The real inno-

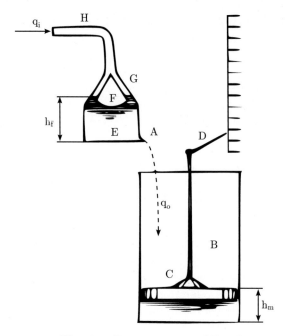

Fig. 3.3. Reconstruction of Ktesibios' clock. Adapted from [311].

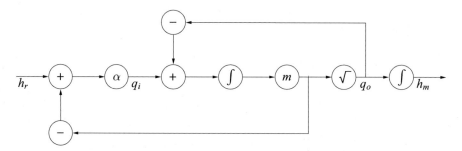

Fig. 3.4. Block diagram from Ktesibios' clock. h_r is the reference level for the canonical float in the feed vessel; h_f is the actual level in the feed vessel; the α block represents the relation between q_i, the flow from the reservoir to the feed vessel and the difference $h_r - h_f$. h_m is the level of the float connected with the indicating element; q_o is the flow from the feed vessel to the measuring vessel, depending on h_f. The *circles* denote operators, + sums its two inputs, − negates its argument, $\int$ integrates, m is a monotonous function of the input. Finally, $\sqrt{}$ represents Torricelli's law: $q_o = k_\beta \sqrt{h_f}$. Modified from Fig. 2 of [311], using a somewhat different convention than in control engineering.

vative element in all mechanical clocks is the appearance of a new regulator, a complicated mechanism called the *escapement*.[4]

The first escapement was the verge and foliot mechanism. The foliot is a horizontal bar with weights on either end. It sits on a vertical rod, called a verge. The verge has pallets to engage and release the main gear. Figure 3.5 shows the principle of the early clock escapement.

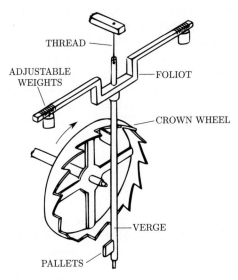

Fig. 3.5. Verge-and-foliot. Adapted from www.elytradesign.com/ari/html/mechanicalclock.htm.

Conceptually, the verge-and-foliot escapement consists of two rotating rigid bodies, which interact with collisions, and causes the rotatory movement of the foliot to one direction, or to the opposite one, depending whether the upper or lower palletes were hit. (The name "foliot" may be associated to its "foolish" motion.) Though the name of the inventor is unknown, there is a consensus that the construction was ingenious. While it was intuitively plausible to use some smooth, continuous movement for measuring the time,

---

[4] Joseph Needham (1900–1995) a legendary British biochemist, deeply interested in Chinese scientific history, suggested [371, 372] that the escapement was invented in China, already in the eighth century, and could be considered as the "missing link" in the technological evolution of clocks between waterclocks and the European mechanical clock. However, the principle behind the European escapement based on centrifugal force of a periodically moving object, seems to be unrelated the Chinese implementation. David Landes, in his wonderful book about the revolutionary role of clocks in making the modern world [304] labels the Chinese contribution as a "magnificent dead end".

the designer of the verge-and-foliot escapement adopted an entirely different approach. He did not try to ensure a constant speed of the falling weight. Alternatively, the speed was increased or decreased due to the collision of the pallet and gear tooth.

Using a somewhat different terminology, the continuous time was decomposed into discrete time intervals defined by time periods between individual collisions between a pallet and a tooth of the crown gear. Escapement makes the gears move forward in small discrete jumps. The measurement of time is reduced to define a unit, and a counter to count the impacts ("ticks and tocks"). From the point of view of dynamics, the speed of the clock depends on the interaction of its components, and it could be considered as a velocity feedback control system [311]. While mechanics is generally considered as a discipline dealing with smooth, continuous motions, the clock itself is a mixture of continuous and discrete models.[5] In any case, feedback control proved to be an invisible thread in the history of technology [57].

Huygens's Pendulum Clock, Feedback Control, Other Escapements

The dynamics of the verge-and-foliot was determined by the interaction of its components. Based on Galileo's discovery Christian Huygens (1629–1695) realized a mechanism, where its period does not depend on interaction, but is given (almost, but not entirely) by a pendulum. Actually it turned out that Galileo's discovery is valid only for small displacements of the pendulum, in this case the approximation leads to harmonic oscillation (see Sect. 3.2.3), while the basic interaction between pallets and gear teeth has been conserved. Since the pendulum is damped, its energy loss should be compensated by this interaction. Feedback control was necessary to establish the appropriate phase of the energy transfer (the phase is important, just like when a child is pushed on a swing).

A general equation of motion for a pendulum is

$$I\ddot{\theta} = -Mg/l \sin\theta - C\dot{\theta} + T(\theta). \qquad (3.1)$$

Here θ is the location of the pendulum (measured as the angular displacement), I is the moment of inertia, l is the given length of the pendulum, g is the acceleration due to gravity, C is the coefficient of friction. The three terms of the rhs of the equation describe the restoring torque due to the gravity, the

---

[5] A more appropriate mathematical framework of modeling the motion of the verge and foliot escapement mechanism is the so-called *impulsive differential equations*. The continuous differential equation describes the motion between impulses, an impulse equation describes the jump to the impulse [449].

loss due to friction, and the escapement torques, respectively [276]. To a fix friction loss C and $T(\theta)$ drive function the $\theta(t)$, the pendulum exhibits an oscillatory motion with constant amplitude and frequency.[6]

The verge and foliot escapement has been substituted by "anchor escapement" by William Clement and Robert Hooke (1635–1703), and later an improved version (deadbeat escapement) by George Graham (1673–1751). A pendulum-driven escape engaged and released gear teeth in the same plane resulting in a reduction in the amplitude of the oscillation, and a strongly improved accuracy ($\sim$ 10 s per day). Anchor escapement established, what is called *recoil*. While the escape wheel turned mostly in one direction, after impact with the lever the escape wheel pushed it backwards (recoil). Graham eliminated this recoil by his deadbeat escapement. Modern control technology uses the deadbeat control, which means that the system is stabilized without overshoot. Mark Headrick[7] labels John Harrison (1693–1776), as the "most brilliant horologist of all time". He adopted grasshopper escapement, (the term was given after the discovery, and characterizes the jump of the lever). The interaction between the wheel and the lever has a minimal friction only. Harrison, an autodidact, spent decades to build several clocks in a competition to determine longitude at sea, which was a serious problem of marine navigation.[8]

Since the local times change in East–West direction, the knowledge of local times in two different points could be used to calculate the longitude distance between them. Sailors wanted to calculate it to navigate more precisely. The Royal Observatory in Greenwich was built in 1675 as reference point. Harrison spent his whole life to build a series of portable clocks with highly precise regulators. Some of his inventions (such as bimetallic strip to compensate the effects of temperature changes, and the caged-roller bearing to reduce friction are still used today). By these and other inventions he proved that longitude could be measured from a watch.

---

[6] For a mathematical analysis of the Huygens's clock see [55].
[7] There is an excellent website (http://www.abbeyclock.com/escapement.html, 11 June 2006), Mark Headrick's Horology Page, what I read to learn the motion of different escapements by looking their computer simulations.
[8] Historical writings also might have self-similar structure. Chap. 9 of David Landes's "Revolution in time" – Clocks and the Making of the Modern World was of my main source about Harrison, his clocks, and his fight to get the Longitude Prize. He writes [304], pp. 150: "...I shall not go into the detail of this mechanism, which interested readers can learn about consulting Gould's classic history..." (Gould restored Harrison's timekeepers).

From the Clock to Clockwork Universe

Lewis Mumford (1895–1990) American historian of technology and science said that the clock (and not the steam engine) was "the key machine of the modern industrial age" [367]. From our perspective the interesting thing is that the most fundamental device, which contributed to making the modern world, and mechanistic world view, is basically a **cybernetic device**. While it measures the continuous irreversible passing time, it is based on periodic motion and on control processes. The mechanical clock contains a component, which generates a time unit by a controlled repetitive process, and a counter, keeping track the time increments. (The result should also be displayed, and the position of clock hand is being used.) The mechanical clock is structurally complex [93] but its behavior (at least in principle) is simple.[9]

Mechanical clocks were intended for signaling (probably used in monasteries to alarm monks to pray). But they soon became models of the solar system, too. The clock became the symbol of regularity and predictability, i.e., of the properties of dynamically simple systems. Nicholas Oresmus (Nicole d' Oresme) (1323–1382) saw the universe as a big mechanical clock created and set by God. While the clockwork universe got its final form in Newtonian mechanics, the mechanical clock became a symbol even earlier, soon after its appearance.[10]

3.2.2 Kepler's Integral Laws

"...My aim is to say that the machinery of the heavens is not like a divine animal but like a clock (and anyone who believes a clock has a soul gives the work the honor due to its maker)"[11]

As it is well known,[12] Kepler searched for simple algebraic relations for the motion of celestial bodies in defense of Copernicus' heliocentric concept, and derived his laws empirically from Tycho Brahe's observations.

---

[9] The somewhat chaotic nature of the clock was studied by [359] See Sect. 3.6.3.

[10] The clockwork analogy was used by Descartes to describe the human body (excluding mental activity), so he was the one of the founders of the monistic mechanistic biology (and also the dualistic brain – mind theory). Darwinian evolution theory sees natural selection as a The Blind Watchmaker, as the title of Dawkins famous book tells [126].

[11] Quotations by Johannes Kepler (1571–1630). http://www.gap-system.org/~history/Quotations/Kepler.html, 12 June 2006.

[12] Arthur Koestler (1905-1983) (one of my heroes) wrote "The Sleepwalkers", a very enjoyable story about the formation of the modern world view of mathematical physics [285].

> **Kepler's laws**
>
> 1. Planets move around the Sun in orbits that are ellipses. The Sun's center of mass is at one of the two foci of the ellipse.
>
> 2. The planets move such that the line between the Sun and the planet (i.e., the radius vector) sweeps out the equal areas in equal intervals of time in any place of the orbit.
>
> 3. The square of the period of the orbit of a planet is proportional to the cube of the mean distance from the Sun (semi-major axis).

In the first law Kepler corrected Copernicus' hypothesis about the circular motion of the planets. While the first two laws refer to the individual motion of the planets, the third law makes connection between the motions of two planets.

> "But the principal thing is that these laws have reference to motion as a whole, and not to the question of how there is developed from one condition of motion of a system that which immediately follows it in time. They are, in our phraseology of today, integral laws and not differential laws."
>
> Einstein: "Isaac Newton",
Smithsonian Annual Report, 1927

In Einstein's remark the term "motion as whole" means that Kepler's laws describe the orbit globally. Kepler did not know the *cause* of the motions, i.e., he was not able yet to predict the motion of a planet from its actual position and velocity. To put it another way, there were no known local rules to calculate the motion. Newton was able to derive (and slightly modify) Kepler's data-driven global, integral laws by deductive, local, differential rules.

3.2.3 Newton's Differential Laws, Hamilton Equations, Conservative Oscillation, Dissipation

Galileo Galilei (1564–1642) initiated (by and large simultaneously with Francis Bacon (1561–1626)) the application of *controlled experiments*. Galileo realized that by combining mathematics and physics the motion of the terrestrial and celestial bodies could be explained by the same theory. He realized that the natural state of an object not influenced by an external force is its stationary motion with a given velocity. (Specifically this velocity might be zero, so the object might be in rest state, too). Objects naturally resist to change their state of motion, and this resistance is called inertia.

Newton second law says that change in the $v := dx/dt$ velocity (i.e., the $a := d^2x^2/dt^2$ acceleration) of a body with mass m is caused by force

$$F := \frac{dp}{dt} = \frac{d(mv)}{dt} = m\frac{d^2x}{dt^2} = ma. \tag{3.2}$$

Here F is the force, m is the mass of the body, x is its position, v is its velocity, p is the momentum, and a is the acceleration. Galileo's law, which is equivalent to Newton's first law, is a corollarium of the second law: $a = 0 \leftrightarrow F = 0$.

If forces are balanced, the acceleration is zero, so there is no change in the velocity: $F = \text{constant} \rightarrow a = 0$. Acceleration $a(t)$ may be zero for a certain type of motion, i.e., if $v(t) \neq 0$ (stationary motion), or in rest state, i.e., $v = 0$.

If there is a linear, (and velocity-independent) relationship between the applied force and the displacement (deformation) of a body, the relationship is expressed by Hooke's law:

$$F(x, \dot{x}) := -kx, \tag{3.3}$$

where k is the spring constant. By combining Newton's second law and Hooke's law and adopting the notation ($\omega_0^2 = k/m$, we have

$$\frac{d^2x}{dt^2} + \omega_0^2 x = 0. \tag{3.4}$$

One form of the general solution is $x(t) = A\sin(\omega_0 t + \phi)$. This describes the motion of a *harmonic oscillation*, with amplitude A and phase ϕ. These two characteristic quantities of the oscillation are determined by the initial conditions. A third quantity, the natural f frequency of the oscillator, is set by the mass and the spring constant: $f = \omega_0/(2\pi)$, and ω_0 is the angular frequency. The total energy (i.e., the sum of the (elastic) potential energy and kinetic energy) has a constant value, it is a *constant of motion*. Harmonic oscillators are conservative oscillators: their properties are NOT determined by the structure of the equation itself, but by the initial values.

$$E_{\text{total}} := E_{\text{kin}}(p) + E_{\text{pot}}(x) = (p^2/2m + Kx^2/2). \qquad (3.5)$$

Here E_{total} is the total energy, i.e., the sum of the kinetic and potential energy. The trajectories lie on the circle defined by (3.6).

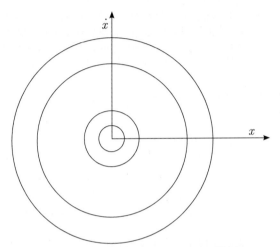

structurally unstable conservative oscillation

Fig. 3.6. Harmonic oscillator is conservative. There is a constant of motion, a closed curve in the state space.

Newton's third law states that for each action there is an opposite and equal action: forces are vectors, i.e., they have directions and magnitudes. Newton supplemented his laws with the law of gravitation. The law of gravitation is based on the assumption that particles interact with each other and the force of interaction is directly proportional the product of their masses and inversely proportional to the square of the distance between them. The gravitational force between two particles with mass m_1 and m_2 respectively is:

$$F = G\frac{m_1 m_2}{r^2}, \quad \text{where } G \text{ is the gravitational constant.} \qquad (3.6)$$

The Newtonian dynamics gave a unified mathematical model to describe the motion of bodies both of the everyday life and of the sky. Orbits of planets should be the result of the inertial and gravitational forces. The Newton equations supplemented with forces specified appropriately are able to predict the motion of any body, knowing its actual coordinates and velocity. Newton was able to calculate the motion of a single planet around the Sun by neglecting the effect of all the other planets, moons etc. The problem of calculating the orbits of two celestial bodies around the Sun (i.e., the "three body problem") remained a central topic of mathematics physics for centuries, and we shall return to this problem in Sect. 3.4.

Hamilton Equations

The power of classical mechanics has been increased by introducing some more convenient formalisms. Both Lagrangian and Hamiltonian formalisms are used extensively . The concept of force does not have explicit role in these formalisms. Here the Hamiltonian formalism is briefly discussed.

The Newton equation is equivalent to the Hamilton equations:

$$\frac{dx}{dt} = \frac{\partial H}{\partial p} \qquad \frac{dp}{dt} = -\frac{\partial H}{\partial x}. \qquad (3.7)$$

If $H(x,p)$ does not depend explicitly on time, then the total energy is conserved. (In general cases, Hamiltonian systems need not be conservative). H is a "constant of motion" in the language of physics, and is a "first integral" using a more mathematical expression.

Newton's (or Hamilton's) equations for N particles in three dimensions involve $3N$ positions and $3N$ velocities. Since it is written in terms of first order DEs we need $6N$ variables. The circular motion is generalized for a higher-dimensional system as motion on a torus.[13] A more general Hamiltonian system contains $2n$ variables, where n is the dimension of the space. Elementary Hamiltonian systems (such as all one-particle systems) have several important properties.

According to the *Liouville theorem* Hamiltonian dynamics *preserves* the *volume* (better saying its *area*), and more generally its *measure* of the $2n$ dimensional phase space. Let $\rho(p,x)$ determine this volume, i.e., the probability $\rho(p,x)\,\mathrm{d}^n x\,\mathrm{d}^n p$ that a particle will be found in the infinitesimal phase space volume $\mathrm{d}^n x\,\mathrm{d}^n p$. The Liouville equation governs the evolution of $\rho(p,x;t)$ in time t:

$$\frac{d\rho}{dt} = \frac{\partial \rho}{\partial t} + \sum_{i=1}^n \left(\frac{\partial \rho}{\partial x^i}\dot{x}^i + \frac{\partial \rho}{\partial p_i}\dot{p}_i \right) = 0. \qquad (3.8)$$

---

[13] Torus is a doughnut-shaped surface, a generalization of the circle. A circle is a one-dimensional torus, a regular torus is two-dimensional, and there is an n-dimensional extension, too.

Another important property is *integrability*. A Hamiltonian equation is said to be integrable, if there exists an $(x,p) \mapsto (X,P)$ transformation where as a function of the new variables H depends on the position coordinates only. For the new variables therefore the Hamiltonian equations are

$$\frac{dP}{dt} = \frac{\partial H}{\partial X} = G \qquad \frac{dX}{dt} = -\frac{\partial H}{\partial P} = 0. \qquad (3.9)$$

This system is obviously integrable, implying

$$P(t) = G(X)t + X(0) \qquad X(t) = X(0). \qquad (3.10)$$

Under this assumption the original equations can be solved, and $2n$ combinations of x_is and p_is, i.e., $2n$ integrals can be obtained. The motion of a general integrable Hamiltonian system is called quasiperiodic, and takes place on an n-dimensional torus.

Integrable systems are very rare among higher-than-two-dimensional systems. It is an important question to ask how the qualitative behavior of an integrable Hamiltonian system changes if the system is slightly perturbed. We shall return to this question briefly in Sect. 3.5.3.

Conservative oscillations are not realistic models of any physical system, since small perturbations destroy the oscillatory behavior. Somewhat more technically speaking they are *structurally unstable*. A system of differential equations is structurally stable if small changes in the equation do not imply a qualitative change in the solution.

Symmetries

Newton's second equation, (3.2) is *invariant* under the transformation $t \to -t$, i.e., when $x(t) \to x(-t)$ and $p(t) \to p(-t)$ the right hand side of the equation remains unchanged, i.e., it is symmetric to the time reversal. Symmetry to time reversal means that a movie running forward and backward is indistinguishable. (Strictly speaking the statement is true only if the forces applied are also time-reversible.)

Mathematical models are intended to describe real phenomena, and in many cases different phenomena can be explained by a common model. It is a natural question to set up criteria in which two phenomena can be qualified as identical, similar, or diverse. The question is in close connection with the *invariance of natural laws* and with the fundamental symmetries being established as a consequence of these invariances.

The term *invariant* is the adjective of a number, function, property etc., which remains unchanged under certain mapping, transformation, or operation. For instance the numbers 0 and 1 are invariants under the operation of squaring, or parallel elements remain parallel after an affine transformation. A mapping, transformation, or operation is *symmetric*, if after its application a certain function, the number of properties remains invariant.

In mechanics, time and space are primary concepts. Not only Einsteinian, but also Newtonian mechanics is relativistic, as the latter complies with Galileo's principle of relativity. Galilean transformation, which introduced the relativity of space, but retained absolute time, does not change the Newtonian equation. Using Hamiltonian formalism it was shown in around 1850 that the invariances under displacement in time, position and angle give rise to conservation of energy, linear momentum and angular momentum respectively.

Invariance principlesinvariance principles are closely connected with transformation groups.[14] In 1918, Emmy Noether (1882–1935) showed the way in which invariance under a continuous group[15] transformation give rise to constraints of motion and proved that the first ten integrals of Newtonian mechanics follow from the invariance properties using the infinitesimal transformation of the Galileo group. (The elements of a Galileo group are Galileo transformations. A Galileo transformation connects a system description from two different coordinate systems which differ from each other only by constant relative motion. Two observers moving at constant speed and direction with respect to one another get the same result. This is Galileo's relativity principle.)

Invariance principles not only are not restricted to classical mechanics, but they were the major players in forming Einstein's relativity theory. The basic equations of classical electrodynamics, the Maxwell equations (dealing with the motion of charged particles and generation of electric and magnetic fields) are not invariant under the Galileo transformation. H. A Lorentz (1853–1928) introduced another transformation which plays a similar role in electromagnetism as the Galileo transformation does in mechanics. The transformation

---

[14] An algebraic structure is a transformation group, if the elements of the structure are transformations satisfying the group axioms. A group $(G, *)$ is composed of a set G, and a binary operation $*$ defined on it: $G * G \to G$. The group axioms: (i) the operation is associative. i.e., for all a, b and c in G, $(a*b)*c = a*(b*c)$; (ii) G contains an element e, which is an identity element: for all a in G, $e*a = a*e = a$; (iii) for each element a there is an inverse element, i.e., an element b in G such that $a * b = b * a = e$, the group is closed in the sense that the result of the operation is also an element of the group: for all a and b in G, $a * b$ belongs to G.

[15] A continuous group contains infinite number of elements, but there are infinite groups (say the integer, or the rational numbers), which don't form continuous groups.

relates the state of the systems by preserving the value of the c velocity for light. The velocity of the light became a parameter of the transformation:

$$x' = (x - vt) \tag{3.11a}$$
$$y' = y \tag{3.11b}$$
$$z' = z \tag{3.11c}$$
$$t' = \gamma(t - \frac{vx}{c^2}), \tag{3.11d}$$

where $\gamma = 1/\sqrt{1 - \frac{v^2}{c^2}}$ is the Lorentz factor.

There was an open question whether the equations of motion or the invariance principles are the more basic "first principles". It was a *choice* of Einstein that he gave the Lorentz transformation primary importance, and so the equations of motion had to be modified for deriving the theory of special relativity.

Invariant quantities

- Acceleration is invariant under the Galilean transformations.

- Energy is invariant for time translation.

- Linear momentum is invariant for space translation.

- Angular momentum is invariant for rotation.

- Speed of light is invariant under the Lorentz transformations of special relativity.

Energy, linear and angular momenta are "conserved".

The mathematical method of group analysis, as a general mathematical method for searching symmetries of differential equations was initiated by the Norwegian mathematician Sophus Lie (1842–1899) at the end of the nineteenth century, and was applied to physics by Emmy Noether.

Dissipation

Newton knew very well that his mathematical model refers to the idealized case, when the body moves under the action of conservative forces, and therefore friction was neglected. Taking friction into account, i.e., a dissipative force (when force really depends on the velocity), e.g.,

$$F(x, \dot{x}) := -kx + c\dot{x}, \qquad (3.12)$$

the resulting motion is damping oscillation: Fig. 3.7 shows the phase plane (the 2D state space) of such a motion.

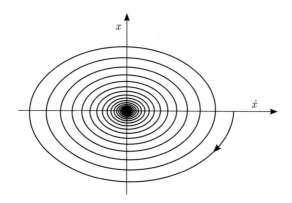

Fig. 3.7. Phase plane representation of a damping oscillator.

Newton's assertion was that the solar system might be unstable without divine interactions. Leibniz accused Newton for not assuming God's ability to design a perfect clock.[16]

Newton's physical results did not support the deistic view, which implies the clockmaker hypothesis. The hypothesis suggests that a watch is too complex to be generated by natural processes. If the Universe works as a watch, there should be a watchmaker to create it, who then stepped aside. Newton pointed out, using Kepler's and his own method that because of the

---

[16] ...Sir Isaac Newton, and his followers, have also a very odd opinion concerning the work of God. According to their doctrine, God Almighty wants to wind up his watch from time to time: otherwise it would cease to move. He had not, it seems, sufficient foresight to create perpetual motion. Nay, the machine of Gods making is so imperfect, according to these gentlemen; that he is obliged to clean it now and then by an extraordinary concourse, and God must mend it as a clockmaker mends his work. Consequently, God must be so much the more unskillful a workman, as he is often obliged to mend his work to set it right.

perturbations due to the interactions among the planets and comets imply irregularities, occasional divine intervention is needed to ensure stability. The theological consideration referring to the stability of the solar system had to be supplemented. Hundred years later Laplace proved that the modification of the orbits due to the perturbations are cyclic and bounded, therefore the solar system remains unstable without divine interaction. This result was the triumph of the Newtonian method of mathematical physics against the Newtonian theology.

Once Again About the Pendulum Clock

The differential equation which describes the motion of a pendulum is given by

$$\frac{d^2\theta}{dt^2} + \frac{g}{\ell}\sin\theta = 0, \qquad (3.13)$$

Galileo's discovery was valid for small displacement i.e., the approximation $\sin\theta \approx \theta$ is sufficiently good if and only if $|\theta| \ll 1$. In this approximation the solution of the equation is a cosine function of time, where the period of the oscillation T_0 does not depend on the displacement (amplitude) θ:

$$T_0 = 2\pi\sqrt{\frac{\ell}{g}}, \qquad |\theta_0| \ll 1. \qquad (3.14)$$

If this harmonic oscillator approximation is not valid, the period is amplitude-dependent, which implies "circular error".

3.3 Mechanics Versus Thermodynamics

3.3.1 Heat Conduction and Irreversibility

The "clockwork universe" concept conceived a world, where all type of changes are causal, the motions periodic, and the equations of mechanics (more or less) express time reversibility.

Jean Baptiste Fourier's (1768–1830) mathematical theory of heat conduction was based on the hypothesis of irreversibility. He derived a partial differential equation for the spatiotemporal distribution of the temperature $T(x,t)$:

$$\frac{\partial T(x,t)}{\partial t} = K \frac{\partial^2 T(x,t)}{\partial x^2}, \tag{3.15}$$

where $K > 0$ is the heat diffusivity. The Fourier equation describes how any initially inhomogeneous spatial distribution of temperature leads to a homogeneous one. Or to put it another way, how a nonequilibrium state leads to an equilibrium.

Figure 3.8 shows a solution of the equation.

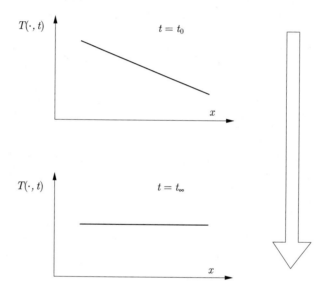

Fig. 3.8. Initially ($t = t_0$), there is a temperature gradient which is eliminated during heat conduction. For $t = t_\infty$ a uniform temperature distribution is generated.

There is an obvious contradiction between time-reversible mechanics and irreversible thermodynamics. However, as Steve Brush remarks [79]: "... It is sometimes asked why Fourier and other scientists in the first half of the nineteenth century did not discuss the apparent contradictions between reversible Newtonian mechanics and irreversible heat conduction, but in fact there is no contradiction at the phenomenological level since Newtonian mechanics had already been successfully applied to problems involving dissipative forces (such as air-resistance or friction) which are not time-reversible. The contradiction arises only when one assumes that all forces at the atomic level must be reversible."

3.3.2 Steam Engine, Feedback control, Irreversibility

Thermodynamics, however, started with industrial applications. James Watt's (1736–1819) steam engine used a feedback principle to control the speed by self-regulation. In 1788 Watt designed a centrifugal flyball governor for regulating the speed of the rotary steam engine. This device was constructed from two rotating flyballs which were flung outward by centrifugal force. When the speed of rotation increased, the flyweights swung further out and up, triggering a steam flow throttling valve which slowed the engine down. Therefore a constant speed was generated by automatic control.

James Clerk Maxwell (1831–1879) in a less known paper (analyzed by Otto Mayr in "Maxwell and the Origins of Cybernetics" [336]) gave a mathematical analysis of Watt's flyball governor. Two equations of motion (i.e., two second-order differential equations leading to a third-order one) were set and combined to describe the dynamic behavior of the control system. Maxwell basically adopted linear stability analysis to calculate the conditions of stability. He studied the effect of the system parameters and showed that the system is stable if the roots of the characteristic equation have negative real parts. This method later has been generalized as became known as the Routh–Hurwitz criterion for the local stability of fixed points.

Sadi Carnot (1796–1832) made an analysis on the theoretical efficiency of steam engines, and introduced a concept, which later became *entropy*, one of the most misunderstood notions of modern science. Carnot was close to postulate the law of irreversibility, what is called now the "second law of thermodynamics". In that time, however, the conservation law for energy ("the first law of thermodynamics") was not known, and without this knowledge the second law could not be formulated.

3.3.3 The First and Second Laws of Thermodynamics

The first law of thermodynamics states the conservation of energy. Different forms of energy can be converted into each other (say mechanical work to heat) but cannot be destroyed. The total energy of an isolated system is constant. The second law tells that while the mechanical work can be converted into heat completely, the reverse transformation cannot be complete. "Heat can never pass from a cooler to a warmer body without some other change..." as Rudolf Clausius (1822–1888) stated in 1854. Heat is not completely convertible to work, and entropy measures this non-convertibility. The entropy of an isolated system has the tendency to increase, and it has a maximal value in equilibrium state. The second law of thermodynamics describes all natural spontaneous processes.

Thermodynamics is not only a theory of heat. It could be held as the general systems theory of the physical world. In a somewhat broader context, the two fundamental laws of thermodynamics suggest constancy and change. The first law reflects nature's constancy, the second one assigns the direction of changes.

Randomness and Irreversibility

Having the fundamental laws of thermodynamics the "kinetic theory of heat" re-emerged. Francis Bacon, Boyle, and Hook already conjectured in the seventeenth century that heat is motion. Clausius, Maxwell, and Ludwig Boltzmann formulated the modern program of the "kinetic theory". Ernst Mach and Wilhelm Ostwald rejected the program of reducing heat to motion of atoms (their existence that time was not yet demonstrated), and suggested that thermodynamics should be independent of mechanics.

While the reduction of the first law of thermodynamics was easy, the experiment to reduce the second law to mechanical principles produced strong debates. Boltzmann defined a quantity (expressed with molecular characteristics) which had a never-decreasing property in time. This quantity is proportional to the macroscopically defined entropy. Boltzmann used statistical assumptions to derive his equations. This was the period, when such kinds of concepts as "probability" and "randomness" became legitimate notions of modern science.

Henry Poincaré's (1854–1912) recurrence theorem states that "almost all" solutions of the equations of mechanics return if not to their initial positions, at least arbitrarily close to them. Based on the recurrence theorem Zermelo stated that it is impossible to explain irreversibility by mechanical concept. Boltzmann acknowledged the mathematical correctness of the theorem, and gave a *statistical interpretation* of the second law. It was a novelty that a law may have a statistical (and not absolute) character. Boltzmann also argued that recurrence time should be extremely long.

Recurrence time

In 1915, M. Smoluchowski calculated the mean recurrence time for a one per cent fluctuation of the average density in a sphere with a radius of 5×10^{-5} cm in an ideal gas under standard conditions would amount to 10^{68} seconds or approximately 3×10^{60} years. The time inter-

> val between two large fluctuations, the so-called "Poincaré cycle" turned out to be $10^{10^{23}}$ ages of the Universe, the age of the universe taken as 10^{10} years (Smoluchowski, 1915).
> http://etext.virginia.edu/cgi-local/DHI/dhi.cgi?id=dv2-12

While the debates did not solve the problem (and Boltzmann committed suicide in 1906), during this time the existence of atoms, the supporting pillar of the mechanistic theory, was almost demonstrated. The debates about randomness and irreversibility helped the scientists to make the transition from classical to quantum physics ([80], p. 637).

3.4 The Birth of the Modern Theory of Dynamical Systems

The dynamical systems theory was born from the generalization of models of mechanics. The *state* of a system at a fixed time point is represented as a point of the *state space*, also called as phase space. If the system is two-dimensional, graphical visualization of the motion helps one get an intuitive interpretation.[17]

Poincaré at the end of the nineteen century reinvestigated the mathematical problem about the stability of the solar system. He won the competition sponsored by King Oscar II of Sweden with the work entitled "The Problem of Three Bodies and the Equations of Equilibrium". Actually he studied a system containing the Sun, Jupiter and an asteroid. First it seemed that Poincaré had proved the stability of the Solar Magnus Gösta Mittag–Leffler,[18] the leading Swedish mathematician and an excellent organizer, who actually ran the competition, published it in Acta Mathematica. Edvard Phragmén,[19] a young mathematician, who helped edit the journal, found a mistake in the

---

[17] Generally coordinate systems consisting of orthogonal axes are used, (this coordinate system is called *Cartesian*) but occasionally *polar coordinates* or *spheric coordinates* are also selected.

[18] There is a recurring gossip about the reason why there is no Nobel prize in mathematics. Different versions of the rumor claim that it is due to an eventual rivalry over a woman between Mittag-Leffler and Nobel. The rumor seems to be unjustified (http://www.cs.uwaterloo.ca/~alopez-o/math-faq/node50.html, 15 February 2006).

[19] Phragmén became a professor and soon the director of the Royal Inspection of Insurance Companies, but he still worked on such "pure" fields of (being continued)

paper. While Poincaré asked to withdraw the paper, it was already under distribution. In any case, Weierstrass icily noted that in his country, Germany, it was axiomatic that prize essays were printed in the form in which they had been judged." ([192], pp. 71–72). Poincaré and Mittag-Leffler exchanged some fifty letters and finally a 270 page long paper was published [409]. This paper proved to be the starting point of the modern theory of dynamical systems. Poincaré introduced new perspectives to analyze nonlinear differential equations, investigated stability, and periodic motions. He used a qualitative, topological[20] approach and searched for the global solution, i.e *all* solutions, as Kepler was doing.

While Poincaré reanalyzed his results, he had to make a big conceptual leap. He noticed, that certain differential equations describing simple mechanical motions are *not integrable* in the classical sense. The three body problem has solutions, which remain within a bounded region while the difference between neighboring solution trajectories grows exponentially with time, so these solutions are what we now call chaotic.

Poincaré found orbits with strange properties, i.e., those which tend to the same point for $t \to \infty$ and $t \to -\infty$. Orbits with such kind of property called now homoclinic, (see Fig. 3.9) while those which go to two different fixed points, called heteroclinic. He discovered the phenomenon of transverse homoclinic intersections, which is a signature of chaos.

Fig. 3.9. Homoclinic orbit.

(continued from Page 79) mathematics, such as complex function theory, besides having publications on such kinds of applications as the theory of voting.

[20] Very non-technically: topology is a branch of geometry, which neglects distances, and angles, and concentrates on neighborhood relationships. Topologically equivalent objects can be mapped into each other by topological transformations, which allow stretching and squeezing, but not tearing and gluing. Graphs are discrete topological structures. A fun theorem of topology is the "hairy ball theorem". If there is a ball with hairs all over it, it is impossible to comb the hairs continuously and have all the hairs lay flat.

3.5 Oscillations

While the thermodynamic arrow of time suggests that processes tend to equilibrium in open systems, other temporal patterns such as regular oscillatory and irregular chaotic behavior may occur; not only in physical, but in chemical, biological and other systems as well. In the next two subsections, the basic models of the two oscillatory behaviors are reviewed.

3.5.1 The Lotka–Volterra Model

The Lotka–Volterra model describes a predator-pray interaction. Prey is assumed to have infinite food resources which would imply Malthusian growth,[21] i.e., exponential increase. In the presence of predator and assuming "binary collisions" between predators and prey there is a bilinear term to describe the decrease of the quantity of prey and the increase of that of the predators. In the absence of prey (so with the absence of food) the quantity of the predator is decreasing proportionally to its own concentration. Based on these assumptions the Lotka-Volterra model is written as:

$$\dot{x} = ax - bxy, \qquad (3.16a)$$
$$\dot{y} = cxy - dy. \qquad (3.16b)$$

Of course it is reasonable to assume that $b \equiv c$ and $a, b, c, d \geq 0$.

The model has two fixed points. There is a trivial fixed point, $(x_{eq}^1 = 0;\ y_{eq}^1 = 0)$, and a nontrivial one $(x_{eq}^2 = c/d;\ y_{eq}^2 = 0)$.

The nature and stability of the fixed points can be determined by linear stability analysis. It starts with the derivation of the Jacobian matrix, which is the coefficient matrix of the linear equation valid around an equilibrium point in question.

The Jacobian matrix of the system is:

$$\begin{bmatrix} a - by & -bx \\ dy & dx - y \end{bmatrix}. \qquad (3.17)$$

The characteristic equation of a matrix is

$$\det A - \lambda I = 0, \qquad (3.18)$$

---

[21] The Malthusian growth model is based on the assumption, that human population increased by a fixed proportion over a given period (specifically doubled every 25 years).

I is the identity matrix and $\det A$ is the determinant of matrix A.

For the trivial fixed point, (i.e., when both populations are subject of extinction), the eigenvalues are $\lambda_1 = a$, $\lambda_2 = -c$. The existence of both a negative and a positive real value means that this fixed point is a saddle point, which is known to be unstable. So, the extinction point is unstable, the dying out of both populations does not imply from this model.

For the non-trivial fixed point the eigenvalues are $\lambda_1 = iac$ and $\lambda_2 = -iac$, $i = \sqrt{-1}$. If both eigenvalues are complex, and the real parts are zeros, the singularity point is called *center*, and it is neutrally stable.[22] The variables show oscillations around the fixed point, and its amplitude is determined by the *initial values*.

The Lotka–Volterra system (i) has a first integral $H(x, y)$, and (ii) leads to closed trajectories in the phase plane implying periodic solutions. The constant H is determined by the initial conditions x_0 and y_0. The Lotka–Volterra model (as the harmonic oscillator) belongs to the class of conservative oscillators, which are not stable (see Fig. 3.10).

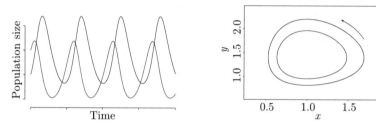

Fig. 3.10. Conservative oscillation. The initial values, not the structure of the equations determine the amplitude of the oscillation.

The Lotka–Volterra model has its roots in chemistry and ecology, since Alfred Lotka (1880–1948) derived it as a model of oscillatory chemical reactions, and Vito Volterra (1860–1940) set it up to explain the oscillatory variation of number of fishes in the Adriatic sea. The model is not a realistic one due to the lack of structural stability. It can be used, however, as a general paradigm of systems with *competitive* and *cooperative* interactions, and was suggested also in socioeconomic context. Goodwin's cyclic growth model [209] is a well known example: it is a model of the Marxian theory of distributive conflict: there is a competition between capital and labor for shares. Specifically, work-

---

[22] Loosely speaking neutral stability means that a small perturbation of the fixed point neither decays back (this would correspond to stability) nor is subject of amplification (this would mean instability), but is maintained.

ers wage plays the role of predator, and the rate of employment the role of prey. Other business cycle models will be reviewed in Sect. 4.8.1.

Generalized Lotka–Volterra Models: Population Ecological Models

The relationship between two species in a community is characterized by the structure of the interactions between them. Table 3.1 summarizes the possible relationships. Symbol 0 means no direct effect on population growth, while + and − are positive and negative effects, respectively.

Table 3.1. Elementary interactions between species in a community

Interaction	direct effect on species X_i	direct effect on species X_j
Neutral relationship	0	0
Commensalism	+	0
Mutualism	+	+
Predation	+	−
Competition	−	−

These model frameworks proved to be very useful for making realistic models of ecological interactions.

3.5.2 Stable Oscillation: Limit Cycles

A limit cycle in a phase space reflects oscillation. It contains periodically recurring states. Technically a limit cycle is an attracting set to which orbits or trajectories converge and upon which trajectories are periodic (Fig.3.11). A stable limit cycle, which attracts neighboring trajectories may imply self-sustained oscillations. A non-damping pendulum and heart beat in resting situation are the most characteristic illustrations, but there are many examples in the literature of applied science from oscillatory chemical reactions via biological clocks and certain electrical circuits to business cycle models.

The van der Pol oscillator, a simple model of both electrical circuits and heart beat was set in 1927. Limit cycles related to the chemical oscillations

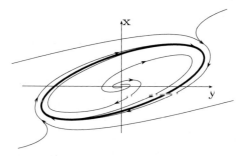

Fig. 3.11. Limit cycle oscillation. The structure of the equations, not the initial values determine the amplitude and frequency of the oscillation.

have been studied extensively in the nineteen seventies. The Brusselator model is an oversimplified model of the Belousov-Zhabotinsky reaction.[23]

There are two components (X and Y) which change their concentrations, and two other components, whose concentrations are kept constant.

The reaction steps:

$$A \to X \quad \text{(zeroth order inflow)}, \tag{3.19a}$$
$$B + X \to Y + C \quad \text{(transformation)}, \tag{3.19b}$$
$$2X + Y \to 3X \quad \text{(cubic autocatalysis)}, \tag{3.19c}$$
$$X \to D \quad \text{(first order outflow)}. \tag{3.19d}$$

Using the *mass action* kinetic rules and some simple transformation the dynamics of the reaction can be described for the dimensionless variables by a two-dimensional ordinary differential equations (2D ODE)

$$\dot{x} = A - (B+1)x + x^2 y, \tag{3.20a}$$
$$\dot{y} = Bx - x^2 y, \tag{3.20b}$$

where x, y are the dimensionless concentrations of the two reactants, and $A, B > 0$

The only fixed point of equation (3.20) is $(x_0, y_0) = (A, B/A)$.

From the characteristic equation of the coefficient matrix

$$\lambda^2 + (A^2 + 1 - B)\lambda + A^2 = 0 \tag{3.21}$$

---

[23] For an excellent brief description of the BZ reaction see http://online.redwoods.cc.ca.us/instruct/darnold/DEProj/Sp98/Gabe/intro.htm.

the eigenvalues can be determined:

$$\lambda_{1,2}(B) = \frac{-(A^2+1-B) \pm [(A^2+1-B)^2 - 4A^2]^{1/2}}{2}. \tag{3.22}$$

If $(A-1)^2 < B < (A+1)^2$, then $\lambda_1(B)$ and $\lambda_2(B)$ are complex conjugates. If their real part is not zero, then (x_0, y_0) is a *focus*. If the real part of the eigenvalues are negative, then it is a stable focus, if not, it is an unstable focus. Stable focus characterizes the damping oscillation.

A periodic solution may emerge when the equilibrium point looses its stability, and the trajectory is still bounded. Bifurcation theory has more power than linear stability analysis. A bifurcation is the qualitative change in the nature of attractor by changing the control parameter(s) of a dynamical system.

The Hopf bifurcations occur when a conjugated complex pair crosses the boundary of stability, and a limit cycle emerges. It has an angular frequency which is given by the imaginary part of the crossing pair (Fig. 3.12).

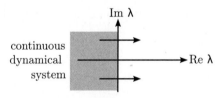

Fig. 3.12. Hopf bifurcation. Pair of complex conjugate of eigenvalues crosses the imaginary axis implying bifurcation to small amplitude limit cycle.

For the Brusselator model the control parameter can be chosen to be B. The real part of the eigenvalues is $\text{Re}(B) = -(A^2+1-B)/2$, and for the critical value $B_c = A^2 + 1$, $\text{Re}(B_c) = 0$ and $\text{Re}'(B_c) = 1/2 \neq 0$ fulfills, so in some neighborhood of B_c a periodic solution emerges. Bifurcation theory also characterizes the emerging oscillation, its amplitude is increasing from zero by increasing the control parameter, while the frequency is finite, and is approximately equal to the imaginary part of the eigenvalue evaluated at the critical point (e.g., [491], pp. 251).

For the Brusselator, the frequency is

$$\omega = \frac{2\pi}{\text{Im}\,\lambda(B_c)} = \frac{2\pi}{A}. \tag{3.23}$$

The Brusselator model is an over-idealized example of chemical oscillations, and chemical oscillations are a kind of building blocks for "biological clocks". A large number of much more realistic models were built and tested by using stability analysis and bifurcation theorems to uncover the *structural assumptions* behind the changing dynamic behaviors.

What are the general conditions of exhibiting limit cycle behavior?[24]

1. *Structural stability.* As opposed to systems showing conservative oscillation, the dynamic behavior should not be destroyed by small perturbation of the system structure or parameters.

2. *Open system.* The system should not be isolated in thermodynamic sense. Generally a system is open, so some chemical component can enter the system.[25]

3. *Feedback* Some kind of feedback is necessary to maintain oscillation. The most direct form of positive feedback is autocatalysis, but indirect effects may be sufficient.

4. *Steady state.* The system should have at least one steady state. An unstable steady state might lead to oscillation, if the trajectory is bounded, confined to a box in the state space.

5. *Limited growth.* If the growth were unlimited, the trajectory would not be bounded, so no oscillation could occur.

3.5.3 Quasiperiodic Motions: A Few Words About the Modern Theory of Dynamical Systems

Andrei Kolmogorov, the leader of the famous Moscow mathematics school developed a general theory of dynamical systems applied to classical mechanics,

---

[24] As many others, I like Edelstein-Keshet's book [143], and am using it for learning and teaching.

[25] Actually, a system may be closed for material flow, but oscillation still may emerge if the system is open to energy flow. The possibility of such kinds of thermokinetic oscillation was published in Russian in 1949, by Salnikov [453] and was reanalyzed several decades later in English [218]. There was a Soviet school on combustion and flame theory labeled by Yakov Zeldovich, who worked on the development of Soviet nuclear and thermonuclear weapons, and Frank-Kamenetsky, who contributed much to rather diverse fields such as chemical engineering and astrophysics.

and combined probabilistic methods with deterministic descriptions. Vladimir Arnold and Jürgen Moser proved a set of theorems that certain quasiperiodic systems were shown to remain stable for small perturbations. The theorems points in the direction that the solar system might be stable. However, orbits of small members of the solar system, such as asteroids and comets, seem to be chaotic. The question is whether or not the major planets' orbits are also chaotic. The answer is not straightforward, but the KAM (Kolmogorov-Arnold-Moser) theory suggests that nearly-integrable Hamiltonian systems (models of the Solar systems might belong to this class) exhibit a remarkable stability. Quasiperiodic systems remain stable for small perturbations.

3.6 The Chaos Paradigm: Then and Now

3.6.1 Defining and Detecting Chaos

Chaotic phenomena are certainly a very important part of the complex systems. In an excellent book [267] Kaneko and Tsuda classified the diversity of dynamical phenomena. Specifically, both low- and high-dimensional chaotic systems were studied. There are other highly recommended books on chaos, again, [491] is suggested for potential students and teachers. This section contains a brief summary, emphasizing several aspects related to the general aspects of complexity issues.

Chaos cannot be confused with randomness. Random phenomena are irreproducible and unpredictable. It is a nonperiodic temporal behavior generated by purely deterministic mechanisms (in reality) and algorithms (in models). In a chaotic region, each trajectory is highly sensitive to initial conditions (i.e., every very small change in initial conditions eventually can yield large divergences in trajectory). The long-term behavior of individual trajectories is practically unpredictable, but the overall behavior of the system – is not! Analogous to stochastic processes in which the probability distributions or density functions can be constructed from random individual realizations, a family of chaotic trajectories can also be analyzed, at least by statistical methods.

Lyapunov exponent

The sensitivity to initial conditions, which is the fundamental property of chaotic systems, is quantified by the Lyapunov exponent. Consider trajectories of a continuous time continuous state space finite dimensional system

$$\dot{x}(t) = F(x(t)), \quad x(t) \in \mathbb{R}^r \tag{3.24}$$

starting near x_0. Prepare a small r-dimensional cube of side ϵ containing x_0, to represent the uncertainty in measurement of initial conditions. Let its volume be $V_r(x_0, \epsilon)$. After time t, let $V_r(x(t), \epsilon)$ be the volume of the smallest (hyper)parallelepiped that contains all the states obtained by system evolution from states in the initial cube. The volume at $x(t)$ divided by the volume at x_0 yields the local r-dimensional Lyapunov exponent by the formula

$$\Lambda_r(x_0, \epsilon, t) = \frac{1}{t} \ln \frac{V_r(x(t), \epsilon)}{V_r(x_0, \epsilon)}. \tag{3.25}$$

Taking the limit as ϵ tends to 0 and t tends to infinity yields

$$\Lambda_r = \lim_{\substack{\epsilon \to 0 \\ t \to \infty}} \Lambda_r(x_0, \epsilon, t) = \lim_{\substack{\epsilon \to 0 \\ t \to \infty}} \frac{1}{t} \ln \frac{V_r x(t)}{V_r(x_0)}. \tag{3.26}$$

Each such number is called a *Lyapunov exponent*. Consider $r = 1$. If the discrepancy $\delta(t)$ in initial conditions were reduced over time with an exponential decay $\delta t \approx \epsilon e^{-\lambda t}$, we would have $\Lambda_1 = -\lambda$, whereas if $\delta(t)$ increased exponentially over time with $\delta(t) \approx \epsilon e^{\lambda t}$, we would have $\Lambda_1 = \lambda$. Thus, the increase of the distance in different directions between adjacent points during motion is characterized by a set of positive Lyapunov exponents (Fig. 3.13).

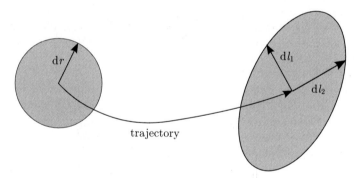

Fig. 3.13. Set of different positive Lyapunov exponents measure the deviation of trajectories into different directions.

A single positive Lyapunov exponent indicates (but does not guarantee) chaos, whereas the presence of more than one positive Lyapunov exponent is associated with hyperchaotic systems [448].

Autocorrelation Function, Power Spectrum

One way of visualizing dynamical properties is to plot the power spectrum of the motion. The power spectrum calculated from a trajectory shows the distribution of frequencies. Specifically, for a discrete dynamical process $x_n = f(x_{n-1})$ the procedure is the following: First the autocorrelation function can be calculated as

$$C_t = \lim_{N \to \infty} \frac{1}{N} \sum_{k=1}^{N} x_k x_{k+t} = \langle x_0 x_t \rangle. \tag{3.27}$$

Second, the power spectrum is derived from the autocorrelation function by the Fourier transform:

$$S(\omega) = c_0 + 2 \sum_{t=1}^{\infty} c_t \cos(\omega t), \tag{3.28}$$

this is called the Wiener–Khintchine theorem. (The whole procedure, mutatis mutandis can be applied to continuous processes, too.)

The spectrum of a simple oscillator is one (vertical) line, while that of quasiperiodic motions can be decomposed into finite number of frequencies (line spectrum). However, chaotic process have a continuous spectrum, which means that all possible frequencies occur. Figure 3.14 shows an example of this illustration.

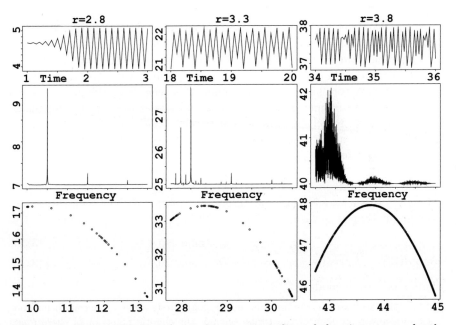

Fig. 3.14. Characterization of periodic, quasiperiodic and chaotic processes by the spectral density and Poincaré plot.

Mixing

It is known from Gibbs' classical statistical mechanics that household mixing is an appropriate analogy for characterizing the approach of conservative systems to statistical equilibrium. Mixing means that not the individual points, but the statistical properties of an ensemble of trajectories starting from random initial values, i.e., the temporal evolution of probability density on the energy surface is investigated. The effect of mixing is that points, located originally nearby, will deviate from each other, the probability density of the states becomes "spread out" over the whole energy surface, i.e., the phase space tends to be characterized by a uniform distribution. Starting from some initial density $\rho(p, x; t = t_0)$ a measure-preserving transformation T_m has the property

$$\rho(T_m^{-1} A) = \rho A, \qquad (3.29)$$

where A is a measurable subset of the phase space. In this sense chaotic processes related to mixing are irreversible. This is true at least in continuous state space systems, as we shall see soon.

Conservative Chaos

There are two types of chaotic processes: conservative and dissipative. In a conservative system there is a transition from *integrability* to *chaos* by changing the numerical value of the control parameter(s). In the KAM system there is a mechanism to collapse the KAM tori implying self-similar structures. While orbits on KAM tori have constant frequencies, chaotic trajectories do not have well defined frequencies, and actions and angles randomly move inside the chaotic zone. Purely integrable and purely chaotic systems occur rarely, the mixture of stable "islands" in the chaotic "sea" form a mixed phase, as Fig. 3.15 illustrates.

One of the most famous models leading to conservative chaos is *Arnold's cat map*.[26] It is a two-dimensional difference equation defined on a torus (i.e., a doughnut-shaped surface) as:

$$x_{t+1} = x_t + p_t \quad \text{mod } N, \qquad (3.30)$$
$$p_{t+1} = x_t + 2p_t \quad \text{mod } N, \qquad (3.31)$$

where mod means modulo.[27]

---

[26] "...There are two reasons for this terminology: first, cat is an acronym for continuous automorphism of the torus; second, the chaotic behavior of these maps is traditionally described by showing the result of their action on the face of the cat..." [45].

[27] Two numbers are said to be congruent modulo N, if divided by N they have the same remainder.

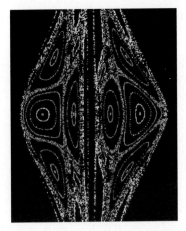

Fig. 3.15. There are periodic, quasi-periodic and chaotic trajectories. The latter can be found in the *fuzzy regions filled up with dots*.

The starting configuration is a two dimensional picture like an image of a cat. Intuitively the mapping materializes the stretching and folding (the second implies contracting) of the system. Loosely speaking if the Arnold map is defined on continuous state space, poor cat will be "smeared out" in the whole state space (and I decided not to include a figure  about the actual state of the cat). However, for the happiness of cats, if N is finite, and the values of x and p might have integer values only, the possible states are also finite. Arnold's cat map for such a finite system is a construction of imitating cyclic world. Such systems have finite (and not long, i.e smaller than $3N$) Poincaré recurrence time. After unfortunate transients states due to a series of stretching and folding, the original form of the cat will been reestablished.

Conservative systems do not really have attractors, since the motion is periodic. For dissipative dynamical systems, however, the measure of the state space shrinks exponentially, and the attractors tend to have zero measure.

3.6.2 Structural and Geometrical Conditions of Chaos: what Is Important and What Is Not?

As (1.2) showed, one variable and a single quadratic nonlinearity is enough to generate (dissipative) chaos, at least in discrete time (but continuous state space) systems. However, in continuous time continuous deterministic systems (i.e., in the world of differential equations) the trajectories of two variable systems (with first order derivatives) tend either to fixed points or closed orbits.[28]

Lorenz derived a three-dimensional ordinary differential equation as an approximation of some partial differential equations known as the equation of motion in hydro- and aerodynamics:

$$\frac{dx}{dt} = a(y - x), \tag{3.32}$$

$$\frac{dy}{dt} = bx - y - xz, \tag{3.33}$$

$$\frac{dz}{dt} = xy - cz, \tag{3.34}$$

with the parameter values $a = 10.0$, $b = 28.0$, $c = 2.66667$. This equation was able to materialize stretching and folding of the points of the state space. However, somewhat surprisingly, I think it is correct to assume that there are no general structural conditions of exhibiting chaotic behavior.

Otto Rössler, a pioneer of chaos research from Tübingen [446, 445, 447], see also [195] used most likely geometrical intuition to design 3D ordinary differential equations (ODEs) systems of chaotic behavior. He constructed a set of equations as a minimal model a of continuous chaotic system:

$$\frac{dx}{dt} = -y - z, \tag{3.35}$$

$$\frac{dy}{dt} = x + ay, \tag{3.36}$$

$$\frac{dz}{dt} = bx - cz + xz. \tag{3.37}$$

The model is minimal in three different senses [195], the number of variables is three, there is only one (quadratic) nonlinear term, and the motion is restricted to one lobe of the state space (as opposed to the Lorenz attractor expanding two lobes).[29] Figure 3.16 shows a Lorenz and a Rössler attractor, respectively.

One way to assign discrete time models for n-dimensional differential equations is to construct a Poincaré map.

---

[28] This comes from the Poincaré-Bendixson theorem. It is based on topological consideration.
[29] The only aesthetic revulsion against this construction is that it was advertised as a model of chemical kinetics, which it is not. For more details see Sect. 4.2.

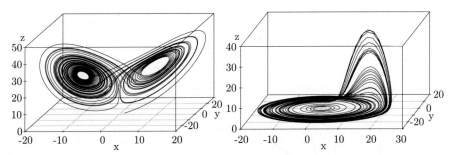

Fig. 3.16. Lorenz attractor has two lobes, the Rössler attractor has one.

Poincaré map

First, the Poincaré sections are generated (Fig. 3.17). Take a hyperplane of dimension $n-1$ transverse to the curve $t \to x(t)$ through x_0. Second, a map $F \mapsto \mathbb{R}^{n-1}$ is induced by associating to t_0 the nearest intersection of the trajectory (with initial condition $x(0) = x_0$) with given hyperplane. The successive intersections generate an $\{x_n\}$ sequence of points, and a $\{t_n\}$ sequence of return times.

If the first such intersection occurs at x_1 we define $F(x_0) \equiv x_1$. The form of F is independent of the index of the series and also of the coordinates, therefore

$$x_{n+1} = F(x_n) \qquad (3.38)$$

and the T return-time function

$$t_{n+1} = t_n + T(x_n) \qquad (3.39)$$

can be specified. From the difference equations associated to the system of differential equations, the vector field of the differential equation, and consequently its properties can be reconstructed.

Another possibility is to derive the next amplitude map (also called Lorenz map) by plotting successive maxima Max_{i+1} vs. Max_i, in the time series, $x(t)$, $y(t)$ or $z(t)$.

There are no clear structural conditions of chaos generation. While a single nonlinearity in the logistic difference equation is sufficient to produce a bifurcation sequence leading to chaos, its continuous time version, i.e., the logistic differential equation, has monotonous trajectories. It is true that to a 3D ODE it is possible to construct 1D discrete time mappings, which capture the essentials of the dynamics, but the procedure is not reversible.

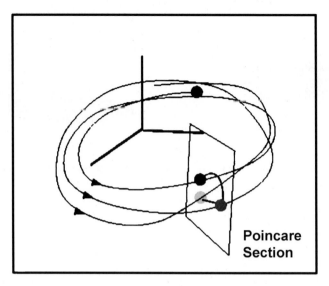

Fig. 3.17. Poincaré section. From http://wwwhome.math.utwente.nl/~geurtsbj/teaching/promo/double_pendulum_background.html.

3.6.3 The Necessity of Being Chaotic

It is nicely said that "a little chaos is essential to the normal operation of machines".[30] Let's return to the mechanical clock, our (not so) simple mechanism. Francis Moon applies nonlinear dynamics [359] to understand machines, even in historical context. Moon set up a theoretical model of clock escapement based on the concept of coupled oscillators leading to chaos. The assumptions of the model:

- Pendulum is modeled by a slightly damping harmonic oscillator.

- A cubic, nonlinear oscillator model coupled linearly to the pendulum equation and describes the impact dynamics in the escapement.

- Driving gear torque and static friction are modeled as a threshold function.

- The driving torque from the weight driven gear train, when released by the nonlinear oscillator, acts to add energy through the escapement pallet when the velocity of the pendulum is positive.

---

[30] http://web.mae.cornell.edu/ugresearch/Moon-Chaotic-Dynamics-in-Clocks.htm, 24 July 2006.

$$\ddot{x}_1 + \beta_1 \dot{x}_1 + \omega_1^2 x_1 + \alpha_1 x_3 = tq(x_3)\,\text{sign}(\dot{x}_1) \tag{3.40}$$

$$\ddot{x}_3 + \beta_2 \dot{x}_3 + \omega_2^2 x_3 + \kappa x_3^3 + \alpha_2 x_1 = 0 \tag{3.41}$$

$$tq(\dot{x}_3) = \begin{cases} T_0 & \text{if } \dot{x}_3^2 > \delta \text{ and } 0 < x_1 < \Delta \\ 0 & \text{otherwise} \end{cases} \tag{3.42}$$

Here $x_1(t)$ is the displacement of the pendulum, $x_3(t)$ denotes the structural connection between the driving train and the escapement. The escapement torque is applied when the amplitude is a in certain region, $0 < x_1 < \Delta$. the noise threshold to release the gears and apply the escapement torque is measured by the constant δ. The model leads to a solution, which exhibits a periodic carrier signal with small high frequency noise riding on the carrier. It was found to be chaotic and "good".

From "An Essay on Clock Repair"

A clock is, in fact, a mechanical device with integral parts that fit together in a predetermined way much like our modern concept of mathematics with its roots in the mechanistic thinking from the time of Sir Francis Bacon, Sir Issac Newton, and Rene Descartes; however an interesting set of circumstances comes into play as a clock runs. The natural world of chaotic events that cannot be fully explained by this mechanistic way of thinking affect the clock as time passes and it runs: bushings wear, springs become fatigued, oil dries out, pivots wear, gear teeth wear, hammer pads wear out and these events do not happen with simplistic predictable regularity because many of them are heavily influenced by human activity which is often quite chaotic. Forces of nature that change air temperature and moisture content in very complex ways also effect the condition of a mechanical timepiece. So, what is in fact a very well organized mechanical device becomes chaotic in nature as it runs and is influenced by the complex array of non-mechanistic events taking place around it – sort of like trees in a windstorm. Consider the question: "If a tree falls in a forest, does it make a sound?". If you are one of those who say immediately, why yes of course, because you can hear it... and we all know you certainly can hear a tree falling in a forest, right? Well you'd better think twice about that one. First, how large is the forest, and how far away are you? Have you ever been in a forest in a huge windstorm? Do you know how much noise there is in a forest in a windstorm? How does a tree fall in a forest? Does someone cut it down? Does it just fall when it is dead? Does it get ripped up by the roots? Have you ever seen a tree fall in a forest in a windstorm? Determining if a tree makes noise when it falls is likely to be the last thing on your mind. Seeing and getting to safety is most likely going to be what you will want to do. Trees often don't hit the ground hard when they fall, because there are other trees around that catch them, and there is so much noise from the windstorm, that there is no way you are going to hear a tree fall unless you are standing right below it. Am I digressing? No. If you can understand what is being said here, then you have one of the main qualities that a clock repair person must absolutely have, and that is the ability to think creatively. The point is, that what seems totally logical and normal does not apply in clocks and clock repair the way we are all used to. One plus one equals two, but not right now, not until the trip cam gets to the 90 degree position and releases the strike gear train.

> http://www.perpetualpc.net/clock_repair_essay.html
> ©2002, David Tarsi

3.6.4 Controlling Chaos: Why and How?

There are several different strategies to control and suppress chaos. The similarities and differences of the approaches can be explained in terms of a few dichotomies, which we list here:

1. *Feedback versus feedforward control.* The processes can be controlled by some feedback mechanism: the controlling signal is determined by the deviation between the actual and the expected behavior. Alternatively, a predetermined strategy can be prescribed to influence the internal operation of the system by some feedforward mechanism.

2. *Model-based versus "data-only" control.* Control is possible with or without the knowledge of the model.

3. *Parameter versus input control.* A system may be controlled internally by adjusting some of its parameter or externally by adding some control function.

4. *Targeting versus control.* In case of targeting (but not in case of control) there is a prescribed goal.

The emergence of the whole field of control of chaos was the result of a single paper [392], which offered a feedback algorithm – now called the OGY algorithm – for stabilizing (not too fast) unstable periodic orbits embedded within a strange attractor which contains an infinite number of unstable periodic orbits – UPO – by making only small time-dependent perturbations in accessible system parameters.

The time-continuous control of chaos by self-controlling feedback (with and without delay) has been offered by Pyragas [416, 417]. The implication of continuous control is that even rapid periodic orbits could also be stabilized, and the level of noise tolerance is also increased.

The adaptive control algorithm (ACA) starts from the model equations of the system exhibiting chaotic behavior

$$\frac{\mathrm{d}x(t)}{\mathrm{d}t} = F(x; p; t). \tag{3.43}$$

The system has a chaotic solution with specific constant p parameter values. The adaptive control algorithm [246] is implemented by specifying the dynamics of the parameter change depending on the difference of the actual and desired states:

$$\frac{\mathrm{d}p(t)}{\mathrm{d}t} = \epsilon G(x(t) - x_s). \tag{3.44}$$

A class of problems of feed-forward control (the term "open-loop" is also used) has been formulated [252, 254]. Having a dynamic system $\mathrm{d}x(t)/\mathrm{d}t = F(x(t))$, where $x(t)$ can be finite-dimensional vector, and there is a prescribed goal dynamics $g(t)$; the problem is to choose an additive control function U to yield the "entrainment"

$$\lim_{t \to \infty} |x(t) - g(t_0)| = 0, \tag{3.45}$$

for all initial values starting from the basin of the entrainment. In the simplest case (e.g., when the goal function is entirely contained in the convergent region of the phase space and some other technically restrictive conditions hold), the control function with the rather simple form

$$U(g(t), g'(t)) = g' - F(g(t)) \tag{3.46}$$

provides the required control.

Another class of feedforward control strategies adopts the (usually weak) periodic perturbation of some parameter [322]. The method starts from the model given by equation (3.43) assuming that it exhibits chaotic dynamics in the range $p' \leq p \leq p''$. Substituting the constant p taken from the inside of the range by some periodic function $p + p\sin(\omega t)$, and giving some restrictive condition for p to force the system to remain within the region of the chaotic attractor, chaotic behavior can easily be suppressed. This method does not need the fixing of some prescribed goal, but the price to be paid is that no guess can be given as to which non-chaotic behavior will emerge as the result of the chaos suppression.

The efficiency of different control algorithms were tested and evaluated in the last ten years. It was noticed with some surprise that [16]: "Interestingly, even now, a decade after the appearance of this area the majority of publications on control of chaos appear in physical journals. On the contrary, the number of papers in the journals devoted to automation and control is small. For example, of more than 1,700 papers presented at the 15th Triennial World Congress of IFAC (Barcelona, 2002) only 10 had the word 'chaos' in their titles".

In Sects. 4.8.2 and 4.9 two examples will demonstrate the applicability of chaos control, even in social systems (in a micro-economic model of competing firms, and in the drug market).

3.6.5 Traveling to High-Dimension Land: Chaotic Itinerancy

Chaotic Itinerancy: The Concept

"Chaotic itinerancy", a sophisticated concept, was proposed as a universal dynamical concept in high-dimensional dynamical systems by Japanese scientists [267, 268].

High (i.e., larger than three) dimensional dynamical systems are not always described by attractors (fixed point attractor, limit cycle, torus, or a strange attractor) known from the studies of low-dimensional dynamical systems. Occasionally high-dimensional (systems with many variables) can reasonably be reduced to low-dimensional systems, but most likely there are phenomena intuitively difficult to perceive.

One of the key related concepts is called an "attractor ruin" (Fig. 3.18). An attractor ruin is said to be "quasi-stable": it attracts in some directions but leads to the onset of instability. The trajectory departs into this unstable direction and can meet another attractor ruin, etc.

Generation of CI: An Example

A well-known model framework of high-dimensional dynamical systems is called the *globally coupled map* (GCM) [265]. In this construction the indi-

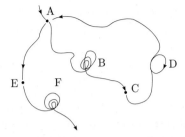

Fig. 3.18. Schematic drawing of chaotic itinerancy. Dynamical orbits are attracted to a certain attractor ruin, but they leave via an unstable manifold after a (short or long) stay around it and move toward another attractor ruin. This successive chaotic transition continues unless a strong input is received. The destabilized Milnor attractor is an attractor ruin, which can be a fixed point, a limit cycle, a torus or a strange attractor that possesses unstable directions. Adapted from [523].

vidual elements are chaotic (say, the function f is the logistic map),

$$x_{n+1} = (1-\epsilon)f(x_n(i)) + \frac{\epsilon}{N}\sum_{j=1}^{N} fx_n(j), \qquad (3.47)$$

where n is the discrete time step, i is the index of an element, and ϵ is the coupling term. The two limiting behaviors are (i) completely synchronized states, which are chaotic, ("coherent phase"), and (ii) completely "desynchronized phase". In "reality", of course there are partially synchronized states, which could be identified as "attractor ruins", and the different clusters are connected by itinerant trajectories.

The Geometry Behind: Milnor Attractors

The Milnor attractor is an extension of the notion of the conventional attractor, which does not exclude the possibility that a trajectory leaves the attractor in the unstable direction. It is defined as a state from which some perturbations of arbitrary small size can kick the system out of orbit although a positive measure of initial points is attracted to it. The Milnor attractor is NOT exceptional! [266] In Milnor's definition [352] both topological and measure-theoretic concepts play roles. Let ρ be a measure equivalent to the Lebesgue measure on a metric space X, on which dynamical flows are defined. A compact invariant set α is called a (minimal) Milnor attractor if the following hold: 1. The basin of attraction $B(\alpha)$ of α has a positive ρ-measure, i.e., $\rho(B(\alpha)) > 0$ and 2. There does not exist a proper closed subset α' satisfying $\rho(B(\alpha) \setminus B(\alpha')) = \emptyset$. Since the Milnor attractor can be connected to unstable orbits that are repelled from the attractor, it provide a mechanism for implementing both transitions from a state and returns to it. This behavior is the feature of chaotic itinerancy.

Chaotic itinerancy could be understood as a sequence of attraction to and escape from a Milnor attractor. Kunihiko Kaneko and Ichiro Tsuda [267, 268] demonstrate and argue that chaotic itinerancy occurring in high-dimensional systems is a very important aspect of complex systems. They extensively studied many details of the underlying complex dynamics, transitions between ordered, partially ordered and truly chaotic phases.

Chaotic Itinerancy: Some Concluding Remarks

While the attractor concept is not applicable in conservative systems, itinerant motion among quasi-stationary states is often observed in a Hamiltonian system with many degrees of freedom.

Itinerant motion over several quasi-stable ordered states offers an alternative interpretation of motions among metastable state, which traditionally explained by random hopping between static states, and were found in optical, electrochemical and biological (including neural) systems. Specifically, the learning capability of neural networks increases in the presence of chaotic itinerancy [522]. For further reading see [267, 268].

3.7 Direction of Evolution

3.7.1 Dollo's Law in Retrospective

The hypothesis, which by-and-large states that evolution is irreversible was formulated by Louis Dollo (1857–1931), a paleontologist. Dollo's law makes the observational statement that new morphological structures established by complex adaptive steps don't resemble an ancestral condition. The law was never understood as an absolute rule, and reversal of isolated characters (such as size) was not excluded. To put it another way, structures or functions discarded during the course of evolution do not reappear in a given line of organisms. If a species becomes extinct, it has disappeared for ever.

There are recent debates about a set of findings, which state the reversal of certain morphological characters, and that way seem to be or are in contradiction with the Dollo's law. Dollo's law, however, is not a strict theorem, and mostly, it was stated much before the theory evolution was integrated with genetics.

Stephen Jay Gould (1941–2002), one of the most influential protagonists of the theory of evolution, liked and analyzed the concepts behind Dollo's law [211]. Dollo (Gould's Dollo) might have used the concept of irreversible evolution in three different senses:

1. An a priori assumption that a whole organism never reverts completely to a prior phylogenetic stage.

2. A testable hypothesis that a complex part of an ancestor never reappears in exactly the same form in a descendant.

3. Certain evolutionary trends are necessarily unidirectional. This interpretation can be attached only to a very few of Dollo's statements. If this thought was in Dollo's mind at all, it played an extremely minor role in his thought on irreversibility.

Günter Wagner originally from Vienna, now Professor of Ecology and Evolutionary Biology at Yale University, formulated a set of hypotheses in the early 1980s which gave a possible explanation of the irreversible evolution [547]. He gave one possible, plausible system of assumptions. Biologically, the prerequisite of the irreversible change is "that the characters of an organism have to be considered as parts of each others environment". In the early 1980's everybody was excited about chaos theory, and it became a fashionable theory for biologists and chemists. Wagner was also motivated by chaotic phenomena, which materializes as a class of irreversibility. Adopting the message (but not the mathematics) of chaos theory he concluded that it was possible to avoid taking a step back.

While probably very few (if any) evolutionary biologists would suggest that long-term evolution would be reversible at all levels of biological hierarchical organization, since a series of steps back would be highly improbable, the reversibility of evolution for particular phenotypes is a controversial issue. Reverse evolution (or re-evolution) can be defined as the reacquisition of the same character states, including fitness, as those of ancestor populations [82] by derived populations. There are a few examples when re-evolution was accomplished under laboratory conditions, and there are few instances where reverse evolution has been observed in nature.

Almost a quarter century later, Wagner saw Dollo's laws somewhat differently [286]. Dollo's law (Wagner's Dollo's law) might have several different versions. First, the whole (evolutionary) trajectory cannot be reversed. Second, the exactly same genotype cannot be re-evolved. Third, assuming the existence of different developmental pathways, in a weak sense the process is re-evolutionary, if an ancestral state is attained by any mechanism. Kohlsdorf and Wagner [286] made a careful analysis of a number of reversals reported recently, and for a number of cases the reversals was confirmed.

There are also debated issues, which need further analysis. "Loss and recovery of wings in stick insects" was reported [556]. Generally it is assumed that once an insect lineage loses its wings, its descendants would remain flightless. It was stated, however, that the wings were lost in the primitive ancestor of stick insects, then reacquired four times during evolution.

What might be the mechanism behind re-evolution? Genetic variation itself does not describe any direction (one might remember that the Newton equations are invariant to time reversal), and the effect of selection might act to any direction. Macroscopic evolutionary reversibility seems to be improbable: "Underlying the assumption of irreversibility of the loss of complex traits is the idea that after a structure had been lost the genes related to its development would degenerate ... and therefore the re-appearance of characters involving complex genetic pathways would be nearly impossible". And the cyclic change of the wings? Trueman et al. are obviously cautious [519], as they titled their article "Did stick insects really regain their wings? Our reappraisal of the evidence convinces us that Whiting et al. have significantly overstated the probability of wing re-evolution in stick insects. When this is taken into account we see no grounds for overturning the traditional view of stick insect evolution... Whiting et al.'s conclusions have already been promulgated in the popular literature (New Scientist, Scientific American, The New York Times)[31] but before this extraordinary evolutionary scenario reaches the entomology textbooks a re-examination is in order." In their answer to the critics Whiting and Whiting [557] state, "Our view of stick insect evolution may change with additional data, but we maintain that the current data and analysis stand as the best supported case for evolutionary recurrence."

3.7.2 Is Something Never-Decreasing During Evolution?

The pseudo-contradiction between entropy-increasing thermodynamics and complexity increasing evolution was not real, since the entropy is increasing in closed, while complexity in open systems.

However, the champions of the new evolutionary theory from the nineteen thirties, who unified evolutionary biology with population genetics by using

---

[31] The original paper was published in Nature.

mathematical models (Ronald Fisher (1890-1962), John B.S. Haldane (1892–1964) and Sewell Wright (1889–1988)) confronted bitterly.

It was certainly common in their way of thinking that they started to accept the view that natural selection acts on genes. Since phenotype traits are generally determined by the interplay of many genes, a single mutation in one gene implies small, and not drastic changes.

Fisher's main contribution is what he called the "Fundamental Theorem of Natural Selection": The rate of increase in fitness of any organism at any time is equal to its genetic variance in fitness at that time.

Wright suggested the "adaptive landscape" concept (Fig. 3.19). The average fitness values are plotted against the allele frequencies, and evolutionary gene frequency changes imply the increase of the average fitness of a population, i.e the system climbs on the hill towards the top during evolution.

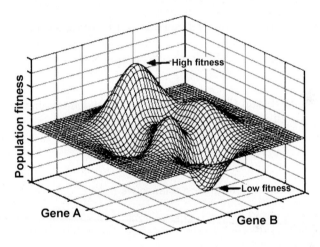

Fig. 3.19. Fitness of different combinations of genes as a hilly landscape, in which the valleys represent less-fit combinations of genes and the peaks represent the fitter one. http://evolution.berkeley.edu/evolibrary/article/_0/history_19.

> It will be noticed that the fundamental theorem proved above bears some remarkable resemblances to the second law of thermodynamics... Both are statistical laws; each requires the constant increase of a measurable quantity...
>
> ... entropy changes lead to a progressive disorganization of the physical world, at least from the human standpoint of the utilization of en-

ergy, while evolutionary changes are generally recognized as producing progressively higher organization in the organic world.

Ronald Fisher

If x_i is the gene frequency of the allele i, $\sum_i x_i = 1$, there is an individual fitness f_i, and the average fitness Φ is defined as $\Phi := \sum_i x_i f_i$, then the variance of Φ is calculated as

$$\mathrm{var}(\Phi) = x_i (f_i - \Phi)^2 . \tag{3.48}$$

Of course, variance measures the "spread" of the individual values. The smaller the variance the more closer are the individual values to each other.

Fisher calculated the time-dependence of the average fitness:

$$\dot{\Phi} = \frac{\mathrm{d}}{\mathrm{d}t} \sum x_i f_i = \sum \dot{x}_i f_i + \sum x_i \dot{f}_i . \tag{3.49}$$

Fisher assumed (and this was a source of misunderstanding for decades)[32] that $\dot{f}_i = 0$, for all i. Fisher derived

$$\dot{\Phi} = \sum \dot{x}_i f_i = \sum x_i (f_i - \Phi) x_i . \tag{3.50}$$

By using the definition of the variance we have

$$\dot{\Phi} = \mathrm{var}(f) . \tag{3.51}$$

Since variance is a never decreasing quantity, any non-zero variance implies the increase of the average fitness.

Fisher knew that his theorem was not about the total change in mean fitness, but rather the change in mean fitness due to natural selection operating in a constant environment. The correct interpretation of Fisher's theorem was given by George Price (1922–1975) [413], see also [182, 144], and led to the formation of evolutionary game theory, what we shall study in Sect. 9.2.2. From ecological perspective Fisher's theorem remained fundamental.

---

[32] For the reevaluation of Fisher's statement see [182, 213]. As the latter says the fundamental theorem may be interpreted as "Non-random changes in gene frequency always change mean fitness positively. No other aspect of the genotypic changes of a population has a systematic optimizing tendency."

3.8 Cyclic Universe: Revisited... and Criticized

The metaphysical idea, that the universe goes through a cycle of Big Bangs followed by Big Crunches, then the cycle repeats, has been reinforced recently by physical arguments. Paul Steinhardt, the Albert Einstein Professor in Science, a physics professor at Princeton University, with his coworker Neil Turok of Cambridge University suggested a new cyclic model of the universe, as a viable alternative hypothesis to the highly successful Big Bang/inflationary scenario.

The Big Bang hypothesis was formulated first in the nineteen twenties, and derived from the discovery that starting from data describing the present state of the universe, and using Einstein's equations of general relativity to extrapolate backward for about 15 billion years, temperature and density become infinite and the space is curled. One of the most important arguments for supporting Big Bang theory came from Hubble, who observed that the universe is continuously expanding, and a galaxy's velocity is proportional to its distance from the Earth. The inflationary theory (Steinhardt himself strongly contributed to its establishment) also explained the changes in the geometry of the universe (stretch, uncurling space-time, and smoothing out its wrinkles.) Inflationary cosmology was suggested by Alan Guth (now the Victor F. Weisskopf Professor of Physics at MIT), and the theory also explains, that superimposed to uniform structure, there are density fluctuations, ripples, found in the cosmic background radiation. The theory states that fluctuations, which had to exist even in the early universe due to the probability character of quantum physics, have been amplified, so they can explain ripples and the large-scale structure of the universe. For one of Guth's popular papers see [224].

An earlier scientific version of the cyclic universe came from Richard Tolman, a mathematical physicist, in the 1930s. He played with the idea of infinite series of oscillation, where each period starts with big bang, and finishes with big crunch (well, "starts" and "finishes" are relative terms). It turned out that the model has a (thermodynamic) time arrow, and cannot avoid the increase of entropy in each cycle, leading ultimately to heat death.

The new cyclic model [484] assumes that the universe is infinite and flat as opposed to Tolman's assumptions (finite and closed), and based on three concepts (as it is written in a "summary of recent progress for students" [485]):

> The intuitive assumptions of the Steinhardt–Turok cyclic universe model
>
> - The Big Bang is not a beginning of time, but rather a transition to an earlier phase of evolution.
> - The evolution of the universe is cyclic.
> - The key events that shaped the large scale structure of the universe occurred during a phase of slow contraction before the bang, rather than a period of rapid expansion (inflation) after the bang.

The details of the cyclic model are much beyond my competence, and I was hesitating whether or not I have the right to include it in this book. Cosmologists use a very sophisticated theory of physics, string theory, "M theory", and the concepts of branes (a short for membranes), which are related to higher-dimensional space-times. According to string theory, not particles, but infinitesimally small strings that vibrate in extra-dimensional space, are the fundamental constituents of matter and energy. The expansion and collapse of the universe does not happen in our normal space, but in an unobserved extra dimension occupied by gravity.[33]

Motivated by the M-theory they assume that the universe consists of two separated four-dimensional space-times (branes). One brane contains the known elementary particles of our world (as leptons, neutrinos, quarks and photons). Other particles lie in the other branes, and influence our particles by gravitational interactions (and not with the so-called weak or strong interactions). So particles on the other brane are so-called a dark form of matter. Two branes might have cyclic motion with a period of many billion years, and they collide occasionally. Functionally, the collision is a big bang. After the big bang, branes moved apart, there was an expansion of matter and energy... and stars and galaxies formed... life evolved, and we are here.

Andrei Linde, a former Soviet physicist, now a professor at Stanford, another founder of inflationary cosmology says that "... instead of being a true

---

[33] I am sure that majority of the readers of this book have read Abbott's Flatland [1]. A Square, an inhabitant of the obviously two-dimensional Flatland describes the geometrical and social organization of life (the latter is a satire of the Victorian society) and his encounters to Lineland, Pointland, and Spaceland.

3.8 Cyclic Universe: Revisited... and Criticized

alternative to inflation, the cyclic scenario is a rather unusual and problematic version of inflationary theory."

The 50-page long paper on the cyclic (also called ekpyrotic) theory appeared on the web in April 2001, "... and ten days later, before any experts could make their judgement, it was already enthusiastically discussed on BBC and CNN as a viable alternative to inflation. The reasons for the enthusiasm can be easily understood. We were told that finally we have a cosmological theory that is based on string theory and that is capable of solving all major cosmological problems without any use of inflation..." [318]

I like Wikipedia's remark on 'Ekpyrotic'

"... Despite these disputes, the ekpyrotic scenario has received considerable attention in the astrophysical and particle physics communities. Regardless of whether it is a correct model of the origin of the universe, it is an excellent indication of the new possibilities opened up by the development of brane cosmology." [34]

---

[34] Note added in the last minute: There is a very recent cyclic model of the universe, the Baum–Frampton model [50].

4
The Dynamic World View in Action

4.1 Causality, Teleology and About the Scope and Limits of the Dynamical Paradigm

The dynamical system is a mathematical concept motivated first by Newtonian mechanics. The state of the system is generally denoted by a point in an appropriately defined geometrical space. A dynamical system operates in time. Typically, we take the time set T to be the real line $\mathbb{R}$ (a continuous-time system) or the set of integers $\mathbb{Z}$ (a discrete-time system). We then formalize an *autonomous* system as a ordered pair (Q, g), where Q is the state space, and $g : T \times Q \to Q$ is a function that assigns to each initial state $x_0 \in Q$ the state $x = g(t, x_0)$, in which the system will be after a time interval t if it started in state x_0. A fundamental property of g, then, is the validity of the identity

$$g(t + s, x_0) \equiv g(s, g(t, x_0)) \tag{4.1}$$

for all states x, and times t, s. Loosely speaking g is a fixed rule which governs the motion of the system.

The behavior of a dynamic system may also depend on the time course of the input applied. Systems that take into account the effect of input belong to the family of *non-autonomous* systems.

Dynamical systems theory became a predominant paradigm, and materialize the philosophical concept of *causality* by mathematical tools. This concept tells that causes imply effects, consequently the present state determines the future. If the fixed rule acting on the actual state is deterministic, there is only "one" future. In case of probabilistic rules, different futures could be predicted with certain probabilities.

For continuous time continuous state deterministic system the equation is given by
$$\dot{x}(t) = f(x(t), k) + I(t); \quad x(0) = x_0. \tag{4.2}$$
It describes the temporal behavior of the system variable x driven by a "forcing function" f, influenced by the parameter k, starting from the initial condition x_0, taking into account that the system is subject of the time dependent external perturbation $(I(t))$ arriving from the environment. The evolution of the system depends on the additive effect of the external perturbation and the internal development.

Robert Rosen, a controversial hero of theoretical biology and of complex systems research suggested [439] that at least three of the four Aristotelian notions of causality can be associated to mathematical objects in equation (4.2):

Aristotelean categories	mathematical objects
material cause	initial values
formal cause	parameters
efficient cause	driving force
final cause	—

"The causality of natural processes may be interpreted as implying that the conditions in a body at time t are determined by the past history of the body, and that no aspect of its future behavior need to be known in order to determine all of them" – wrote Truesdell and Noll in 1965 in their rigorous encyclopedia on the nonlinear mechanics of continua.

4.1.1 Causal Versus Teleological Description

While differential equations (and dynamical systems) describe motions locally, for certain systems there is an equivalent, global description by using integral principles for the *whole* motion. Specifically, a quantity is defined, which should have an extreme (i.e., maximum or minimum) value for the *whole* motion from the beginning to the end. The concept of *optimal processes* reached physics at least four hundred years ago.

Pierre de Fermat (1601–1665), a French lawyer and mathematician (probably more famous for the latter profession) posed and solved a well-defined physical (specifically geometrical optical) problem by mathematical methods.

4.1 Causality and Teleology

How does a light ray move from an initial point to a given endpoint through an optical medium, where the optical density may vary from point to point? There is a set of possible solutions, and Fermat's "principle of least time" gives an answer. While there are infinite number of possible trajectories which connect the two fixed points, there is a selection criterion to choose one of them. The answer suggests that the *time* of the traveling (and nothing else, e.g *not* the geometrical *distance* between the two points) should be minimal.

The "principle of least action" (most likely formulated by Pierre-Louis Moreau de Maupertuis (1698–1759)) is the mechanical analogue of Fermat's principle. There is a mechanical quantity called "action" which should be minimum for the path taken by a particle among the all possible trajectories which are compatible with the conservation of the energy. Maupertuis principle is also an optimality principle. About a hundred years later Hamilton's principle, the most effective optimality principle in physics, has been formulated. The principle of William Rowan Hamilton (1805–1865) states that for mechanical systems there is a quantity, called the Lagrange function, and the integral of this function will be stationary for the actual mechanical process. This function is defined as the difference between the kinetic and potential energy, $L := E_{\text{kin}} - E_{\text{pot}}$. According to the Hamilton's principle:

$$\delta \int_{t_0}^{t_1} L \mathrm{d}t = 0. \tag{4.3}$$

The quantity $\int_{t_0}^{t_1} L$ is the action integral. For the classical mechanical systems the local and global ("teleological") descriptions are equivalent. The mathematics behind the extremal principles of mechanics (and a few other branches of physics) is based on the mathematical disciplines called variational calculus. The Lagrange function satisfies the equation

$$\frac{\mathrm{d}}{\mathrm{d}t}\frac{\partial L}{\partial \dot{x}_i} - \frac{\partial L}{\partial x_i} = 0. \tag{4.4}$$

For the harmonic oscillator $E_{\text{kin}} := 1/2 m\dot{x}^2$ and $E_{\text{pot}} := 1/2 kx^2$, and the action integral is $J = 1/2 \int_0^T (m\dot{x}^2 - kx^2)\mathrm{d}t$, and $\delta J(x) = 0$ is fulfilled when $m\ddot{x} + kx = 0$. This is exactly the Newton equation for the harmonic oscillator.

The equation for the pendulum could be also rewritten in the form of Lagrange equation.

The fact that many (but far from all) dynamical equations of natural process based on mechanisms localized to a single time point are mathematically equivalent to integral description referring to a time interval implies that for this (restricted, but important) class the "causal" description is equivalent with the "teleological" one. The variational principle is not only an equivalent formulation of the local one in the Newtonian mechanics, but has been used in relativity theory and quantum physics, as well.

4.1.2 Causality, Networks, Emergent Novelty

According to the mechanistic worldview, Science, Technology and Metaphysics seemed to be unified by the Newtonian principles. The motion of mechanical machines as well as celestial bodies were thought to be determined by the (same) Laws of Nature. The clockwork world view of Kepler, Galileo and Newton, characterized by causality, determinism, continuity and reversibility, promised to *reduce* all kinds of dynamic phenomena to mechanical motions. At the end of the 18th century chemistry and medicine started to challenge the view that material nature is *nothing but* inert mass and motion. As we know, the invention of the steam engine contributed very much to the disorganization of the cyclic reversible mechanistic world concept and to the birth of the theory of irreversibility.

One of the supporting pillar of the Newtonian world view – paradigm, if you like – is the strict principle of causality, which can be stated as follows:"Every event is caused by some other event" ([83], pp. XXII). The concept of the *linear* causality, which separates 'cause' and 'effect' by a simple temporal sequence, has been found to be appropriate for describing **simple** systems, and could not be considered as a universal concept. *Circular* causality relies on the suggestion that in a feedback loop there is no meaning in separating cause and effect, since they are mixed together. "Circular causality is not just a subcategory of causality, but a concept that supersedes the traditional notions of cause and effect. Hence, these traditional notions no longer apply" ([455]. p. 129). The term "network causality" means the ability to take into account interactions among structural loops. The models of linear causal systems are appropriate subsets of single level dynamic systems.

We assume here that the models of these generalized causal phenomena can be given within the framework of dynamic system theory. In a really excellent, thought provoking book [263] George Kampis identified material causation as a creative agent which causes self-modifications of systems. Dynamical system theory can try to tackle the emergent complex structures and creative processes.

> "Material causation is just a word. It is clear ... that what it means is that *we do not know anything about the causes* that determine things by an invisible and unapproachable 'actor'. The actions of this actor bring forth something new. It has to be taken very seriously that molecules, thoughts, artefacts, and other qualities never existed in the Universe before they were first produced by a material causation. Therefore, irreducible material causation is creation per se: *free construction of new existence, with new properties, i.e., that in no conceivable form pre-exist either physically or logically.* Before they are already there, absolutely no hint can be gained about their possibility, about their properties, about how they come into being, what they will look like and what will happen to them next..."
>
> Kampis, G:
Self-modifying systems in biology and cognitive science.
pp. 257–258.

We will not go further into the metaphysical issues of causality and dynamics. Our real concern in this chapter is that by using the concepts of dynamical systems spatiotemporal patterns observed in natural and socioeconomical systems are supposed to be subject of *causal explanations*. A set of model frameworks and models will be shown to illustrate how structural conditions, interactions among qualities induce predictable or unpredictable temporal, spatiotemporal patterns.

4.2 Chemical Kinetics: A Prototype of Nonlinear Science

The intention of the theory of chemical kinetics is to describe the interactions among the components (species) of a chemical system [164]. The *state* of a chemical system can generally be characterized by a finite-dimensional vector. The dimension of the vector is the number of interacting qualities (i.e., components); the value of the vector describes the quantities of the qualities. A chemical reaction is traditionally conceived as a process during which chemical components are transformed into other chemical components. Stoichiometry investigates the static, algebraic relation of the network of reactions.

Rules of reaction kinetics govern the velocity of the composition change. The most often used rule to prescribe the dynamics of the system is the "mass action law".

The velocity of the reaction

$$aA + bB \xrightarrow{k} cC + dD \tag{4.5}$$

is given by this law as

$$\begin{aligned}
r(t) &= 1/a \frac{dc_A}{dt} \\
&= -1/b \frac{dc_B}{dt} \\
&= 1/c \frac{dc_C}{dt} \\
&= 1/d \frac{dc_D}{dt} \\
&= k(c_A(t))^a (c_B(t))^b ,
\end{aligned} \tag{4.6}$$

where the scalar k is the rate constant characterizing the velocity of the process.

The kinetic behavior of a chemical reactions is traditionally described by a system of (generally) nonlinear differential equations:

$$\dot{\mathbf{c}}(t) = \mathbf{f}(\mathbf{c}(t); \mathbf{k}); \quad \mathbf{c}(0) = \mathbf{c}_0 , \tag{4.7}$$

where $\mathbf{c}$, an m-dimensional concentration vector (m is the number of the components), is the *state* of the system, and the f function, which determines the temporal evolution of the system is determined by the stoichiometry, k is the vector of the parameters, i.e., of the rate constants, and $\mathbf{c}_0$ is the initial value vector of the components.

However, not all kinds of differential equations, not even the special class of ODEs with polynomial right-hand side, can be considered as reaction kinetic equations. Trivially, the term $-kc_2(t)c_3(t)$ cannot occur in a rate equations for referring to the velocity of c_1, since the quantity of a component cannot be reduced in a reaction in which the components in question does not take place. Putting it another way, the *negative cross effect* is excluded.[1]

The deterministic models of classical kinetics are appropriate only when the system can be considered as macroscopic, and even then the deviations from the "average values" remain negligible. A number of situations can be given, when the *fluctuations* are relevant:

- The size of the chemical system is small. In this case the state space is discrete, the continuous approximation is very bad. Discrete state space,

---

[1] Please note that equations (3.34) and (3.37) are not kinetic equations, since they contain a term with negative cross-effect. Even more, orthogonal transformations cannot transform these equations into kinetic ones [516].

4.2 Chemical Kinetics: A Prototype of Nonlinear Science

but deterministic models are also out of question, since fluctuations cannot be neglected even in the "zeroth" approximation, because they are not superimposed upon the phenomenon, but they represent the phenomenon itself.

- The system operates near instability point of a deterministic model. In this case small fluctuations may be amplified and produce observable, even macroscopic effects.

- Fluctuations[2] can be a source of information. The fluctuation-dissipation theorem connect the spontaneous fluctuations around an equilibrium state, and the dissipative process leading to equilibrium. Using this theorem applied to chemical kinetics, rate constants can be calculated from equilibrium fluctuations.

A continuous time discrete state space stochastic model is defined to describe chemical fluctuation phenomena. The concentration, (a continuous variable) should be transformed into a function of the number of components (discrete variable): $c(t) \to V^{-1}\mathbf{n}(t)$, where V is the volume of the system, $\mathbf{c}(t)$ is a concentration vector and $\mathbf{n}(t)$ is the number of components at a fixed time t.

Introducing a stochastic description let ξ be a stochastic vector process, the dimension of which is equal to the dimension of the concentration vector. The *state* of a system is described by the $P_n(t) \equiv \mathcal{P}(\xi(\sqcup) = \mathbf{n})$ probability distribution function. The *temporal evolution* of the distribution is determined by the assumption that the process is considered as a *Markovian jump process*. Markovian character of a stochastic process means that the change of state does not depend explicitly on the past values of the process, so the process does not have memory. the future depends only on the current time step.

The structure of a stochastic model is more complicated than of its deterministic counterpart. The equation for the temporal evolution of the absolute

---

[2] The Scottish botanist Robert Brown discovered the existence of fluctuations when he studied microscopic living phenomena. However, the physical nature of the motion, which was named after its discoverer, was not known for a long time. As Darwin wrote in 1876: "I called on him [Brown] two or three time before the voyage of the Beagle (1831), and on the occasion he asked me to look through a microscope and describe what I saw. This I did, and believe now that it was the marvelous currents of protoplasm in some vegetable cell. I then asked him what I had seen; but he answered me: This is my little secret." The theory of Brownian motion was given by Einstein (1905) [148] and Smoluchowski (1906) [545], who calculated the temporal change of the expectation of the square displacement of the Brownian particle, and the connection between the mobility of the particle and the – macroscopic – diffusion constant.

distribution function (called the *master equation*) is derived from taking into account two types of elementary reactions. The effect of the first class of reaction is that the state j is available from l (l can denote different possible states). The second class describes all of the possible transitions from state j. Therefore we can write:

$$\dot{P}_j(t) = \sum_i \text{transition to } j \text{ from } i \text{ - transition from } j \text{ to } i.$$

Stochastic Chemical Reaction: An Example

Here is a simple example to show the importance of fluctuations. Let us consider the reaction ([164], Sect. 5.6.1):

$$A + X \xrightarrow{\lambda'} 2X$$
$$X \xrightarrow{\mu} 0 \ . \tag{4.8}$$

where A is the external and X is the internal component, and 0 denotes a so-called zero complex. This reaction can be associated with a simple birth-and-death process. The deterministic model is the following:

$$dx(t)/dt = (\lambda - \mu)x(t); \quad x(0) = x_0, \tag{4.9}$$

where $\lambda = \lambda'[A]$. The solution is

$$x(t) = x_0 exp(\lambda - \mu)t. \tag{4.10}$$

If $\lambda > \mu$, i.e., the birth rate constant is greater than the death rate constant, x is exponentially increasing function of the time. For the case of $\lambda = \mu$

$$x(t) = x_0. \tag{4.11}$$

The stochastic model of the reaction is

$$dP_k(t)/dt = -k(\lambda + \mu)P_k(t) + \lambda(k-1)P_{k-1}(t) + \mu(k+1)P_{k+1}(t) \tag{4.12}$$
$$P_k(0) = \delta_{kx_0}; k = 1, 2, ...N. \tag{4.13}$$

There are two consequences of the model:

(1) the expectation coincides with the process coming from the deterministic theory, i.e.,

$$E[\xi(t)] = x_0 exp(\lambda - \mu)t, \quad (4.14)$$

which in case of $\lambda = \mu$ reduces to the form

$$E[\xi(t)] = x_0.$$

(2) the variance of the process is

$$D^2[\xi(t)] = (\lambda + \mu)t. \quad (4.15)$$

For the case of $\lambda = \mu$

$$D^2[\xi(t)] = 2D\lambda t,$$

i.e., progressing with time, larger and larger fluctuations around the expectation occur) Fig. 4.1.

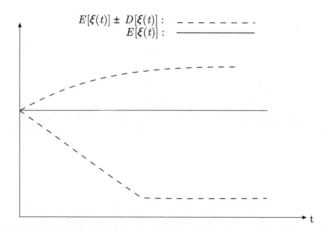

Fig. 4.1. Amplifications of fluctuations might imply instability. While the expectation is constant, the variance increases in time.

It is quite obvious that in this situation it is very important to take the fluctuations into considerations. Such kinds of formal reactions are used to describe the chain reactions in nuclear reactors. In this context it is clear that the fluctuations have to be limited since their increase could imply undesirable instability phenomena.

4.2.1 On the Structure – Dynamics Relationship for Chemical Reactions

A complex chemical reaction (i.e., a *mechanism*) is a set of elementary reactions. Stoichiometry describes the static, algebraic relationship among the chemical components. Since the kinetic differential equations of the chemical reactions have some special structure, it is possible to derive relationship between the structure of the reactions and its dynamics *without solving* the system of differential equations itself. This is the spirit of the "qualitative theory of differential equations". The question was at least to exclude chemical reactions to show "exotic" behavior, which in chemical context means periodicity, chaos or multistationarity. Horn [241] and Feinberg [172, 171] gave negative criteria for the existence of exotic behavior. Here one of their result is described by using their notation.

Zero Deficiency Theorem

Let the chemical components or species of a mechanism be $A(1), \ldots, A(M)$, the *complexes*

$$C(n) = \sum_{m=1}^{M} y^m(n) A(m) \quad (n = 1, 2, \ldots, N). \tag{4.16}$$

Therefore the *complex vectors* of stoichiometric coefficients are

$$y(n) = (y^1(n), \ldots, y^M(n))^T \quad (n = 1, 2, \ldots, N), \tag{4.17}$$

T denotes the transposed vector.

The *elementary reactions* of the mechanism are

$$C(i) \to C(j) \quad (i \neq j; i, j = 1, 2, \ldots, N). \tag{4.18}$$

Let L be the number of the connected subgraphs of the directed graph formed by the complexes as vertices and reactions as edges, i.e., L is the number of *linkage classes*.

The reaction is *weakly reversible*, if the transitive closure of the reaction determined by the above defined directed graph is a symmetric relation.

Let s be the dimension of the *stoichiometric space* S, where

$$S := \mathrm{span}\{y(j) - y(i); C(i) \to C(j)\} \tag{4.19}$$

4.2 On the Structure – Dynamics Relationship for Chemical Reactions

The *deficiency* of the mechanism is

$$\delta := N - L - s. \tag{4.20}$$

According to one of the assertions of the zero deficiency theorem if a chemical system with $\delta = 0$ is weakly reversible, then for mass action kinetics with any choice of positive rate constants the existence, uniqueness and asymptotic stability of positive equilibrium point follows, i.e., the exotic behavior of these systems is excluded.

Examples

Let's see two, slightly different mechanisms, this is mechanism A:

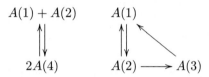

In this mechanism there are four components, five complexes, and the number of linkage classes is two. This is mechanism B:

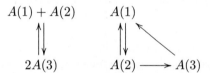

Complex $2A(3)$ replaces complex $2A(4)$ of mechanism A. Both mechanisms are weakly reversible, $N = 5$ and $L = 2$.

However, for mechanism A, $s = 3$, and for mechanism B, $s = 2$. Therefore the deficiency δ is zero for mechanism A, and one for B. The zero deficiency theorem ensures certain stability properties of mechanism A, which is not guaranteed for mechanism B.

4.2.2 Chemical Kinetics as a Metalanguage

The structure and models describing chemical reactions are almost trivial. Chemical kinetics generally takes into consideration binary, and rarely, ternary interactions among the molecules. It is an extensively used procedure to decompose (not only chemical) complex phenomena into binary, or perhaps ternary interactions. Therefore the formal theory of chemical kinetics can be extended to describe transformation phenomena in other-than-molecule populations.

Chemical kinetics is able to describe competition, cooperation and selection among the constituents. A number of biomathematical models at different hierarchical levels, such as prebiotic chemical [146, 147], genetic, population dynamic and evolutionary models lead to the same type of differential equations. The replicator equations [461] have the form

$$\dot{x}_i(t) = x_i(t)\Big(f_i(x(t)) - \sum_{j=1}^{n} x_j(t) f_j(x(t))\Big). \qquad (4.21)$$

The replicator equation and its variations are important mathematical models of evolution, and we shall return to them in Sect. 4.5.

4.2.3 Spatiotemporal Patterns in Chemistry and Biology

A variety of spatial structures such as chemical fronts or waves, periodic precipitates and stable spatial patterns have been subject of studies, both experimental and theoretical, for decades. It is widely accepted that mechanisms responsible for patterns and order in a reaction-diffusion system may also play a fundamental role in understanding certain aspects of biological pattern formation. Though it is evident that chemical reactions and diffusion processes are everywhere and always present in biological systems, one question seems to be difficult to answer completely: are the pattern forming reaction-diffusion systems the real basis of biological morphogenesis or do they offer some analogy to obtain insights into the mechanism of pattern formation? How are patterns of sea-shell forms and of animal coasts (zebras, giraffes, lions etc.) formed?

Since chemical reactions and diffusion processes are the "consequences" of the interactions among chemical constituents, chemical patterns may be interpreted at the molecular level. In the first example, stationary spatial chemical patterns (i.e., the so-called Turing structures) are reviewed. Then we follow with a general theory of biological pattern formation applied among others to the regeneration of Hydra's head after its removal. Finally a very important phenomenon, somitogenesis, a challenging problem of biological morphogenesis is discussed.

Turing Structures

> Alan Turing(1912–1954) From T. machines to T. structures. Computer scientists remember Alan Turing for his fundamental contribution around 1936 to theory of computability. Cognitive scientists celebrate his paper of 1950 "Computing machinery and intelligence". For the biologists (I guess, mostly for mathematical biologists), Turing's main achievement is the 1952 Royal Society paper "On the chemical basis of morphogenesis". And of course, he contributed very much to break the secret codes of the Germans during the war. Probably the different fields he worked can be interconnected: that algorithms corresponding to local mechanism of interacting components produce ordered structures.

Turing wanted to show the possibility of the emergence of spatially inhomogeneous (but temporally stationary) stable structures starting from the perturbation of (spatially) homogeneous structures. He constructed a model which he thought to be a reaction diffusion system in which there exists a spatially stable temporally homogeneous stationary state which loses its stability as a result of inhomogeneous perturbations.

Turing's example was

$$\dot{x} = 5x - 6y + 1 + D_x \Delta x, \qquad (4.22)$$
$$\dot{y} = 6x - 7y + 1 + D_y \Delta y. \qquad (4.23)$$

This model, where D_x and D_y are diffusion constants, and Δ is the second spatial derivation (Laplacian) operator, was able to produce the formation of stable spatial structures and influenced the way of thinking on chemical morphogenesis, though main stream biology neglected it for decades. The importance of this paper, and the relevance of *diffusion-driven instability* has been recognized by theoreticians. It influenced people in all the three schools (catastrophe theory, dissipative structures, synergetics) very much. It is obviously an abstract model, such numbers as 5, and 6, are not realistic in terms of elementary mechanisms. Furthermore, and more importantly, a term $-6y$ on the right hand side of the first equation, i.e describing the change of x, materializes "negative cross effect". Equation (4.23) is not a system of differential equations of chemical kinetic models.

It was shown a few decade later [502, 503] that the presence of cross-inhibition (i.e., $(\partial f_i / \partial x_j)(c) < 0$) is a necessary condition of Turing instability.

This result implies that the presence of higher than first order reactions is a necessary condition of Turing instability.

However, no well defined experiments had been made until the late eighties. All the systems exhibiting spatial patterns either contained convection or surface effects, thus the origin of pattern formation has never been pure Turing instability. It was putting the CIMA reaction into a carefully designed 'gel ring reactor' [384] (in Patrick DeKepper's group in Bordeaux) [92] which is generally considered to have produced the long-sought-for result first: the emergence of stationary patterns as a result of diffusion driven instability. Lengyel and Epstein [308] were able to illuminate an important condition of generation of Turing structures: it should be a great difference between the diffusion constants of the various species, and it is provided by the starch indicator present in the CIMA system.[3] Detailed analysis showed that the same dynamical system may lead to different spatial patterns, such as hexagonal blobs and stripes, see Fig. 4.2.

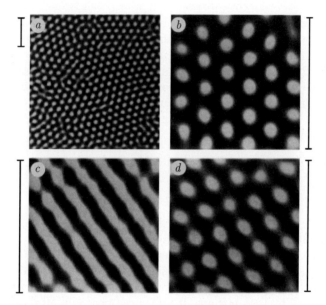

Fig. 4.2. Different spatial patterns, such as blobs an stripes occur in the same system for different parameter values. Adapted from [393].

---

[3] For a thorough mathematical analysis of the CIMA system, see [383].

A Generative Principle of Biological Pattern Formation: The Gierer–Meinhardt Model Framework

A family of models was developed by Hans Meinhardt and Alfred Gierer based on the interaction of two types of (partially hypothetical) morphogen molecules (activators and inhibitors) and the assumption (which was already assumed by Max Delbrück, whom the reader already knows well from previous chapters). It was assumed that the interaction for different initial arrangements and parameter values may lead to different patterns [349].

Suppose that there are two types of diffusible chemicals, an activator (a) and an inhibitor (h), that are produced at the same place in the animal. Both depend on the coordinate x and time t. The rate of change of a is:

$$\frac{\partial a}{\partial t} = \rho + k\frac{a^2}{h} - \mu a + D_a \frac{\partial^2 a}{\partial x^2}, \tag{4.24}$$

where ρ is the production rate, ka^2/h expresses that the generation of a is an autocatalytic process which is hindered by the inhibitor h, μ is a decay constant and D_a is diffusion constant.

The inhibitor has a decay in time and it can diffuse as well, but its generation is triggered by the activator:

$$\frac{\partial h}{\partial t} = ca^2 - \nu h + D_a \frac{\partial^2 h}{\partial x^2}. \tag{4.25}$$

Of course, the details might be subject of modifications, the important thing is that the system should be locally unstable and globally stable.

Models and Reality. A Case Study: Somitogenesis

Somitogenesis is the segmentation process in vertebrates and cephalochordates[4] which produces a periodic pattern along the head-tail axis of the embryo. Somites are formed by the successive segmentation of the presomitic mesoderm so that the first somite is generated at the most cranial end of the embryo and segmentation propagates caudally, to the direction of the tail (Fig. 4.3a, b). Somites are than divided by a fissure into anterior (A, front) and posterior (P, back) halves that differ in their gene expression and differentiation (Fig. 4.3c). Further differentiation of somitic cells leads then to the formation of bones, musculature and connective tissue of the skin.

---
[4] Fishlike animals having a flexible, rod-shaped body instead of true spinal column present in vertebrates. Cephalochordates are probably the closest living relatives of the vertebrates.

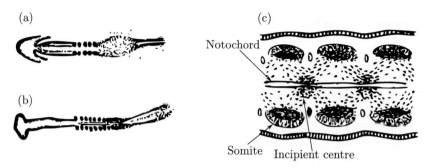

Fig. 4.3. Somitogenesis in chicken embryo. Formation of somites starts at the cranial side of the embryo (**a**) and proceeds in caudal direction (**b**). (**c**) For the formation of vertebrae, cells of the anterior part of the somite migrate in cranial while cells of the posterior part migrate in caudal direction. Adapted from [401].

The above mentioned mechanism of pattern generation in somitogenesis is a strongly investigated but still an unresolved problem in developmental biology. Two of the existing models will be reviewed in the following paragraphs.

Reaction–Diffusion Model

Meinhardt's somitogenesis model is derived from his former reaction–diffusion models of morphogenesis [349]. These models are able to produce structures with spatial peridocity that are reminiscent of the periodic anterior-posterior half-somite pattern in somitogenesis. We introduce such a reaction-diffusion model first.

Suppose that there are two types of diffusible chemicals, an activator (a) and an inhibitor (h), that are produced at the same place in the animal. Both depend on the coordinate x and time t. The rate of change of a is:

$$\frac{\partial a}{\partial t} = \rho + k\frac{a^2}{h} - \mu a + D_a \frac{\partial^2 a}{\partial x^2},$$

where ρ is the production rate, $k\frac{a^2}{h}$ expresses that the generation of a in an autocatalytic process which is hindered by the inhibitor h, μ is decay constant and D_a is diffusion constant.

The inhibitor has a decay in time and it can diffuse as well, but its generation is triggered by the activator:

$$\frac{\partial h}{\partial t} = ca^2 - \nu h + D_a \frac{\partial^2 h}{\partial x^2}.$$

4.2 On the Structure – Dynamics Relationship for Chemical Reactions

Figure 4.4 shows the periodic structure that this interplay between activator and inhibitor leads to.

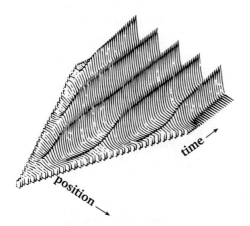

Fig. 4.4. Spatial structure generated as the result of the interplay between activator and inhibitor. Adapted from [228].

In Meinhardt's model for somitogenesis a similar periodic pattern is generated first which models the generation of anterior-posterior (A-P) half-somites. Somites are formed only after this by the separation of subsequent A-P pairs (so that one somite will consist of one A-P pair). This is in agreement with experimental observations in the so called bithorax (Drosophila) mutant where it was found that the segmental specification of the insect can be disturbed without changes in the A-P pattern. This suggests that the formation of A-P pattern is the primary and somite formation is the secondary event indeed.

According to Meinhardt's model [347, 348, 346], a presomitic cell can be in two different states: either gene A or gene P is active which leads to the production of substance A or P, respectively. In the former case the cell will become an anterior while in the latter case a posterior somitic cell. The state of a cell can be influenced by the neighboring cells: if they produce substance A, P production will be reinforced and vice verse. The mechanism of this mutual reinforcement can be imagined so that the substances diffuse to the neighboring cells where they function as transcription factors: A activates the gene of P whereas P triggers transcription from the gene of A. Under these conditions small random initial fluctuation in the level of A or P gene activation would be sufficient to generate the periodic A-P pattern (Fig. 4.5a).

However, the resulting pattern is somewhat irregular and the emergence of A and P stripes occurs parallel along the whole axis of the embryo instead of being subsequently generated from cranial to caudal direction. In order to get around these problems a further feature of the cells is introduced, the

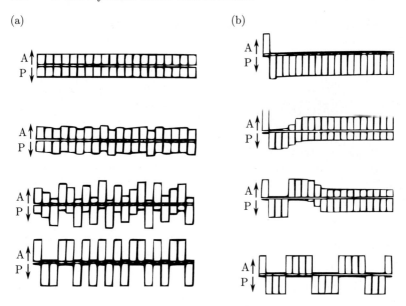

Fig. 4.5. Reaction-diffusion model of segmentation. (**a**) A small random initial difference in the A and P gene activation in the cells automatically leads to the formation of a more or less regular A-P pattern. (**b**) Regularity of the pattern can be increased by a slight modification of the model. At the beginning all the cells, except for the most anterior ones, switch from state A to P because of the assumed morphogen concentration gradient. Cells near to the A-P border are mutually stabilized while cells far from it switch back to state A. This process continues until a highly regular A-P pattern is generated throughout the whole axis of the embryo. Adapted from [348].

ability to oscillate: if a cell is surrounded mainly by cells of the same state, it switches to the other state as explained above. If the neighboring cells switch their states as well, the process can be repeated and cells will oscillate between states A and P.

According to the improved model, at the beginning of the segmentation process all presomitic cells are in state A and there is a morphogen concentration gradient in the embryo increasing from cranial to caudal direction. Cells exposed to a morphogen concentration above a certain (low) threshold switch from state A to P by which the first A-P border is born (Fig. 4.5b). A and P cells near the border stabilize each other while all the farther P cells switch back to state A, because of their oscillating tendency, generating a new P-A border. This A-P stripe forming process continues until it reaches the caudal end of the embryo. By this mechanism a caudally propagating and highly regular A-P pattern can be generated (Fig. 4.5b).

4.2 On the Structure – Dynamics Relationship for Chemical Reactions

The next step in somitogenesis is the generation of sequential pattern of somites: although somites are similar, they also differ from each other, so a mechanism is needed that distinguishes between subsequent somites. The borders of somites has to be exactly at the P-A borders (each somite containing one A-P pair), that is the mechanism has to ensure the precise superposition of the periodic (A-P) and the sequential pattern. These features can be easily explained by Meinhardt's model: it is a property of the model that the number of A-P oscillations a cell has made correlates with its position along the anterio-posterior axis (Fig. 4.5b). Assuming that every switching from state P to A activates a new gene, different sets of genes will be active in different AP pairs, which can cause the differences between somites. By this mechanism somite borders would automatically coincide with P-A borders.

Cell Cycle Model

In the late 1980s observations from single heat shock experiments suggested that cell cycle correlates to somite segmentation [415]: heat shock to chick embryos resulted in anomalies separated by constant distances of six to seven somites, the number of somites that develop during 9 hours which is the duration of the cell cycle. As presomitic cells leave the Hensen's node (where they are derived from) and settle along the anterioposterior axis of the embryo strictly in the order in which they were formed (Fig. 4.6), the observation was explained by that heat shock affects an oscillatory process (the cell cycle) within the presomitic cells. This view is supported by further observations as well: similar periodic anomalies can be caused by drugs inhibiting cell cycle progression [415], and there is some degree of cell cycle synchrony between cells that are destined to segment together to form a somite [490].

Based on these experiments Stern et al developed the "cell cycle model" [490]: P1 and P2 denote two time points in the cell cycle, about 90 min apart (Fig. 4.6). Cells that reach P2 are assumed to start to produce a signal to which cells, whose cell cycle is between P1 and P2, will respond by increasing their adhesion to each other (after this they become unable to signal). As mentioned above, cells along the embryonic axis are located according to their time of formation so neighboring cells are in similar cell cycle state. Therefore cells between P1 and P2 are close to each other and due to the adhesion molecules they produced they will aggregate and start to form a somite.

Collier et al. [102] proposed a mathematical formulation of this model: the system is described by two state variables, the concentration of the adhesion molecule ($g(x,t)$) and the signal molecule ($s(x,t)$). The model equations are:

$$\frac{\partial g}{\partial t} = \frac{(g+\mu s)^2}{\gamma + \rho g^2}\theta(x,t) - \eta g,$$

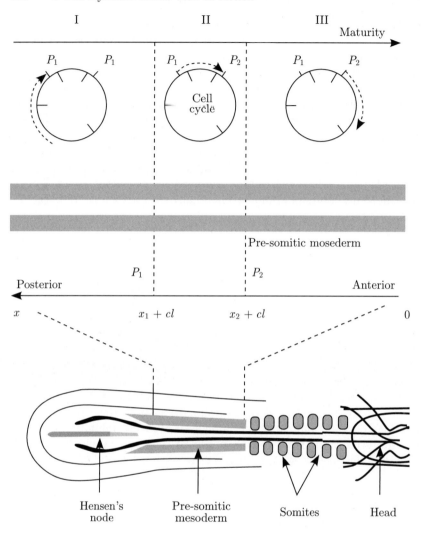

Fig. 4.6. Cells cycle model of somitogenesis. Cells, whose cell cycle is between P1 and P2, aggregate to form a somite. Adapted from [102].

$$\frac{\partial s}{\partial t} = \frac{\kappa}{\epsilon + g}\chi(x,t) - \lambda s + D\frac{\partial^2 s}{\partial x^2},$$

where

$$\theta(x,t) = H(ct - x + x_1),$$
$$\chi(x,t) = H(ct - x + x_2).$$

H is the Heaviside function, x_i is the position of P_i on the x axis when $t = 0$, for $i = 1, 2$, thus g and s production can only occur in cells that passed P_1 and P_2 state, respectively. The $(g + \mu s)^2/(\gamma + \rho g^2)$ term in equation X expresses that production of g is autocatalytic, it is enhanced by s and saturates for large g whereas $-\eta g$ represents linear degradation. The production of s is inhibited by g ($\kappa/(\epsilon + g)$), its concentration also decays linearly (λs) and it diffuses along the axis (last term).

Figure 4.7 shows the numerical solutions of the equations. At the position of $P2$, a peak in s occurs because cells start to produce the signal molecule s here (Fig. 4.7b). s diffuses rapidly so there is a decreasing level of s in the neighboring positions too. s triggers the g production in cells between $P1$ and $P2$ which in turn inhibits s production in these cells. Due to this mechanism a wavefront of g is propagating down the axis in jumps which ensures the aggregation of groups of neighboring cells which in turn leads to somite formation.

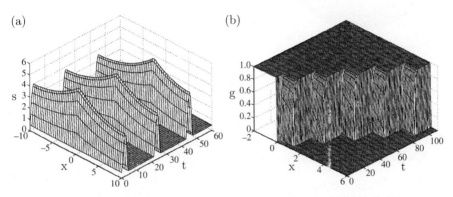

Fig. 4.7. Numerical solution of the cell cycle model. Peaks of signal molecule production (**b**) trigger an abrupt increase in the production of the adhesion molecule (**a**) which leads to cell aggregation and somite formation. Adapted from [102].

Chemical Waves

Zaikin and Zhabotinsky [578] found not only the spontaneous appearance of periodic temporal and spatial patterns in an initially homogeneous chemical systems, but wave phenomena were also demonstrated. Arthur Winfree (1942–2002) [564] discovered the existence of spiral waves experimentally and basically explained them mathematically, based on the partial differential equations of the reaction-diffusion systems. Figure 4.8 shows a typical spiral wave emerging in chemical systems. Spiral waves in an other excitable media, actually in cardiac muscle, was studied even by founders of the cybernetics Wiener and Rosenbluth in 1946, [559] Wave propagation in chemical systems now seems to be well-controllable [277]. Interestingly, it was shown [364] that by the aid of in excitable media logical gates are emulate, which may serve, as element of information processing.

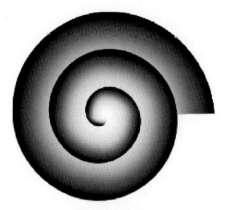

Fig. 4.8. The emergence of spiral waves in chemical medium is a prototype of nonlinear spatiotemporal phenomena.

4.3 Systems Biology: The Half Admitted Renaissance of Cybernetics and Systems Theory

4.3.1 Life itself

Systems biology is an emergent movement to combine system-level description with microscopic details. It might be interpreted as the renaissance of cybernetics and of system theory, materialized in the works of Robert Rosen (1934–1998) [438, 440]. In an excellent review Olaf Wolkenhauer [567] explained how the concepts of systems theory, and of cybernetics were applied by Rosen to biology, and how his ideas returned now under the name of systems biology.

Rosen gave a formalism, which connected *phenotype* (i.e., what we can observe directly about an organism) and *genotype* (the genetic makeup). In particular, phenotype is interpreted as being "caused" by genotype. He also argued that to understand biological phenotype, in addition to the Newtonian paradigm, the organizational principles should be uncovered. He realized that a crucial property of living systems, that while they are thermodynamically open systems, organizationally they should be closed. To put it in another way, all components, which are subject of degradation due to ordinary wear and tear, should be repaired or resynthesized within the cell. Rosen gave a mathematical framework to show how it is possible to do. The original treatment use a branch of mathematics called *category theory* and will not repeated here. We restrict ourselves here to discuss briefly the question why cells might be considered as self-referential systems.

4.3.2 Cells As Self-Referential Systems

Robert Rosen, by analyzing machines and organisms noticed that a main difference is that organisms not only make and reproduce themselves but are also able to repair themselves. Rosen gave the formal framework of what he called metabolism-repair or (M,R)-system. Recently, Athel Cornish-Bowden, a British biochemist in Marseilles (yes, we live in global world, don't we?) and his coworkers reanalyzed Rosen's results [312, 112], and attempted to transfer to the community of biochemists. What is the message to be transferred? Traditionally, in cell biology, enzymes are considered as proteins which catalyzes the metabolic conversion of substrates to proteins. These (and any other) proteins, however, have also finite life-times, so they should be resynthesized. But this synthesis also needs enzymes, so there is a infinite regress. Rosen's central result was to show that logically it is possible to avoid this infinite regress, and the essence of cell is the existence of *organizational closure*, see Fig. 4.9. The operation of the cell is controlled *internally*.

> About Rosen:
> His work is almost totally unknown to biologists, but it is essential for placing in a broader context the idea of understanding the parts of a system in terms of the whole. He tried to analyze metabolism in terms of what he called metabolism-repair systems, or (M,R)-systems. "Repair" was an unfortunate choice of term, and what he meant by it was not repair but resynthesis. In other words, his (M,R)-systems were an attempt to give mathematical expression to the ideas ... where enzymes are explicitly considered as products of metabolism.
> From [312].

I belong to that camp, whose members believe that Rosen was way ahead of his time. However, he offered a purely functional theory, and did not try to give any structural realization of the (M,R) system. Therefore biologists less prone to abstract thinking could not prove or falsify his hypothesis. In any case, some people in the systems biology community now give credit Rosen's pioneer work.

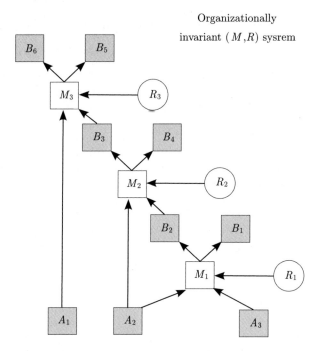

Fig. 4.9. Logical structure of the (M,R) system. An M is a general enzyme which converts A substances to B products. The M enzymes are subject of natural degradation. They could be repaired (resynthesized) internally. Figure based on [312].

Biological Complexity (To Be) Explained

Biological complexity, as we have already seen, has different roots. The fundamental components of living beings, i.e., cells are organizationally closed, as Rosen discussed. Biological systems, however, are open from the perspective of material, energetic and information flow.

Robert Rosen [441] recalls the physicist's approach, which denies that the mind can be the object (or subject) of legitimate scientific study, since it cannot be identified with objective reality. Rosen's analysis points out that this kind of objectivity is narrowly understood and based on mechanistic notions. He also remarks that biologists adopt a more narrow concept of objectivity: it should be independent not only from perceptive agents, but also from the environment: to explain wholes from parts, that is "objective", but parts in terms of wholes, that is not. To put it another way: closed causal loops are forbidden in the "objective" world. Rosen's conclusion is that the world of systems determined by linear (and only linear) causal relationships belongs to the class of "simple systems" or mechanisms. The alternative is not a "subjective" world, immune to science, but a world of complex systems, i.e., one which contains closed causal loops.

Leaving the clear and well-organized world of linear causal systems we find ourselves in the jungle of the second order cybernetics (Sect. 2.2.2). Biological systems contain their own descriptions, and therefore they need special methods or "at least" special language.[5] It is rather obvious that, despite the methodological success of the analytic sciences, the marvelous complexity of life cannot be explained completely in terms of physics. I think, the main methodological problem what we should confront is to understand what dynamical systems can and cannot do. Whether or not the framework of dynamical systems is sufficiently rich to tackle novelty generations, emergence of complexity. I have to tell the kind Linear Reader that the intention of the whole book is to show how well dynamical systems can be applied to understand the emergence of complexity, but its limit also will be analyzed.

4.3.3 The Old–New Systems Biology

As opposed (better yet, complementary) to molecular biology, the systems biological approach emphasizes the *integration* of components (mostly proteins and genes) by dynamical models. Slightly modifying Kitano's approach [283] I believe that systems level understanding requires the integration of five basic properties.[6]

Architecture. The structure (i.e., units and relations among these units) of the system from network of gene interactions via cellular networks to the modular architecture of the brain are the basis of any system level investigations.

Dynamics. Spatio-temporal patterns (i.e., concentrations of biochemical components, cellular activity, global dynamical activities such as measured by electroencephalogram, EEG) characterize a dynamical system. To describe these patterns dynamical systems theory offers a conceptual and mathematical framework. Bifurcation analysis and sensitivity analysis reveal the qualitative and quantitative changes in the behavior of the system.

---

[5] Such kinds of languages were offered by Maturana and Varela [331] speaking about *autopoiesis*. Autopoiesis is term to describe the basic properties of living systems, as their ability to self-created.

[6] While nowadays systems biology generally not incorporates the brain, we don't see any reason to exclude it [162]. In Sect. 8.5.2 the application of the systems biological perspective to neuropharmacology will be briefly presented.

Function. This is the role that units (from proteins via genes, cells and cellular networks) play to the functioning of a system (e.g., our body and mind).

Control. There are internal control mechanisms which maintain the function of the system, while external control (such as chemical, electrical or mechanical perturbation) of an impaired system may help to recover its function.

Design. There are strategies to modify the system architecture and dynamics to get a desired behavior at functional level. A desired function may be related to some "optimal temporal pattern".

I see systems biology as a relation of these five properties as Fig. 4.10 shows.

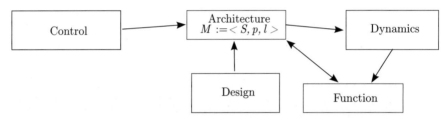

Fig. 4.10. Interdependence of the key properties of biological systems. Architecture is defined in a general sense. Structure is supplemented with parameters and initial conditions to specify a model. The role of control is to shift the system dynamical state to a desired one.

Systems biology adopts a set of different modeling techniques from deterministic and mostly stochastic chemical kinetics to more phenomenological description of both genetic and biochemical networks [68, 135]. For the phenomenological description the prototypical example is the random Boolean network suggested by Stuart Kaufman. While the original idea was suggested in 1969 [269], the model framework has been subject of extensive studies from the mid-eighties, when Kaufman moved to the Santa Fe Institute (and in any case, computers with large capacity appeared).

4.3.4 Random Boolean Networks: Model Framework and Applications for Genetic Networks

The Structure of the Model

Kaufman introduced the so-called "NK model" in the context of genetics and evolution [269, 270]. Its generalization became a quite generic dynamical system, which is able to show regular and irregular behaviors, and was applied not only in biology, but also in many discipline from computer science to social dynamics. There is an excellent review of the model framework[8]. The model is called NK, since there are N elements, each of them interacting with K others. The model is Boolean, since the N variables are Boolean, i.e., they take values 0 and 1. Formally it is a discrete time binary state model, the dynamics is defined by the updating rule

$$\sigma_i(t+1) = f_i(\sigma_{j1(i)}(t), \sigma_{j2(i)}(t), ..., \sigma_{jKi(i)}(t)), \qquad (4.26)$$

Where $\sigma_i(t)$ is the value of an element.

To specify the dynamics we should know the number of elements which influence all the ith variables, (connectivity and linkages, respectively) and the updating rule f. While in the generalized model framework. the number of influencing variables may be different for each node, The Kaufman model assumes that the number of connections are the same, K. For variable connectivities it is possible to define the average connectivity of the system $<K> = 1/N \sum_{i=1}^{N} K_i$.

The dynamic behavior of the system strongly depends on the structure of the coupling. It is possible to assume some topology among the variables, say, they form a hyper-cubic lattice, and one may assume that the state of the elements depend on their somehow defined neighbors. Alternatively, the dependence may be chosen from a uniform distribution, so each element has equivalent chance to influence the others. This random uniform Boolean network is called as the Kaufman net. (One may define coupling, which is intermediate between the lattice-like and the purely random, i.e., "small world" networks, see Sect- 7.4.1).

The number of coupling functions is 2^{2^K}, so for $K = 3$ there are 256 functions. The set of functions can be classified into groups. E.g., there is a set of *canalizing functions*, in which the value of the functions is determined by one of its argument only. Say, the value of the $f(\sigma_1, \sigma_2, \sigma_3)$ might be 1, for $\sigma_1 = 0$, and 0 for $\sigma_1 = 1$, independently of the values of σ_2 and σ_3.

Another possibility is to use weighted f_is, i.e., the functions are weighted with probabilities p and $1-p$.

Characterization of the Dynamic Behavior

Information flow and phases

One characteristic property of the dynamics is its sensitivity to initial state. Assume two initial states:

$$S(t) := [\sigma_1(0), \sigma_2(0), ...\sigma_N(0)] \quad (4.27)$$

$$S^+(t) := [\sigma_1^+(0), \sigma_2^+(0), ...\sigma_N^+(0)]. \quad (4.28)$$

A time-dependent measure for the (of course time-dependent) distance is defined as

$$D(t) := \sum_{i=1}^{N} \left(\sigma_i(t) - \sigma_i^+(t)\right)^2. \quad (4.29)$$

Let's assume that the distance between the two initial state is small. If the information flow is localized, the distance remain relatively small (the system is said to be in *frozen* state). In certain situations $D(t)$ can diverge for large times, so the information can be transferred to the whole system (the system is said in *chaotic phase*).

Simple assumptions lead to

$$D(t) = D(0)exp[t\ln(0.5K)]. \quad (4.30)$$

The distance is measured by the Hamming distance.[7]

The analysis of this equation implies that three different phases occur by varying the value of the connectivity K:

- *frozen*: if $K < 2$ (so $K = 0$ or $K = 1$), the Hamming distance is exponentially decaying function of time.

- *chaotic*: if $K > 2$, the Hamming distance increases exponentially with time.

- *critical*: $K = 2$, the system's behavior critically depends on fluctuations.

---

[7] Hamming distance between two strings of equal length is the number of positions for which the corresponding symbols are different.

4.3 Systems Biology

This critical behavior became famous (and as it often happens with too successful notions, a little bit infamous), as the "edge of chaos". But first let's speak about an other characterization of the dynamics.

Cycles

The state space is finite, and contains 2^N configurations. Starting the system from any initial condition there is a cyclic attractor, where the system tends. The different cyclic attractors are characterized by their length and the size of the basin of attraction. The different phases might have very different cycles. The frozen phase has relatively short cycles, and short transients. The chaotic phase contains long cycles and long transients. The cycle length growth as a power of the size. In the critical phase these dependencies are algebraic, and it was found that specifically it is proportional to the square root of N. This result seemed to support the applicability of the whole model framework, since such kinds of behavior was found in biological experiments, what we will review soon. It might be the case, that the dependency is linear, as it was suggested several years ago [62], and the situation is under dispute.

> The structure of the nodes is very important for the dynamics of Random Boolean Networks. The descendants of a node are the nodes that it affects, while the ancestors of a node are those that affect it. To have cycle attractors, i.e., of period greater than one, there should be at least one node that will be its own ancestor. A circuit of auto-activating nodes is called a linkage loop, and when there is no feedback, linkage trees are formed. Note that loops spread activation through trees, but not vice verse. The relevant elements of a network are those nodes that form linkage loops, and do not have constant functions, for these cause instabilities in the network, which might or not propagate. Note that as there are more connections in a network (higher K), the probability of having loops increases. Therefore, finding less stable dynamics for high values of K is natural. From [86].

Edge of Chaos?

The critical phase separates the frozen (ordered) and the chaotic (disordered) phase. The verbal hypothesis is that complex systems evolve that way that their parameters will be in the ordered phase, close to the border of the chaotic phase.

It seems to be plausible, in general terms, that somehow, complexity lies between the regular determinism and randomness/chaos. Specific calculations [355] with cellular automata (a method, offered by John von Neumann, see Sect. 5.5) suggested some cautiousness to avoid over-interpretation of the notion. In any case, Kaufman managed to show, that the model framework was very useful to got insight about the operation of genetic networks. Every model frameworks may be subject of criticism (and actually, they have been), including the random Boolean NK model. No doubt that the model framework is beautiful, illuminating, and helps to understand many aspects of genetic networks. OK, not all, but it never was expected.

NK Modeling of Genetic Networks

The development of cells within the developing embryo into different cell types happens in a process regulated by certain DNA sequences, i.e "regulator genes". These genes produce ("express') proteins that regulate other genes. Interaction of genes can be modeled by networks. Genes are the nodes, and two genes are connected by an edge if the product protein of one gene influences the expression of the other one. While it was obvious that the genetic network is not inherently random, due to the fact the linkages were complicated, and not well-known, the random network approximation was reasonable.

In the model the *state of a cell* is characterized

- by the state of its genes denoted by the vector $S(t) := \sigma_1, \sigma_2, \ldots \sigma_N$.

- the state of each gene is binary, i.e., a gene is either "on", or "off".

- each gene is connected to the same number K other genes, (and canalizing functions are used in some cases).

- the linkages among the genes are selected randomly.

The dynamics of the gene network in the NK model is specified by the following assumptions:

- The updating rule, or evolution function f_i, related to σ_i gets value 1 with probability p, and value 0 with probability $1-p$.

- The network updates is *synchronous*, the state of each gene is calculated in the same time.

Under these assumptions, as it was mentioned, cycles, i.e., periodic attractors appear, but the properties are very different in the frozen and chaotic phase. The biological interpretation is that the attractors represent the resulting cell type. Biological plausibility exert constraints on the properties of the attractors. The length of an attractor, (and the transient leading to it) must be not too long, since a cell should reach it stable state during a reasonable time. Since the chaotic phase is characterized by cycles, which show exponential increase with system size, genetic networks are not supposed to be in the chaotic phase.

While in the frozen phase the cycles are short, they have an other property, which does not support their biological relevance. The number of *relevant elements* in the frozen phase is close to be zero. Therefore the system is very resistant for point mutations (i.e., when the value of a single gene is subject to change). A real genetic network, which is able to evolve, should have some degree of sensitivity to mutations.

Kaufman's suggestion, that biologically realistic genetic networks should operate at the edge of chaos, i.e., the systems evolve that way that they are at or close to the critical phase. In this phase the size-dependence of the number and length of attractors is linear, it ensures a relatively quick convergence to stable, and not very long periodic attractors.

One family of experimental data supported the view, that genetic networks perform at the edge of chaos: the number of different cell types is more or less proportional to the square root of its DNA content. Disputes about the specific form of the size dependence are weakening this argument. An other problem is that in real genetic network the connectivity is much larger than two. To keep the system at the critical phase weighted or canalized, updating rules should be used.

Concluding Remarks

The random Boolean network model framework has been very popular, since it was intuitively well understandable, biologically plausible. It was able to produce rich dynamic behavior, and specifically offered a concept, which reached the "popular culture". The view, that "interesting" things happen between the regions of order and randomness is appealing. Intellectually it grew from the John von Neumann's cellular automata concept. Many extensions of the model framework exist. Among others, instead of assuming constant K for all genes, the input distribution may be taken from some distribution and the network might be so-called *scale-free*. The concept of "scale-free" will be explained later in Sect. 7.4.1. (Of course, many mathematical results obtained in the last 20+ years are restricted to constant connectivity). Kaufman sees clearly the scope and limits of the original framework, and participates in developing it. For his own analysis, see [271].

4.4 Population Dynamic and Epidemic Models: Biological and Social

4.4.1 Connectivity, Stability, Diversity

The fundamental question to be answered is how does the stability of an ecological system change if there is a change in the size and/or connectivity of the network of interacting elements, and/or in the strength of the interactions.

The May–Wigner Theorem

While ecologists believed for a while that diversity and stronger interactions among species enhance stability, Robert May [332, 333] proved for certain system sizes connectivity cannot exceed a threshold to ensure stability. The story goes back to cybernetics. Ashby [28] has found and Gardner [193] is said to have found that the probability of stability exponentially decreases with system size (i.e., with diversity). The same authors [194] reported the above mentioned threshold effect.

May [332, 333] published theoretical results, for randomly assembled deterministic systems by using linear stability analysis. The mathematics is a corollary of Wigner's theorem [560] on the eigenvalues of certain random matrices. May's argument follows.

Let B_n be a random $n \times n$ matrix with connectivity c_n, i.e., B_n has $c_n n^2$ non-zero elements. The non-zero elements are chosen independently from a fixed symmetric distribution with mean 0 and variance α_n^2. The interaction matrix is defined as $A_n := B_n - I_n$, and the trivial solution of the differential equations $\dot{x} = A_n x$ is for large n almost surely stable if $\alpha_n^2 n c_n < 1$ and almost surely unstable if $\alpha_n^2 n c_n > 1$. (The number α_n measures the range of the distribution, or it controls the magnitude of the elements of the matrix A_n, i.e., it characterizes the strength of connections).

The problem of connecting the structure and dynamics of ecological systems has been discussed as the *diversity-stability* debate [338, 337].

4.4 Population Dynamic and Epidemic Models: Biological and Social

May's results suggested that both diversity, and too strong connections tend to destabilize the equilibrium state of a community. The state of a community, in a more general ecological sense can be characterized by different measures. *Richness* is the total number of species, while *diversity* takes into account, somehow, both richness (the *dimension* of a vector), and the quantity of each species (the *value* of each vector component). *Complexity* of a community is determined by the richness, level of connectedness (number of non-zero elements in the interaction matrix), and the strength of the interactions.

Measures of Ecological Stability

The stability of complex ecosystems cannot be restricted to the stability of the equilibrium point [337], and several measures of stability can be defined. There are two classes of stability measures, the first is related to the usual stability concepts of dynamical systems, and the second measures the system's ability to preserve its function after disturbance ("resilience", "resistance").

1. Measures of system dynamics

   - Equilibrium stability: external perturbation implying deviation from the equilibrium state decays and the system returns to the vicinity if the state.

   - General stability: the quantities of the each species vary under limits, the lower limit is far from zero, so extinctions is not a threat.

   - Variability: The variance of the quantities of species is the usual experimental measure of stability. Larger general stability implies smaller variation.

2. Measures of resilience and resistance

   - Equilibrium resilience: is proportional with the inverse of the time necessary for the system to return to its equilibrium state.

   - General resilience: is proportional with the inverse of time necessary for the system to reach its original (not necessarily equilibrium) state.

   - Resistance: e.g the ability of the community to survive invasion of new species.

Weak Interactions Enhance Stability

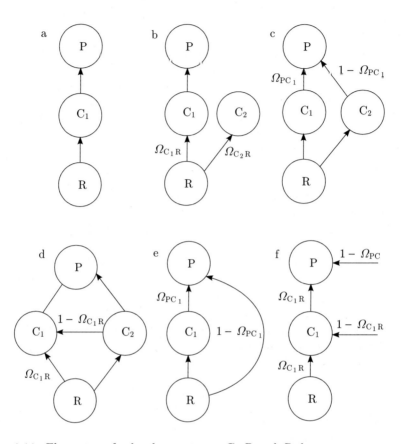

Fig. 4.11. Elementary food-web structures. C, P and R denote consumers, top predator and resource densities. Ω_{ij} denotes a parameter expressing the preference of a species i to consume species j. (**a**) Simple food chain; (**b**) exploitative competition (multiple intermediate consumers); (**c**) apparent competition (top predator feeding on two intermediate consumers); (**d**) intraguild competition (i.e., the killing and eating of species that use similar resources and are thus potential competitors); (**e**) omnivorous predator ("eats everything"); (**f**) food chain with external input. Based on McCann et al. [338].

McCann et al. [338] used a variation of the Rosenzweig-MacArthur model [444], where (as opposed to the assumption of the Lotka-Volterra model) predators' kill rate will approach an upper bound as the density of prey increases. Here V stands for the victim component (may be R or C), and P is

4.4 Population Dynamic and Epidemic Models: Biological and Social

the predator (P or C) in Fig. 4.11.

$$\dot{V} = rV - \delta V^2 - \frac{kPV}{\chi + V} \tag{4.31}$$

$$\dot{P} = \frac{\beta k PV}{\chi + V} - mP. \tag{4.32}$$

Here r is the victim species' Malthusian rate of increase, k the prey-predator kill rate, β a conversion factor from victims to baby predators, and m the predator's death rate, δ expresses the effects of intraspecific competition, χ is the density of prey at which the predators' kill rate reaches half its maximum.

Model studies showed the crucial dependence of the dynamic behavior on the strength of interactions. Many weak,[8] i.e., non-zero, and not too high, (it is difficult to give much more quantitative statements) in addition of a few strong connections tend to stabilize ecological communities. Obviously there is not to much experimental data about interaction strength, but based on the small amount available, and the model calculations McCann states:

> It seems ... that the weak interactions may be the glue that binds natural communities together. [338]

Further Reading

For a review on modeling food webs, see [136]. They evaluated the available food web data, discussed several classes of models. Static models assign links between species according to some simple rule. The second class uses general kinetic equations for interacting populations. In the third model new species may arrive through "invasion". The third model framework may lead to large stable network.

---

[8] Somewhat (somewhat!) analogously a very important concept about social networks [214, 215], Granovetter's "The Strength of Weak Ties..." argues that not only strong connections among people ("close friends") but weak connections ("acquaintances") also have an indispensable structural role in social organization. Weak ties play the role of the glue between densely connected friendship networks.

4.4.2 The Epidemic Propagation of Infections and Ideas

The Basic Kermack–McKendrick Model

In an excellent, very clearly written book Joshua Epstein [153] demonstrated that a set of sociodynamical phenomena could be interpreted by using nonlinear dynamical models elaborated in biological context. Epidemics is a fascinating topic in his books, since he uses the analogy of "revolution as epidemics", and models of describing and controlling infectious disease for modeling the propagation of ideas between population of people.

In the simplest situation there are two populations, $I(t)$ denotes the number of already infected (either with biological objects capable of transmit infection or with revolutionary ideas to be transferred to others) individuals, and $S(t)$ is the number of susceptible individuals. It is assumed that who is infected is also infective.

What is now the simplest assumption? The encounter between an infective and a susceptible may imply the transformation of a susceptible to an infective one. Chemical kinetics here also serves as a metalanguage to encode the process:

$$I + S \xrightarrow{r} 2I,$$

where r is the effectivity of the encounter, i.e., the infection rate: the larger this rate the more contagious the infection. By adopting (i) the mass action kinetic assumption and (ii) "prefect mixing" (i.e., the space is homogeneous), the kinetic equation is:

$$\dot{S} = -rSI, \tag{4.33a}$$
$$\dot{I} = rSI. \tag{4.33b}$$

Perfect mixing in social applications should be interpreted, as the lack of the existence of any (spatial or social) organization among the participants.

If the total population is constant (no birth and death, no immigration from and to the community), i.e., $I(t) + S(t) = N$, then the temporal change of the infected population is described by the equation:

$$\dot{I} = rI(N - I). \tag{4.34}$$

This is the logistic differential equation. The equilibrium solution $I_{eq} = 0$ is unstable. Instability here means that its slightest perturbation implies that $I(t)$ tends to N, the whole population will be infected.

The oversimplified model just presented consists of two subpopulations and it is called as the SI model. The classical strategy to defend the susceptibles against infection is to remove infectives, so to apply SIR models. Kermack and McKendrick defined a model, by adding a first order removal to (4.33) and so with the structure $S \to I \to R$:

$$\dot{S} = -rSI \tag{4.35}$$

$$\dot{I} = rSI - \gamma I \tag{4.36}$$

$$\dot{R} = \gamma I, \tag{4.37}$$

where γ is the removal rate. (Of course, this is the simplest assumption, the removal is proportional with the actual number of the infective ones). The relative removal rate is defined as $\rho := \gamma/r$. A plausible initial condition is that $S(0) > 0$, $I(0) > 0$ and $R(0) = 0$, nobody is in a quarantine before the epidemic breaks out.

The kinetic condition of the outbreak of epidemic is that the infectious population increases, i.e., $\dot{I} > 0$, i.e., $S(t) > \rho$. The number of susceptibles should exceed a *threshold* to have an epidemic outbreak. Since epidemics is threshold phenomena, one implication is that there is no necessity for full vaccination of the population to avoid the outbreak.[9]

Modeling More Realistic Epidemics

In the original model (4.37) the assumption has been made that the contact rate does not depend on the size of the population. For a set of sexually transmitted diseases it seemed to be a more realistic assumption that the contact rate is a non-increasing function of the population size. Another extension is to take into account the change of the total population size. A simple assumption is that a fraction of f of the infective class recover, and the remaining fraction $1-f$ die of disease. The extended model based on these assumptions is written [49] as

$$\dot{S} = -r(N)SI, \tag{4.38}$$

$$\dot{I} = r(N)SI - \gamma I, \tag{4.39}$$

$$\dot{N} = -(1-f)\gamma I, \tag{4.40}$$

More realistic models take into account age-dependence, delay, spatial heterogeneity, etc. In Sect. 7.4.3 instead of assuming "perfect mixing" among the

---

[9] Well, I leave to social scientists to analyze which subpopulation of a society will be immunized and which will not...

participant, some (network) organization will be assumed. In the first case the probability of the encounter of any two members of the population is the same. For organized communities some people may encounter with a subset of people more frequently than with others. Actually this organization makes a difference for the spreading of epidemics, as you will see.

Recently the investigation of the effects of infective immigrants to epidemic dynamics seems to be important, in context of HIV, severe acute respiratory syndrome (SARS) and avian influenza and measles (say [464]. From the perspective of public health policy the possibility of providing access to health care for populations of illegal immigrants should be seriously assessed. While deterministic models proved to be efficient, stochastic models are more faithful [15].

4.4.3 Modeling Social Epidemics

Epstein's witty perspective is to use models of (biological) epidemics to interpret social revolutions [153]. To do so, the variables, parameters, and each term in the equations should be reinterpreted. This reinterpretation requires to assume that everybody, who is not infected is susceptible (so there are no neutral, inert individuals), and if the contact between a susceptible and an infected has an effect, always the susceptible changes her attitude. Removal can be identified with imprisonment. Since in the equation (4.37) there are two parameters, the revolutionary and counterrevolutionary tactics is to modify these parameters.

The reduction of the r, i.e the contact rate between infectives and susceptibles can be obtained, say, by restricting the right of assembly, and the increase of the γ removal rate can be attained by increasing the rate of imprisonment. Ambush help to reduce γ, and the underground literature "Samizdat" and Radio Free Europe (a certain geographically and demographically defined (challenged?) subset of the Readers remember very well) increased r.[10]

Supplementing the basic model with the birth of susceptible population with a rate $\mu > 0$, the following equation is obtained:

$$\dot{S} = -rSI + \mu S, \qquad (4.41)$$
$$\dot{I} = rSI - \gamma I. \qquad (4.42)$$

---

[10] The author decided not to discuss in this book what is revolutionary idea and what is not.

The system of equations is identical to the Lotka–Volterra model. The susceptibles play the role of the preys, and the infectives do that of the predators. The interaction between the populations can be viewed as predator-prey interaction in a political ecological field.

In all these models it was implicitly assumed that the population is completely "mixed", and the result is that the outbreak of the epidemics is a threshold phenomenon. In Sect. 7.4.3 we shall discuss that assuming certain topological structure of the individuals, the propagation is not threshold. Propagation of computer viruses on network belong to this category. Consequently, the slightest infection of the Internet may propagate through the whole network.

4.5 Evolutionary Dynamics

The replicator equation (originally offered by Taylor and Jonker in 1978 [505]) was already mentioned, in the context of using, as a metalanguage:

$$\dot{x}_i(t) = x_i(t)\left(f_i(x(t)) - \sum_{j=1}^{n} x_j(t) f_j(x(t))\right). \tag{4.43}$$

In the context of evolution, x_i is identified with the relative abundance of a genetic sequence. This equation expresses that the selection is frequency-dependent, (or abundance-dependent); the fitness of a single individual depends on the abundance of the other sequences. $f_i(x(t))$ is the fitness of type i, and $\sum_{j=1}^{n} x_j(t) f_j(x(t))$ is the average fitness, also denoted by

$$\Phi := \sum_{j=1}^{n} x_j(t) f_j(x(t)). \tag{4.44}$$

This equation is both broader and narrower than the *quasi-species* equation offered to describe molecular evolution by Manfred Eigen and Peter Schuster [146, 147]:

$$\dot{x}_i(t) = \sum_{j=1}^{n} x_j(t) f_j q_{ji} - x_i(t) \Phi. \tag{4.45}$$

Equation (4.45) is narrower than the replicator equation (4.43), since it does not contain frequency selection. However, there is a factor to take into account the probability of mutations in the replication; q_{ji} is the probability of that sequence i will be copied as sequence j. The trivial unification of the two equations is the replicator-mutation equation:

$$\dot{x}_i(t) = \sum_{j=1}^{n} x_j(t) f_j(x(t)) q_{ji} - x_i(t)\Phi. \tag{4.46}$$

The relationship among the different equations were analyzed by Martin Nowak and his coworkers [394, 385].

Further Remarks

The replicator equation has stochastic version, too [181]. Such kinds of concepts, such as stability, survival, extinction have some different meanings in stochastic context. There is another school of modeling evolutionary systems, called *adaptive dynamics*. It connects phenomena with different time-scales, specifically the more rapid population and ecological dynamics and the slow evolutionary changes. It is a promising field, and one fundamental paper is [198].

We shall return to evolutionary dynamics in Chap. 9, related to *digital evolution* and *evolutionary game theory*.

4.6 Dynamic Models of War and Love

4.6.1 Lanchaster's Combat Model and Its Variations

The Basic Model

R.W. Lanchester, a British engineer derived and analyzed a mathematical model of warfare (e.g., [153]). For a combat between two sides, the state of the system is a two-dimensional vector, the number of soldiers on both sides. The interaction between the two sides implies the reduction of the number of soldiers, i.e., the attrition of forces. The simplest possible assumption is that there is a linear relationship between the attrition of a population of soldiers, and their number of the other side, the $k_b > 0$, $k_r > 0$ rate constants are the firing efficiencies. By denoting the number of "red" and "blue" combatants with R and B, the Lanchester equation is written as:

$$\frac{dR}{dt} = -k_B B, \qquad (4.47)$$

$$\frac{dB}{dt} = -k_R R, \qquad (4.48)$$

with the initial values $R(0) = R_0$ and $B(0) = B_0$. Instead of solving the equation, but eliminating its time-dependence, and adopting some stalemate assumption, the Lanchester "square law" is derived:

$$B_0 = \sqrt{\frac{k_R}{k_B}} R_0. \qquad (4.49)$$

The interpretation of the square law has a dramatic military implications: if your initial adversary twice as numerous as that of your enemy, it is not sufficient for your enemy to double her firing efficiency. To attain stalemate she must be four time as effective. Under this assumption numbers count more than effectiveness. Advantage of numbers strongly influenced the Pentagon strategy, maybe even today.

Density-Dependence

The assumption of the Lanchester equation is that the battle field is "spatially homogeneous", i.e., completely mixed. Also, it assumes that the increase of the numbers of combatants has benefits only. But a soldier is not only predator, but also a prey. Increasing numbers implies increasing the risk of being incapacitated. The *density dependent* Lanchester equation

$$\frac{dR}{dt} = (-k_B B)R, \qquad (4.50)$$

$$\frac{dB}{dt} = (-k_R R)B, \qquad (4.51)$$

takes into account both the predation benefit (in parentheses) and the prey cost (proportional with its own number). Under this assumption the stalemate condition is $k_b B_0 = k_r R_0$. Here linear increase in technological efficiency is sufficient to compensate the increase of adversary.

Ambush and Asymmetry

It is reasonable to assume some asymmetry between the troops: one side is able to ambush, the other is not. The model is defined as:

$$\frac{dB}{dt} = -k_B R, \tag{4.52}$$

$$\frac{dR}{dt} = -k_R RB. \tag{4.53}$$

The stalemate condition is

$$b(B_0^2 - B(t)^2) = r(R_0 - R(t)).$$

Assuming equal firing effectiveness $r = b$, B_0 can stalemate Red force with number B_0^2. That was the case in the Battle of Thermopylae.

Reinforcement

A further extension of the basic model is to incorporate some flow of reinforcements. It is reasonable to set some saturation to this flow, so the equation for the logistic population growth is added to the (density dependent) model:

$$\frac{dR}{dt} = -k_B RB + \alpha R\left(1 - \frac{R}{K}\right), \tag{4.54}$$

$$\frac{dB}{dt} = -k_R BR + \beta B\left(1 - \frac{B}{L}\right), \tag{4.55}$$

where α, β, K and L are positive constants. This equations is a generalized Lotka-Volterra equation, a famous model of competition of "the struggle for existence", given by Georgyi Frantsevitch Gause (1910-1986), a biologist from Moscow. There are four basic solutions of (4.55), and they have different interpretations, as Fig. 4.12 shows. Two cases implements the "principle of competitive exclusion". In these cases one side wins, since the other population becomes extinct. In the third case there is a stable fixed point (technically a so-called *node*), with finite coordinates. This scenario implements "coexistence" by using ecological terminology, but in this "combat dynamical" model it means "permanent war". The fourth case is an unstable (so-called *saddle*) equilibrium. Any small perturbation implies the win of the one or the other side.

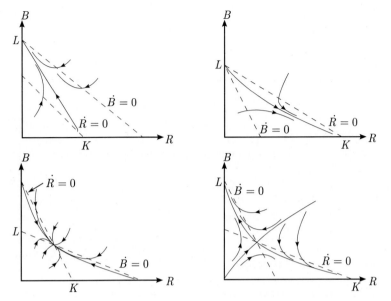

Fig. 4.12. Phase plane analysis of the Gause model. The *upper two plots* implements the "principle of competitive exclusion". The *left lower figure* visualizes the "permanent war" situation (the equilibrium point is stable focus). The *lower right plot* shows an unstable saddle point.

4.6.2 Is Love Different from War?

"Love is what we do." There are several models to use dynamical models to describe the temporal changes in love. The models are semi-serious, and help motivate students to learn in thinking with dynamical models, but they give an insight of possible scenarios. Based on some previous studies Julien Sprott gave a series of models to describe the dynamics of romantic relationshp, say, between Romeo and Juliet [480].

The state of the system is characterized by a two-dimensional vector $(R(t), J(t))$. $R(t)$ is Romeo's love (or hate if negative) for Juliet at time t and $J(t)$ is Juliet's love for Romeo. A simple assumption to model the interaction is that their change of their emotional state depends on their own and the other's state, and the dependence is linear and additive. These assumptions lead to the linear model:

$$\frac{dR}{dt} = aR + bJ, \qquad (4.56)$$

$$\frac{dJ}{dt} = cR + dJ. \qquad (4.57)$$

The constants a, b, c, d determine the romantic style of the relationship. If $a > 0$, than Romeo's feeling is subject of self-amplification (positive feedback), $b > 0$ means that his feelings is encouraged by Juliet's feeling.

From emotional attitude to love dynamics

1. Eager beaver: $a > 0, b > 0$ (Romeo is encouraged by his own feelings as well as Juliet's.)

2. Narcissistic nerd: $a > 0, b < 0$ (Romeo wants more of what he feels but retreats from Juliet's feelings.)

3. Cautious (or secure) lover: $a < 0, b > 0$ (Romeo retreats from his own feelings but is encouraged by Juliet's.)

4. Hermit: $a < 0, b < 0$ (Romeo retreats from his own feelings as well as Juliet's.)

5. out of ouch with one own's feeling: $a = 0$ or $d = 0$

6. oblivious to the other feeling: $b = 0$ or $c = 0$

The terminology goes back to Strogatz and his students [491], Chap. 5.3.

There are $6 \times 6 = 36$ possible dynamics, but the R/J symmetry reduces this number to 21. Specifically, the case of two nerds (i.e., $b < 0$ and $c < 0$) with "out of ouch with their owns feelings" is equivalent with the original Lanchester equation. Of course, it not surprising that occasionally love models and war models coincide. The case of $b > 0$ and $c > 0$), i.e mutual activation, also may be interpreted in combat dynamic context.

The model framework has obvious difficulties. It is not trivial to assign numbers to our love state. Furthermore, more realistic parameters should be time-dependent, etc. Still, it is worth to take a look to some possible outcomes of this simple model of love affair dynamics.

Equation (4.57) has a single equilibrium at $R = J = 0$. The nature of the dynamics around an equilibrium is determined by the eigenvalues of the coefficient matrix $A := \begin{bmatrix} a & b \\ c & d \end{bmatrix}$.

The eigenvalues are the solutions of the characteristic equation (3.18) in this case are

$$\lambda = \frac{a+d}{2} \pm +\frac{1}{2}\sqrt{(a+d)^2 - 4(ad-bc)}. \qquad (4.58)$$

As is well known, nature of the equilibrium points depend on whether the eigenvalues are real, complex conjugate, or purely complex, and we get three classes (focus, node, saddle point), as Fig. 4.13 shows.

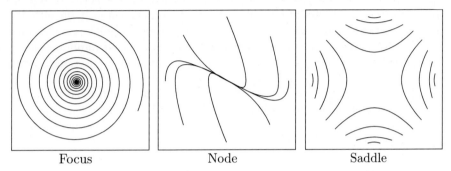

Focus   Node   Saddle

Fig. 4.13. Dynamics in the vicinity of an equilibrium point. From http://sprott.physics.wisc.edu/pubs/paper277.htm.

The stability depends on the sign of the real parts of the eigenvalues. Some specific examples for the love dynamics follow.

Fire and Ice

This situation id defined, if the two lovers are opposites, so $b = -c$ and $a = -d$, and the eigenvalues are $\lambda = \pm\sqrt{a^2 - b^2}$. Since the dynamics depends on a and b, there are two possibilities:

- $ab > 0$: *eager beaver plus hermit*

- $ab < 0$: *narcissistic nerd plus cautious lover*

There are two cases, again. If $a > b$, i.e., the responsiveness to one's own state is larger than to the other's, both λ is real, one is positive, and one is negative. The dynamics can end up in any of the four quadrant, so all four combination of love and hate may emerge depending on the initial values. If $a < b$, i.e., the self-responsiveness is smaller that to responsiveness the other's state), the equilibrium point is a *center*, as it was explained in Sect. 3.5.1. The dynamics is oscillatory.

Other Possibilities

The *peas in a pod* situations is characterized by the interaction of two equivalently romantic characters: $b = c$, and $a = d$. If $|a| > |b|$, the singularity point (i.e., the $0, 0$) is a stable node, the relationship will end with mutual apathy (not the worst case scenario). In the *Romeo the robot* scenario, Romeo's feeling is unchanged ($dR/dt = 0$). Juliet feeling is determined by the sign of Romeo's (constant, so constant sign) feeling and her own romantic style. The equilibrium point is $J_{eq} = -cR/d$. What if Romeo loves Juliet, i.e., $R > 0$? Juliet will love him back if $cd < 0$. This can be implemented if she either is a cautious lover or narcissistic nerd. But the singularity point is stable only if she is cautious ($d < 0$). Narcissism ($d > 0$) will lead either unbound love or hate. Her feeling never goes to zero, or and will not show oscillation between love and hate.

Further most exciting possibilities nonlinear models, love triangles (*ménage à trois*) are also analyzed by Sprott. But even simple linear models produced dynamics with sufficiently interesting insights.

4.7 Social Dynamics: Some Examples

4.7.1 Segregation dynamics

Thomas Schelling published a very influential paper in 1971 with the title "Dynamic Models of Segregation" [457]. This is a paradigmatic paper to demonstrate how local rules (micromotives in Schelling's terminology) imply global ordered social structures (macrobehaviors). Technically it is a *cellular automata* model, space can be one- or two-dimensional. Each player (or agent or turtle, the terminology may differ) sits on a grid point surrounded by others. (You may imagine that the points are people living in a house encircled by their neighbors). The players may have a predominant parameter. This parameter in the USA is a visible color, and it is correlated to the race of the players, and the model describes *racial segregation*.

The model was elaborated in the context of residential segregation, but of course the phenomena cannot be restricted to it, as Beverly Daniel Tatum's book, "Why Are All The Black Kids Sitting Together in the Cafeteria?" explains [504].[11]

---

[11] Of course, other than racial segregation exists, such as gender segregation in the labor market, distribution of different nationalities in an international meeting, etc.

The *state* of the each player is now characterized by the degree of satisfaction. Here this degree depends on the number of neighbors having the same parameter only. The *dynamics* of the system is determined by a local rule, which tend to eliminate the individual dissatisfaction by reallocating the player into a suitable place. The "degree of satisfaction" maybe player-dependent. Some players may feel themselves unsatisfied by having a single different neighbor, others are more "tolerant" to "race inhomogeneity". Obviously, it would be easy to explain segregation, if the individual players would be inherently racists. What the simulation results suggested is that *slight* preference to live "among their owns" imply global segregation, and the formation of ghettos.

In an extensive analysis Pancs and Vriend studied both 1D and 2D model [398], with a supplement [399], and Schelling's basics results proved to be robust. Four types of utility functions were defined. (This is what theoretical social models can do: to define classes of idealized behavior types and analyze the different outcomes due to different structural conditions). See Fig. 4.14.

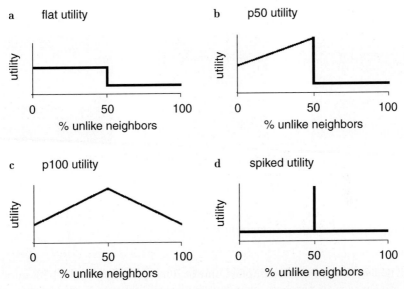

Fig. 4.14. Utility functions used by [398]. For the details see the text.

The functions define the degree of satisfaction in the function of unlike neighbors. All the figures have a special value at 50% ("perfect integration"). There is no other percentage what the player better prefer than to live in a perfect integration, at least in some sense. The *flat utility* reflects Schelling's original perspective. This player is indifferent for having unlike neighbors (or not to have any) while they does not exceeded the 50% tolerance threshold.

For this player, complete segregation is as acceptable as perfect integration, and unacceptable to live in any minority.

The 'p50' utility function has a clear preference for perfect integration, and the larger the better for having unlike neighbors at the 50% point, where there is a cut-off. The nature of this function reflects that this player also has aversion being in minority.

The 'p100' function is symmetric with maximum at 50%. Her preference is to live in perfect integration, and any deviation from it is symmetrically (un)acceptable.

The spiked utility function defined the case when any deviation from the perfect integration is highly non-preferred.

A simulation result with 4000 players of each type on a 100 × 100 board is presented in Fig. 4.15.

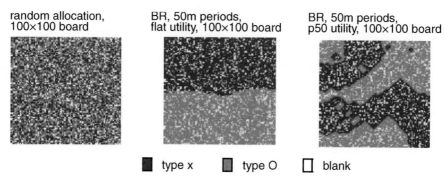

Fig. 4.15. Simulation results from [398]. The left board is the control set randomly. The middle board illustrates that if the utility functions is set to be flat, the separation is close to be complete. The "p50" utility function implies the formation of patches, stripes. Occasionally embedded islands occur.

The flat utility function implied almost full separation, while 'p50' implied the appearance of patches, islands. Occasionally islands in the islands also emerge.

Schelling's segregation became popular, since it attacked a socially and intellectually challenging problem by simple conceptual tools, produced easily interpretable, really interesting, somewhat paradoxical results. The emerging macroscopic structure is not designed, it is not in accordance (but occasionally also not against) to the intention of the players. The principle of "Think globally, act locally!" is violated.

4.7.2 Opinion Dynamics

Interactions of people in a group (in extreme case this may be the whole society) imply changes in their opinion about different issues and may lead to *consensus, fragmentation* or *polarization*. Consensus means that all players share the same opinion, fragmentation occurs when several opinions emerge, polarization is a special case of fragmentation for having two parties of people with two different opinions.

Mathematical models help to understand the dynamics of the interactions among players. A family of models were set and investigated by Hegselmann and Krause [234]. There are n number of players. The opinion of a player at a certain time point t is a real number, $x_i(t)$ and the complete state of the model is described by a **continuous** opinion vector. There is a simpler situation, if there are two possible opinions the state space is binary. The n-dimensional $\mathbf{x}(t) = (x_1(t), \ldots, x_n(t))$ giving the **state** of the system is called the *opinion profile*.

The change of the state, i.e., the opinion dynamics, is governed by the local interactions among people through their opinions, and the $\mathbf{A}$ matrix specifies this influence; an element a_{ij} denotes the influence of i on j. For simplicity $\sum_{j=1}^{n} a_{ij} = 1$. In the following we assume that $a_{ij} \geq 0$, this expresses that a player is never influenced negatively by another player (but of course the influence may be 0). The time may be discrete. The simplest dynamic model (the "classical model") is based on the assumption that player i adjusts her opinion by taking a weighted average of all other j players' opinion, and the influence matrix is constant:

$$\mathbf{x}(t+1) = \mathbf{A}\mathbf{x}(t). \qquad (4.59)$$

Assuming that the weights change with time or opinion, a quasi-linear (basically nonlinear) model can be defined:

$$\mathbf{x}(t+1) = \mathbf{A}(t, x(t))\mathbf{x}(t). \qquad (4.60)$$

The main questions to be answered are related to the dynamic change of opinions starting from some initial opinion profile $x(0)$. A specific problem is to find out the conditions, which imply that the group of players approach a consensus c, i.e., $\lim_{x \to \infty} x_i(t) = c$, for *all* players i. (Structural conditions leading to fragmentation and polarization are also analyzed).

There is another variation of the classical model (called the Friedkin and Johnsen model). While the assumption behind the classical model is that the only driving force of the dynamics is the influence, the Friedkin and Johnsen model takes into account the *nonfidelity* of the player to an original idea. The degree of fidelity is denoted by g_i, and the $1 - g_i$ is the measure of a player's susceptibility for social influence. The model is defined as

$$\mathbf{x}(t+1) = \mathbf{G}\mathbf{x}(0) + (\mathbf{I} - \mathbf{G})\mathbf{A}\mathbf{x}(t). \tag{4.61}$$

Here $\mathbf{G}$ is a diagonal matrix with elements g_i, $0 \leq g_i \leq 1$ in the diagonal and $\mathbf{I}$ is the identity matrix.

There are some analytical results for certain models of opinion dynamics. These results connect structural conditions of the models (such as the structure of the influence matrix), and the qualitative dynamic behavior of the system.

1. There are sufficient and necessary conditions for the model (4.59). If for any two players i, j there is a third one k, with positive weights $a_{ik} > 0$ and $a_{jk} > 0$ then the consensus property holds for all initial value. (Of course, the *value* of the consensus depends on the initial value.)

2. For the Friedkin-Johnsen model there is a sufficient condition to converge a fixed vector of fragmented opinions. The existence of at least one positive degree g_i under some technical condition implies that any initial profile will converge to a stationary opinion pattern. This pattern has the consensus property if and only if there were preexisting consensus among all players with positive degree.

Hegselmann and Krause [234] extended these models by introducing the concept of *bounded confidence*. This concept expresses the assumption that players with too different opinion don't influence each other. The opinions of two players should be closer than a given ϵ confidence bound. Only the opinion of those other players will modify the opinion of player i, who are within this confidence interval.

The model is a special case of the nonlinear model (4.60), (for time-independent influence matrix). One (maybe too simplifying) assumption is that the confidence level is the same for all players: $\epsilon_i = \epsilon$ for all i. Obviously, ϵ is a parameter of the model, and as always, there is the question: "How robust are the results of changing the parameters?"

The bounded confidence (BC) model is defined as:

$$x_i(t+1) = |I(i, x(t))|^{-1} \sum_{j \in I(i,x(t))} x_j(t), \tag{4.62}$$

where $|I|$ is the cardinality of the set I, and the set of players whose opinions are taken into account is: $I(i, x) = \{1 \leq j \leq n; |x_i - x_j| \leq \epsilon_i\}$, and $\epsilon_i > 0$ is the confidence level of player i.

Analytical results were obtained for specific cases. ϵ_i may be uniform for all players, so $\epsilon_i = \epsilon$ for all i. A confidence interval maybe symmetric or asymmetric. Symmetric confidence interval $[-\epsilon, +\epsilon]$ here means that a player equivalently tolerant for a "less" and a "more" opinion. Asymmetric confidence interval is denoted by $[-\epsilon_l, +\epsilon_r]$, where the indices l and r denotes "left" and "right". ϵ_l and ϵ_r are positive, but they may be different.

An asymmetric confidence interval may imply *one sided split*. The term means that while one player (i, if $\epsilon_l < \epsilon_r$) takes into account the opinion of j, but not vice verse. In an opinion profile **x** there is a *split* between players i and j, if $|x_i - x_j| > \epsilon$. The dynamical model has some characteristic properties:

1. The dynamics is "order preserving", i.e., if $x_i(t) \leq x_j(t)$ for all $i \leq j$ then $x_i(t+1) \leq x_j(t+1)$ for all i, j.

2. The dynamics is split preserving. If there is a split between two players, this property is persistent.

The general take home message is that these models help to get an insight about structural conditions of different qualitative outcomes, such as *consensus*, *polarization* and *fragmentation*. Of course, the assumptions of the models may vary, other effects (say repulsion) could be included. Propagation of extreme views are not very different for social epidemics. Specifically, a model resembling for epidemic spread was introduced to analyze the interaction among four subpopulations, such as general, susceptible, excited, and fanatic ones [483]. In a certain parameter regime the excited and fanatic subpopulations survive, in another one die out. The model, (as usually model do) give hint about the conditions of theses different qualitative outcomes.

4.8 Nonlinear Dynamics in Economics: Some Examples

4.8.1 Business Cycles

Kaldor Model

The Goodwin model, an application and extension of the Lotka-Volterra model was briefly mentioned in Sect. 3.5.1. Nicolas Kaldor (1908–1986) gave a mechanism for the generation of temporal oscillatory dynamics in *income* (y) and *capital* (k). Kaldor assumed nonlinear dependence of investment (I) and saving (S) on income.

Assuming time-independent investment and saving the dynamics is given as:

$$\dot{y}(t) = a[I(k(t), y(t)) - S(k(t), y(t))], \quad (4.63a)$$
$$\dot{k}(t) = I(k(t), y(t)) - \delta k(t) \quad (1.63b)$$

Equation 4.63a describes income dynamics. If the rate parameter $a > 0$, the direction of change of the income depends on the sign of the difference between investment and saving. Equation 4.63b specifies that the increase in the capital is equal to the investment, and is reduced by depreciation controlled by the parameter δ. Chang and Smyth [95] proved (actually for the case of $\delta = 0$) the conditions of the existence of limit cycle. Kaldor made assumptions for the form of the functions I and S. Investment, savings, and their difference, dy, as functions of income, y, for capital at $k = k_e$, with $a = 4$ was shown in the left side of Fig. 4.16. With some other restrictions, the model may lead to limit cycle, as the right-hand side of Fig. 4.16 shows.

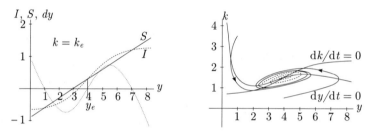

Fig. 4.16. Non-linear investment and savings and Kaldor's trade cycle. From http://www.egwald.com/nonlineardynamics/limitcycles.php\#kaldormodel.

The Kaldor–Kalecki Model

In a version of the model another mechanism to induce periodicity, i.e., some delay in investment, is also incorporated. The modified model is a time-delay differential equation:

$$\dot{y}(t) = a[I(k(t), y(t)) - S(k(t), y(t))], \quad (4.64a)$$
$$\dot{k}(t) = I(k(t), y(t - \tau)) - \delta k(t) \quad (4.64b)$$

where τ is the time delay. It was shown [299] that τ is bifurcation parameter, and there is a Hopf bifurcation mechanism for the onset of limit cycle behavior.

4.8.2 Controlling Chaos in Economic Models

Controlling of some economic process might be very important. While traditional economics used linear models, nonlinear dynamics helps to understand the mechanism of chaos generation and control in micro- and macroeconomics, too. The use of a chaos control technique (discussed in Sect. 3.6.4) in a microeconomic model was demonstrated [240] Chaos occurs in a simple microeconomical model of two competing firms.

Two firms X and Y competing on the same market of goods. The firms perform active investment strategies, i.e., their temporary investments depend on their relative position on the market. The strategies can be asymmetric:

- Firm X invests more when it has an advantage over the firm Y;

- Firm Y invests more if it is in a disadvantageous position compared to the firm X.

A discrete-time dynamical model was defined for the temporal change of the sales x_n, y_n of the two firms as:

$$x_{n+1} = F^x(x_n, y_n) = (1-\alpha)x_n + \frac{a}{1+exp[-c(x_n-y_n)]}, \quad (4.65a)$$

$$y_{n+1} = F^y(x_n, y_n) = (1-\beta)y_n + \frac{b}{1+exp[-c(x_n-y_n)]}. \quad (4.65b)$$

The parameters $0 < \alpha < 1$ and $0 < \beta < 1$ control the velocity of sales reduction without investment, while a and b denote the investment effectiveness, This is the Behrens–Feichtinger model, which also well describes the control of the drug market Sect. 4.9. The analysis showed that there is a parameter region, where chaotic behavior occurs.

Chaos suppression might be advantageous, since the unpredictable dynamics will be at least predictable. Different control strategies could be applied:

1. Change in the forcing function. A time-delay feedback is given to the right-hands site of the governing equation: To ensure periodic (say, period-1, or period-2 behavior), the equation for sales dynamics can be written as

$$y_{n+1} = F^y(x_n, y_n) + F_K, \quad (4.66)$$

where the form of the F_K control force is by an additive term.

$$F_K = K_{yy}(y_n - y_{n-m}), \quad (4.67)$$

where $m = 1$ or $m = 2$ for period-1 or period-2 orbits, and K_{yy} is a control constant.

2. Change in the parameters. A time-delay feedback term is added to a system parameter. It is installed by

$$y_{n+1} = (1-\beta)y_n + \frac{b + K_{yy}(y_n - y_{n-m})}{1 + exp[-c(x_n - y_n)]}. \qquad (4.68)$$

Time-delay feedback is a plausible control mechanism in this context. The policy of firms depend on the difference between actual and past values of sales. There are realistic mechanisms (strategies adopted) to suppress chaos.

4.9 Drug Market: Controlling Chaos

A set of models were established to study the dynamics of the number of addicts ($A(t)$) and sellers ($D(t)$) in a city's illegal drug market [53].

The assumptions of the model describe the increase and decrease of the size of the two populations. The decrease term is described by a first-order decay term. Increase of both population size depend on their relative numbers. There is a positive feedback effect, since relative large number of users will increase the number of sellers (since the market offers good profit opportunities). The increase in the number of new users depends, however, mostly on the number of the actual users, since their habit propagates through their social network. The number of sellers influence the number of new users only indirectly. The structure of the model is shown in Fig. 4.17.

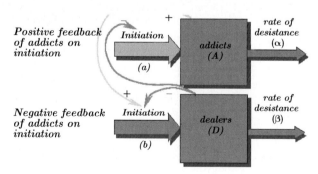

Fig. 4.17. Flow diagram of the drug market model. Adapted from [53].

4.9 Drug Market: Controlling Chaos

The discrete-time dynamical model equations are specified, as

$$A_{t+1} = (1-\alpha)A_t + af(A_t, D_t), \quad A_0 = A(0), \tag{4.69a}$$
$$D_{t+1} = (1-\beta)D_t + bf(A_t, D_t), \quad D_0 = D(0), \tag{4.69b}$$

$a, b > 0$, and $\alpha, \beta \in [0, 1]$. The positive and negative feedback effects should be expressed in the form of the function f. A rather general form for the function is

$$f(A, D) = \frac{1}{1 + e^{-c(A-D)}}, \quad c > 0. \tag{4.70}$$

Statistical data suggest that the outflow rate of the dealers (due to death, law enforcement, policy) is larger than the outflow rate of the addicts, i.e., $\beta > \alpha$. It was shown that for a reasonable region of parameters, chaos may occur, as the bifurcation diagram Fig. 4.18 shows.

Since a chaotic drug market is interpreted as unpredictable, there is a natural question whether it is possible to control the system to stable cycles or even (low level) equilibrium.

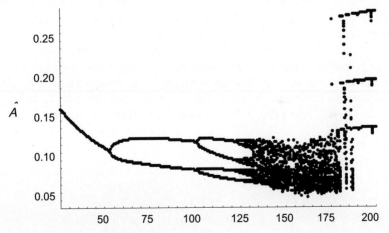

Fig. 4.18. Bifurcation diagram of the model of the drug market. Chaos may occur in a certain parameter region. Adapted from [53].

5

The Search for Laws: Deductive Versus Inductive

5.1 Deductive Versus Inductive Arguments

In this chapter we will take two steps back and have a look at the big picture of the methodological issues of complex systems from a somewhat philosophical perspective. The understanding of the behavior of complex systems requires some thoughts about the epistemological basis of our methods.

Epistemology traditionally deals with three questions (say, [77]):

1. What is knowledge?

2. What can we know?

3. How is knowledge acquired?

Our point is here that while there is an age-old debate between the fact-based *inductive* and the axiom-based *deductive* approaches, a healthy combination of the two is best for understanding complexity.

Deductive Arguments

The deductive approach can be summarized in a simple statement: "If the premises are true the conclusions must be true".

Though they are not always phrased in syllogistic form, deductive arguments can usually be phrased as "syllogisms", or as brief, mathematical statements in which the premises lead to the conclusion. Deduction is truth preserving.

Sherlock Holmes - Arthur Conan Doyle

... the man who most impressed and influenced him, was without a doubt, one of his teachers, Dr. Joseph Bell. The good doctor was a master at observation, logic, deduction, and diagnosis. All these qualities were later to be found in the persona of the celebrated detective Sherlock Holmes...

from Sir Arthur Conan Doyle Biography:
http://www.sherlockholmesonline.org/Biography/index.htm

In the language of logic *modus ponens* is used for deduction:

If P, then Q.
P.
Therefore, Q.

The conclusion must be true if all the premises are true. Aristotle is credited to be the first person to use these kind of arguments although he never coined the term *modus ponens*. His incredible accomplishment is that he studied the *form* and *not* the *content* of the statements. P might be Peter, or Paul, or "it is raining", or anything else, the STRUCTURE of the reasoning is the same.

Modus tollens is expressed by the argument form:

If P, then Q.
Q is false.
Therefore, P is false.

The argument has two premises, (P implies Q; and Q is false).

Inductive Arguments

Inductive arguments use sentences with this form: If the premises are true than the conclusion is *likely* to be true. What we should notice is that it is not guaranteed to be true.

*P*1. There are heavy black clouds in the sky.
*P*2. The humidity is very high.
It will rain soon.

Inductive arguments are never valid in the logician's sense of the term, because their premises do not imply their conclusion.

Francis Bacon (1561–1626) suggested that the only way of acquiring scientific knowledge is "true and perfect induction". However successful this program was (nobody denies the role of thorough observation and data collection), Bacon underestimated the role of the creative imagination, and the formation of testable (i.e *falsifiable*, using Karl Popper's terminology) hypotheses.

5.2 Principia Mathematica and the Deductive Approach: From Newton to Russell and Whitehead

Newton's *Philosophiae Naturalis Principia Mathematica*, often called *Principia*, gives the basis of classical mechanics. While the theory was in accordance with data, Newton

- broke (Francis Bacon's) purely inductive method,

- used minimal experimental data and

- everything in his theory was deduced from a few observation-based conclusions ("mathematical principles of philosophy").

Based on the theory described in *Principia* it was possible to make predictions which led to the discovery of Neptune and Pluto, still it was never proved that the theory is *absolutely* true.

Principia Mathematica (Whitehead and Russell) was a very big enterprise aimed to deduce mathematical truths from logic starting from clearly defined axioms and using logical inferences Even though whole program was not successful, in the end, it showed the power of deduction.

> "I think Whitehead and Russell probably win the prize for the most notation-intensive non-machine-generated piece of work that's ever been done..."
>
> S. Wolfram
> For supporting Wolfram's remark, see Fig. 5.1.
>
> http://www.stephenwolfram.com/publications/talks/mathml/mathml2.html

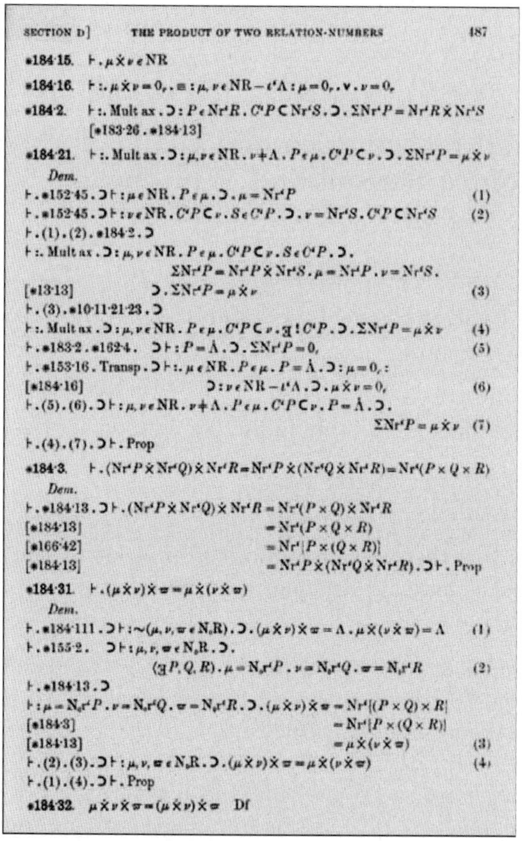

Fig. 5.1. An almost arbitrarily chosen page from the Prinicipia Mathematica.

5.3 Karl Popper and the Problem of Induction

The method of induction is based on observations, and theory is formed by generalization. While logical positivism, which originated in the Vienna Circle, emphasized the exclusive role of experience and logical analysis. They expected that statements based on empirical observations could be verified. However, Karl Popper challenged the view that general empirical statements can be verified. He stated that statements cannot be verified only falsified. His book "The Logic of Scientific Discovery" [411] is a very important milestone in the philosophy of science. This is the scientific credo of many naive scientist, as well. There is an obvious asymmetry between verification or confirmation and the falsification of a theory. Even if a scientist believes in a theory, it is never legitimate to say: "The theory is proved by this experiment." The correct terminology is to say that "The experiment supports the view."

Modus tollens is in the center of Popper's argument against the use of the inductive method, and offering the method of falsification. Scientists should not use inductive inference, but should offer conjectures subject to falsification. This is the way of the Popperian progress of science.[1]

5.4 Cybernetics: Bridge Between Natural and Artificial

While Newtonian mechanics deals with such kinds of concepts as position, velocity, motion, dynamics, and philosophically led to the clockwork world view, cybernetics is the science of information, control and communication.[2] Cybernetics hoped to be a (or *the*) bridge between the *natural* and *artificial*, or in other words, organisms and mechanics. The legendary mathematician John von Neumann attended the first several Macy conferences. His posthumous book The Computer and the Brain [543] is a controversial masterpiece of the brain-computer analogy/disanalogy problem.

---

[1] In a down-to-earth way, scientists generally accept Popper's method. For the debates about the progress of science among such philosophers as Karl Popper, Tomas Kuhn, Imre Lakatos and Paul Feyerabend see e.g., [14].

[2] Interestingly, there is a well edited website worth visiting, using the ambitious term "Principia" in its title (Principia Cybernetica: http://pcp.lanl.gov), but seemingly it has not been updated in the last several years.

5.5 John von Neumann: The Real Pioneer of Complex Systems Studies

While we are speaking about cybernetics, we should briefly review John von Neumann's incredible contributions to science and technology. He is certainly a pioneer of complex systems studies, and most likely (who knows how his brain worked?) combined deductive and inductive reasoning.

Short Biography

John von Neumann (1903–1957) was born in Budapest. In Budapest there were several academically very strong high schools (gymnasiums), where the boys studied together between the age of ten to eighteen, and had the chance to form lifelong friendships. Neumann (he is called just Neumann in Hungarian) was with a close friendship with Eugene Wigner (Nobel prize in physics, 1963). Both of them attended the Lutheran Gymnasium. (Gymnasium has nothing to do with the English term "gym". Gymnasiums gave a high level secondary education which prepares students for elite universities.) After his studies (almost simultaneously, in Budapest and in the ETH in Zurich) he worked in Berlin. He moved to the U.S in 1930 to join the faculty of the Institute for Advanced Study at Princeton. From 1955 he served as the Commissioner of the Atomic Energy Commission. He died of cancer, which might have developed from attending A-bomb tests. Von Neumann worked in many fields, or it is better to say he created a set of new fields. I belong to the community which considers him as the father of the computers.

Foundations of Mathematics

The existence of logical paradoxes, including Russel's one, pointed out the crisis of mathematics. Von Neumann participated in David Hilbert's program to formalize mathematics based on the axiomatization of set theory. While von Neumann had some very important results (which later became the von Neumann–Bernays–Gödel set theory which was based on a finite number of axioms), the whole program was abandoned after the appearance of Kurt Gödel's theorem. Gödel proved that "no formal system is powerful enough to formulate arithmetic that could be both complete and consistent".

Quantum Mechanics

Quantum mechanics started with two independent formulations of Heisenberg's matrix mechanics and Schrödinger's wave mechanics, as it was discussed in Sect. 2.1.1. Von Neumann showed [541] the equivalence of the two descriptions by representing the state of a quantum system in a Hilbert space. (Very loosely speaking Hilbert space is a infinite-dimensional extension of the finite-dimensional Euclidean space. The distance between points in this space is defined by the scalar product of the vectors.) He also concluded that the inclusion of any "hidden parameters" does not help to make quantum mechanics deterministic, as it is inherently nondeterministic. While the approach was not perfect, it finally led to Bell's theorem. It turned out that the quantum theoretical version of the concept of *reality* is strongly different from that of classical physics.

Game Theory

Von Neumann formed game theory as a set of tools to analyze economical behavior by mathematical methods. First, in 1928, he analyzed the zero sum games of two players. A zero-sum game is characterized by the exact balance of gain and loss. For this two person game, one players loss is equal to the other players loss. Of course, say in chess, there could not be more than one winner and one looser. A draw implies the equal split of the score. Von Neumann claimed and proved a *minimax* theorem, where the best of the "worst case scenarios" can be found.

> The fundamental theorem of game theory states that every finite, zero-sum, two-person game has optimal (mixed) strategies.

The thick monograph "The Theory of Games and Economic Behavior" which Von Neumann wrote with Oscar Morgenstern [544] was a landmark of social sciences, and gave a formal framework to consider economics as a strategy game among players. Game theory became a method in applied social sciences, from political to military applications. Actually, thinking about the induction-deduction dichotomy, the concept of "backward induction" was applied here.

> Backward induction: A paradox of backward induction.
>
> The unexpected hanging paradox is a paradox which arises with backward induction. Suppose a prisoner is told that she will be executed sometime between Monday and Friday of next week. However, the exact day will be a surprise (i.e., she will not know the eve of her execution). The prisoner, interested in outsmarting her executioner, attempts to determine which day the execution will occur. She reasons that it cannot occur on Friday, since if it had not occurred by the end of Thursday, she would know the execution would be on Friday. Therefore she can eliminate Friday as a possibility. With Friday eliminated, she decides that it cannot occur on Thursday, since if it had not occurred on Wednesday, she would know that it had to be on Thursday. Therefore she can eliminate Thursday. This reasoning proceeds until she has eliminated all possibilities. She concludes that she will not be hung next week. To her surprise, she is hung on Wednesday. Here the prisoner reasons by backward induction, but seems to come to a false conclusion. This paradox has received some substantial discussion by philosophers.
>
> (From Wikipedia)

Game theory has earned five Nobel prizes. The most publicized being John Nash's award, which was made popular by the movie "A Beautiful Mind." He discriminated between cooperative and non-cooperative games. Nash found an equilibrium point for non-cooperative game (Nash equilibrium), and Richard Selten refined the concept. The third Nobel prize in 1994 went to John C. Harsanyi (from the same gymnasium where von Neumann and Wigner were educated). He gave a method for analyzing games when the players don't have complete information. We already know that Thomas Schelling was awarded for his works on applying dynamical models to social sciences, and Robert Aumann, among others, analyzed old dilemmas of the Talmud (say, [32]).

> Nash equilibrium
>
> If there is a set of strategies with the property that no player can benefit by changing her strategy while the other players keep their strategies unchanged, then that set of strategies and the corresponding payoffs constitute a Nash Equilibrium.

Computer Architectures

By far, the largest family of computers (still used today) have what is called Von Neumann architecture [30]:

- The machines have three basic hardware subsystems: a CPU, a main memory system and an I/O system.

- They are a stored-program computer. Maybe this was the most important step. Programs are just stored as data.

- They carry out instructions sequentially.

- They have, or at least appear to have, a single path between the main memory system and the control unit of the CPU; this is often referred to as the von Neumann bottleneck.

The Computer and the Brain in Broader Context

Von Neumann was motivated to think on the possibility of constructing a machine to dramatically accelerate the large-scale calculations necessary to solve shock wave equations related to the problem of developing the implosion bomb. "The Computer and the Brain" is a posthumously published book of a lecture series at Yale which was never actually delivered.

1. McCulloch–Pitts and the Cybernetic Movement

As we discussed in Sect. 2.2.2, von Neumann was certainly motivated by the method that McCulloch and Pitts defined to describe the logical operation of the brain. He understood what neurobiologists today often still haven't, that McCulloch consciously neglected many even then known details of the internal neuronal mechanism, and constructed a simple model. He created a binary threshold device to model the fundamental logical properties of a neuron. The all-or-none character of the neurons was isomorph with the that of the elementary computing units. The logical design of computers, and the techniques of switching theory used by von Neumann grew out from the MCP model. We may assume that von Neumann also saw that an analogy exists between computers and the brain, not only at the elementary hardware level, but at the level of mathematics, as well. Obviously he saw the similarity between the mathematical model of the brain (i.e., the MCP network model), and of the computer (i.e., the Turing machine). The difference lies in the fact that while neural networks contain finite number of neurons, the Turing machine contains an infinite number of elements (i.e., the length of the tape).

2. Self-Replicating Automation: The Machine and Its Description

The general interest in logical automata led von Neumann to the questions of how biological organisms have the ability to self-replicate themselves, and how to increase complexity by evolution.

Von Neumann was able to clarify the logical relationship between *description* (the instruction tape in a computational model, and genotype for biological organisms), and *construction* (the execution of the instructions to build a new individual, or phenotype) in self-replicating systems. Among others he designed the "cellular automata", CA, which is a two-dimensional arrays of cells. His construction was far from simple, each cell had 29 possible states. While the construction was later subject of simplification, von Neumann not only constructed the first CA, but started a new way of thinking about the algorithmizability of biology, which led to the emergence of the Artificial Life (Alife) movement, as we shall discuss it in Sect. 9.1.2. Cellular automata implements parallel computing. Basically, von Neumann offered the first non-Neumann computing architecture as well.

3. Reliable Calculation with Unreliable Elements

Michael Arbib's (slightly edited) story well explains the problem [19]. "One of the issues that John von Neumann and McCulloch discussed was reliability in the brain. One version of the story was that McCulloch got a 3 : 00 AM phone call from von Neumann to say, "I have just finished a bottle of creme de menthe. The thresholds of all my neurons are shot to hell. How is it I can still think?" (In other versions of the story, von Neumann was called by McCulloch, and the drink was whisky.) Three answers came to that question. First, von Neumann devised explicit ways of building redundancy into model nervous systems, with one reliable neuron replaced by a bank of many unreliable neurons; the strategy was to keep taking majority votes after each bank of similar neurons so that even if many neurons were unreliable, the overall ensemble would be reliable [542]. Second, McCulloch's idea was to build circuits using neurons whose function would not change with moderate shifts in threshold [339]. The diagrams here are very similar to those of the 1943 classic model, but now the "logical calculus" has the additional twist that network function must be relatively stable in the face of threshold fluctuations. Finally, Jack Cowan and Shmuel Winograd, working in McCulloch's group, took Shannon's theory of reliable *communication* in the presence of noise [463] and showed how to use the redundancy that Shannon had come up with for the encoding message to recode neural networks to provide sufficient redundancy for reliable *computation* in the presence of noise [565]. "

Cybernetics, Artificial Intelligence, Cognitive Science

While there seemed to be an analogy between the brain and the computer at the elementary hardware level and at the level of mathematical (quasi)-equivalence, the *organization principles*, however, are very different. It became clear that the actual biological substrate is very important. In particular, the synaptic organization of neural networks has a fundamental role in the implementation of neural functioning. In the same period when AI developed, John Eccles (1903–1997) worked on and understood the physiological mechanism of excitatory and inhibitory synaptic transmission and their interplay in neural networks. These works have been summarized in [139].

5.6 Artificial Intelligence, Herbert Simon and the Bounded Rationality

The birth of the formal AI was the Dartmouth Conference held in the summer of 1956 (an important year, in many respects) and organized by John McCarthy. The goal was to discuss the possibilities to simulate human intelligent activities (use of language, concept formation, problem solving).[3] Among ten or so others, Herbert Simon (1916–2001), who became a hero of complex systems science, attended the meeting.

Herbert Simon

1. From Mechanism to Function: The First Pillar of AI Is Symbol Manipulation

One of the (not immediate) consequences of the meeting was the break with the tradition of cybernetics. The challenge of the AI was to write computer

---

[3] The perspectives of the cyberneticians and AI researchers have not been separated immediately. Some of McCulloch's papers also belong to the early AI works, as the article titles reflect: "Toward some circuitry of ethical robots or an observational science of the genesis of social evaluation in the mind-like behavior of artifacts". Or this: "Machines that think and want".

programs for showing intelligent behavior without trying to make any connection between the structure of the program and the structure of the brain, and/or the computational algorithm and the neural mechanisms.

The transition from Cybernetics to AI was also a transition from binary symbols to more general ones: neural nets at that time used zeros and ones, AI programs wanted to be able to manipulate general symbols. As Newell and Simon remarked in their 1975 ACM Turing Award Lecture "To put the scientific question, we may paraphrase the title of a famous paper by Warren McCulloch: ...What is a symbol, that intelligence may use it, and intelligence, that it may use a symbol" [375]. Basically, the AI program was based on the assumption that manipulations on general symbols are capable of intelligent behavior.

The Logic Theorist (written by Allen Newell, J.C. Shaw and Herbert Simon) was a simple program, with the intention to mimic the human mind (and has nothing to do with the human brain). It was a *deductive machine*, which was able to prove a set of theorems found of the Principia Mathematica. The program started from a set of axioms, and by using rules of inference lead to a conclusion.

Russell learned what about the achievement of he Logic Theorist and as Simon stated: "...he wrote back that if we'd told him this earlier, he and Whitehead could have saved ten years of their lives. He seemed amused, and I think, pleased."

2. The Second Pillar of AI: Heuristic Search

A heuristic search is an inductive strategy to find solutions to problems.

> Heuristic Search Hypothesis. The solutions to problems are represented as symbol structures. A physical symbol system exercises its intelligence in problem solving by searching – that is, by generating and progressively modifying symbol structures until it produces a solution structure...

3. How Do People Make Inferences?

AI adopts several strategies to make inferences.

There are three types of inference strategies:

- Deduction, finding the effect with the cause and the rule.

- Abduction, finding the cause with the rule and the effect.

- Induction, finding the rule with the cause and the effect.

Automated inference systems were very popular in AI, but did not prove to be the ultimate method of high level problem solving.

4. The Architecture of Complexity

Simon concluded (as we know from the example of the two watchmakers whom we met in Chap. 3), that complex systems (organisms, societies, ecologies) are hierarchical systems. The principle is based on the belief in the existence of intermediate stable organizational levels.

5. Bounded Rationality

Neoclassical economics assumes that people behave rationally, they have preferences among their options, and they try to optimize their utilities. Should the players in an economy not be rational, we are subject of "irrational" emotions. At the level of organizations, the firms tend to maximize their profit. However, with too much information, and not sufficient power to process it, a "sufficiently good" solution instead of a single "best one" could be chosen. Simon's main interest was the understanding of decision making. What are the sources of bounded rationality? Players (I simply don't like the word "agent", even if (s)he is not secret) don't have perfect information about their environment and have a limited computational capacity, therefore they cannot execute all the calculations from the knowledge they have to get an implication, and might have other finite resources as well (such as memory).

6. Inductive Artificial Intelligence

Simon was involved in a famous "rediscovery" of Kepler's third law. He wrote a program called BACON (since it was inductive, Sir Francis got the credit) and the same data was fed which Kepler knew (distances of planets from the sun and the period of revolution). The program was taught to search for patterns by using simple heuristics, and found that the square of a planet's period divided by the cube of its distance is a constant quantity. This was a clear, simple, but conceptually important example for demonstrating what *machine learning* can do by induction. Now, having large-scale databases, new methods of data mining are used in what is now called *Knowledge-Discovery in Databases*. The challenge is to develop and use methods in mining both structured and unstructured data. Structured data may be represented by graphs. Chemical molecules are represented (both in 2D and 3D) by graphs, and there are big corporate databases to support pharmaceutical research. *Chemoinformatics* is a discipline to provide tools for the storage, retrieval and processing databases of chemical structures, see e.g., [205] in an excellently edited book by Sean Ekins. *Texts* and *Images* are stored in an unstructured database as their analysis is different from the structured ones. Searching the database of chemical patent abstracts and published research papers is still important in drug discovery. *Content analysis* has been used much earlier in social sciences to analyze human communication, pragmatically speaking, speeches of politicians. Content analysis helped generate appropriate keywords for texts, and conceptually it was one of the methods which were incorporated by the development of search engines. Inductive reasoning based on large data sets is able to induce conjectures from these databases and tries to verify them, generally by statistical methods.

5.7 Inductive Reasoning and Bounded Rationality: from Herbert Simon to Brian Arthur

Simon clearly stated that bounded rationality better describes the behavior of economic players than "optimal rationality". Popper criticized the inductive method. Somewhat surprisingly, Brian Arthur, who has been the main contributor in applying the complex systems perspective to economics in the Santa Fe Institute, cited neither Popper nor Simon in his papers [26, 27].

5.7 Inductive Reasoning and its Limits

As we have already seen, Arthur adopted/rediscovered the concept of positive feedback (i.e., "increasing return") in economics, in particular its role in magnifying small, random events.

Arthur argues that the typical situation in economics is when there are players, generally heterogeneous (i.e at least two players of different types) and they form *mental models*. Players interact with each other (either directly by communication, or indirectly through information). For the latter an example is the stock market, players interact via the price information.

The El Farol bar problem

"Consider now a problem I will construct to illustrate inductive reasoning and how it might be modeled. N people decide independently each week whether to go to a bar that offers entertainment on a certain night. For concreteness, let us set N at 100. Space is limited, and the evening is enjoyable if things are not too crowded – specifically, if fewer than 60 of the possible 100 are present. There is no way to tell the numbers coming for sure in advance, therefore a person or agent goes – deems it worth going – if he expects fewer than 60 to show up, or stays home if he expects more than 60 to go. (There is no need that utility differs much above and below 60.) Choices are unaffected by previous visits; there is no collusion or prior communication among the agents and the only information available is the numbers who came in past weeks. (The problem was inspired by the El Farol bar in Santa Fe which offers Irish music on Thursday nights; but the reader may recognize it as applying to noontime lunch-room crowding, and to other coordination problems with limits to desired coordination.) The point of interest is the dynamics of the numbers attending from week to week." From [26].

There is one element of the game I don't like. There is no penalty for the situation when the number of guests is too small. I think the majority of the readers also would find it boring to be the only guest in the El Farol Bar.

The analysis of the El Farol bar problem led to the birth of the Minority Game.

5.8 Minority Game

A minority game (MG) is a repeated game where N (odd) players have to choose one out of two alternatives (say A and B) at each time step. Those who happen to be in the minority win. Although being rather simple at first glance this game is subtle in the sense that if all players analyze the situation in the same way, they all will choose the same alternative and will lose. Therefore, players have to be *heterogeneous*. Moreover, it is frustrating since not all the players can win at the same time: this is an essential mechanism for modeling competition.

MG is an abstraction of the El-Farol's bar problem [26]. It is simply a game with artificial players with *partial information* and *bounded rationality*. (Herbert Simon is cited very rarely in the econophysics literature.) *Decisions* are brought based on the knowledge of the M last winning alternatives, called *histories*. Take all possible histories and fix a choice (A or B) for each of them, such an assignment is called a *strategy*, which is like a theory of the world.

Players should have some inductive strategies to predict the number of attendees. Strategies may be subject of change, as the game is *adaptive*.

Let us denote the history of the game by $\mu(t)$ and let the N interacting players bring their $a_i(s_i(t), \mu(t))$ $(-1; +1)$ decisions based on this and their $s_i(t)$ strategy.

Let us suppose that we make simulations for this game. In all simulations each agent has a *binary strategy*, i.e., she can choose from one of two strategies which are randomly chosen for each agent at the beginning of the game. The strategy is a look-up table which assigns the actual choice to the M component vector $\mu(t)$ of the previous M outcome signs, $sgn(A(t))$. $A(t)$ is defined as $A(t) = \sum_i a_i(s_i(t), \mu(t))$. An *exponential learning* rule is adopted: each player chooses the sth strategy from the possible ones with probability

$$P[s_i(t) = s] = Z_i e^{\Gamma_i U_{is}(t)}, \qquad (5.1)$$

where

$$Z_i^{-1} = \sum_{s'} e^{\Gamma_i U_{is'}(t)}, \qquad (5.2)$$

Γ_i is constant (analogous to inverse temperature), and the performance of the player i's s strategy is evaluated by a cumulative score $U_{is}(t)$. The updating rule for the evaluation is

$$U_{is}(t+1) = U_{is}(t) - a_i(s_i(t)\mu(t))A(t), \qquad (5.3)$$

where the second term is the gain of individual players.

5.8 Minority Game

MG can be considered as very simple toy model of minority mechanisms in financial markets. The stock market can be considered as a complex adaptive system, and in Sect. 9.1.4 the Santa-Fe Artificial Stock Market model will be mentioned, as a famous, elaborated example.

As a generalization of the original game, one of my former students, Csaba Földy, defined a hierarchically organized minority game, and studied its properties [179]. In this hierarchical extension of the model we imagine N_g games each with N players being run simultaneously. The structure of the model is shown in Fig. 5.2.

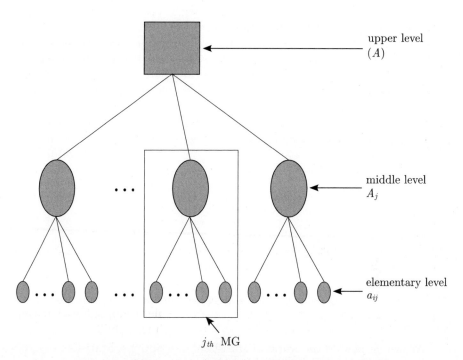

Fig. 5.2. Hierarchical organization of the minority games.

The performance of a strategy now depends not just on the results of the player's own game, but also on the results of the other games. In this way there is a link to the other games. The magnitude of this link is a control parameter in the game. The results show that if the link with other games is strong enough, then the efficiency of the games changes significantly. Also, there are a number of measures of efficiency and the symmetry between the two possible decisions, which relate to the interaction between the different levels consisting of the individual players, the individual games, and the global average over all games. These different measures give different insights into the behavior of these games, and the details can be found in the original paper [179]. The general message is, in any case, that inductive reasoning is an applicable strategy in the real life. But be careful and remember the anecdote!

From Russell to B. Arthur, a famous Bertrand Russell story cited by B. Arthur:

A schoolboy, a parson and a mathematician are crossing from England into Scotland in a train. The schoolboy looks out and sees a black sheep and says, "Oh! Look! Sheep in Scotland are black!" The parson, who is educated, says, "No. Strictly speaking, all we can say is there is one sheep in Scotland that is black." The mathematician says, "No, still not correct. All we can really say is that we know that in Scotland there exists at least one sheep, at least one side of which is black."

From [27].

5.9 Summary and "What Next?"

For the logical interdependence of deductive and inductive approaches to complex systems studies see Fig. 5.3.

We switch now to the world of statistical descriptions. Statistics suggests that mass phenomena are characterized by means, and by the deviations form these means. Gaussian distribution, the most important statistical distribution suggests that there IS a mean, and the deviation from it is symmetric. While it is extremely useful, many complex phenomena are different. The next chapter is about the properties and generation of these distributions.

5.9 Summary and "What Next?"

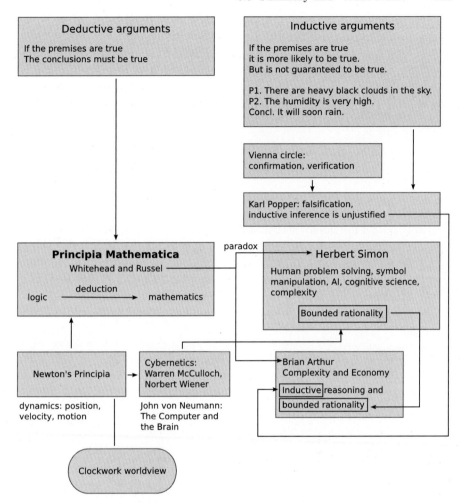

Fig. 5.3. Logical relationship among the deductive and inductive aspects of complex systems research.

6

Statistical Laws: From Symmetric to Asymmetric

6.1 Normal Distribution

6.1.1 General Remarks

Generally, (continuous) biological variables (from heights and weights to IQ) are characterized by the Normal (or Gaussian) distribution. The Gaussian distribution is symmetric, so deviation from the "average" to both directions has similar properties. In addition, the distribution is unimodal, i.e., the probability density function (PDF) has one maximum; furthermore the expectation value (or mean) coincides with the value, where the probability density function takes its maximum value. The plot of a normal probability density function has the form of a *bell curve*, and seen in Fig. 6.1.

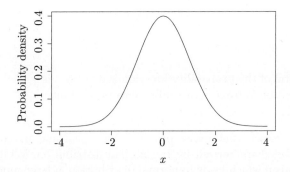

Fig. 6.1. The normal distribution is symmetric and unimodal.

In applications, the notions of *location parameter* and *scale parameter* are very useful. A location parameter is identified by the location of the maximal value of the PDF, so in Fig. 6.1 it equals zero, and the scale parameter characterizes the width of the distribution; here it is one. Change in the location parameter means a shift in the horizontal axis, while an increase in the scale parameter expresses a stretch of the PDF. Although for Gaussian the location and scale parameters coincide with the mean and standard deviation, this coincidence is not valid for the majority of distributions.

There is an empirical rule called "The 68–95–99.7 Rule For Normal Distributions", which tells:

- Approximately 68% of the observations fall within one standard deviation of the mean.

- Approximately 95% of the observations fall within two standard deviations of the mean.

- Approximately 99.7% of the observations fall within three standard deviations of the mean.

So it is *highly* improbable to get experimental data, which has a *much larger* deviation from the mean. We never see people who are four meters tall, not to mention 30 meters. (A 30 meters tall man would be a data point 20 standard deviations from the mean.)

The probability density function $f(x)$ for a continuous random variable is defined by these properties:

- The probability that x is between two points a and b is $p[a \leq x \leq b] = \int f(x)\mathrm{d}x$ where the integration is from a to b.

- It is non-negative for all real x.

- The integral of the probability function is one, that is $\int f(x)\mathrm{d}x = 1$ where the integration is from minus infinity to plus infinity.

It is well known that many physical, biological and psychological phenomena can be characterized by the normal distribution. While there is no single mechanism which leads to normal distribution, a large number of small, independent, *additive* actions tends to imply normal distribution. *Brownian motion* is a physical model of these assumptions.

6.1.2 Generation of Normal Distribution: Brownian Motion

Physical Model and Some More

We already learned in Sect. 4.2 that the motion of certain (not too small, not too large) particles were seen by Robert Brown. Intuitively a Brownian particle (say, a pollen) moves due to the impact of much smaller (say, water) molecules. The Brownian framework is a special case of *random walk* models, and was used in other contexts as well.

Louis Bachelier (1870–1946) in a paper in 1900 [35] defined Brownian motion and applied it as a model for *asset price* movements. While the paper (actually a dissertation) did not gain a very high reputation after its preparation, recently it is qualified now as the starting point of financial mathematics. Independently of Bachalier, of course, in context of molecular physics, Einstein (1905) [148] and Smoluchowski (1916) gave a sophisticated explanation of Brownian motion, by calculating the temporal change of the mean square displacement of the Brownian particle, and the connection between the mobility of the particle and the – macroscopic – diffusion constant.

Activity of a single neuron was considered as an abstract Brownian particle, and its motion to the effect of excitatory and inhibitory actions was given first in a famous paper by Gerstein and Mandelbrot in 1964 [199], which opened the research of mathematical analysis of neural noise processes. Stochastic neuron models will be reviewed in Sect. 8.4.2.

Mathematical Model

The Brownian motion belongs to the class of continuous time continuous state space stochastic processes. The classical derivation assumed that the forcing function has a "systematic" or "deterministic" part, and an additive term due to the "rapidly varying, highly irregular" random effects:

$$\text{"rate of change of state"} = \text{"deterministic rate"} + \text{"random rate"}. \qquad (6.1)$$

According to a standard further assumption the random term is a linear function of "white noise". White noise is considered as a ξ stationary Gaussian process. It is characterized by the mean $E[xi_i] = 0$ and autocorrelation function $E[\xi_i \xi_{i'}] = \delta_{ii'}|t - t'|$; δ is the delta function.

A stochastic process X_t obeys an (Ito-type) stochastic differential equation (SDE) which has the form (in the autonomous case)

$$dX_t = a(X_t)dt + b(X_t)dW_t, \qquad (6.2)$$

where dW_t is the Wiener process.

The introduction of the Wiener process was motivated by its connection to white noise. Accordingly

$$dW_t = \xi_t dt. \qquad (6.3)$$

Both from theoretical and practical points of view the connection between the SDE and the evolution equation for the PDF (called the Fokker-Planck equation in physics) is very important. The solution of (6.2) under a rather general condition is a diffusion process (a special case of Markovian stochastic processes) defined by the infinitesimal generator A (for scalar case):

$$A = a\frac{\partial}{\partial x} + \frac{1}{2}b\frac{\partial^2}{\partial x^2}. \qquad (6.4)$$

Here $a(x,t)$ is the velocity of the conditional expectation (called "drift"), and $b(x,t)$ is a the velocity of the conditional variance (called a "diffusion constant").

The general form of a Fokker–Planck equation is:

$$\frac{df}{dt} = Af. \qquad (6.5)$$

6.1.3 Liouville Process, Wiener and Special Wiener Process, Ornstein–Uhlenbeck Process

Three important continuous Markov processes are briefly discussed here. (For educational purposes Gillespie's book [204] is recommended.)

1. $b = 0$ defines the Liouville process (Fig. 6.2). The Liouville process is characterized by the lack of the diffusion term. Since the effect of the diffusion term is to spread out the pdf, the Liouville process is visualized as a drift of the PDF in time without changing its shape. Starting from a degenerate density function (i.e., when the initial PDF is concentrated to a point), the point will be drifted.

2. $a(x,t) = a$, $b(x,t) = b$ defines the Wiener process (Fig. 6.3), i.e., $X(t) \equiv W(t)$.

$a = 0$, $b = 1$ define special Wiener processes.

The Wiener process is a Gaussian process, but is not stationary. It is used in modeling the Brownian process.

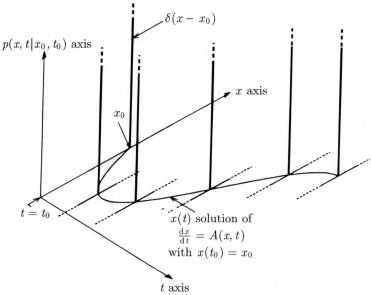

Fig. 6.2. Temporal evolution of a Liouville process. From [204].

For finite $t > t_0$ the PDF is Gaussian, with mean $x_0 + a(t - t_0)$, and the standard deviation $|b(t-t_0)|^{1/2}$ is flattening out as time tends to infinite. For a special Wiener process the center of the spreading curve remains unchanged.

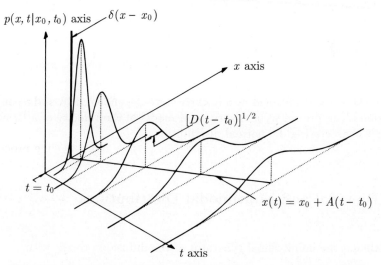

Fig. 6.3. Temporal evolution of special Wiener process. From [204].

3. The Ornstein–Uhlenbeck process (OU, Fig. 6.4) is defined as

$$a(x,t) = -kx, b(x,t) = b, \quad (k > 0, D \geq 0). \tag{6.6}$$

OU has a stationary PDF, and it is the normal density function. For any $t > 0$ the PDF is the normal density function with an exponentially moving mean. The temporal evolution of the standard deviation describes spreading out, but the width of the curve is finite even for infinite time.

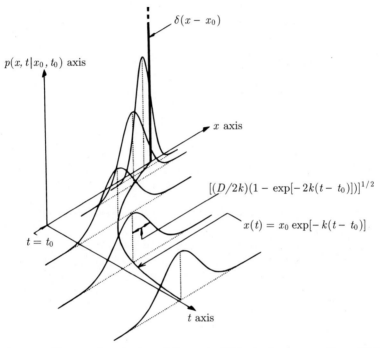

Fig. 6.4. Temporal evolution of Ornstein–Uhlenbeck process. From [204].

The OU process, defined by a linearly state-dependent drift and a constant diffusion term, proved to be a very good model of the velocity of a Brownian particle characterized by normal distribution.

6.2 Bimodal and Multimodal Distributions

By defining the infinitesimal generator of the diffusion process with

$$a(x,t) = Ax^3 + Bx + c, \quad b(x) = b > 0, \tag{6.7}$$

bimodal distribution may emerge. A bimodal distribution is characterized by two peaks (see Fig. 6.5). For such kinds of distribution, the mean does not bear too much information. The location (of the maximum values, also called as mode) is not unique but there are two values.

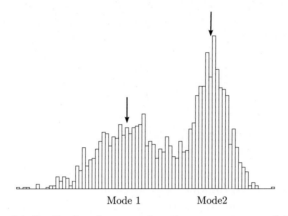

Fig. 6.5. Bimodal distributions have two location parameters, and the mean is not a characteristic value.

Multimodality of the stationary distribution might be associated, at least loosely speaking, with multistationarity of deterministic models. The maxima correspond to the stable stationary points and the minima to the unstable ones. Stochastic catastrophe, an extension of the deterministic catastrophe theory, deals with the sudden transition between distributions characterized by different modes [101, 546].

6.3 Long Tail Distributions

6.3.1 Lognormal and Power Law Distributions: Phenomenology

The family of "long tail" or "heavy tail" distributions is well known in statistics. These distributions are *skewed*. Skewness is a measure of the asymmetry of a distribution. The positively and negatively skewed distributions have a longer tail to the right and to the left, respectively. Occasionally there are tails on both sides (say, the double Pareto distribution, see Fig. 6.6).

It is far from being a rule, but biology is dominated by Gaussian distributions, while some social systems show striking skew distributions (such as the

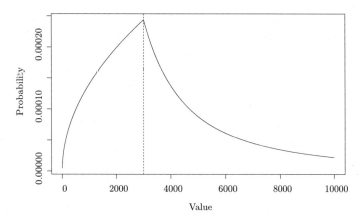

Fig. 6.6. Double Pareto distribution.

lognormal distribution or power law distribution). A catalogue of phenomena showing power law distributions was listed and analyzed by Mark Newman [381]. More or less we follow his list, but not rigorously his analysis.

1. The *frequency of words* is more or less inversely proportional with their ranks in a frequency table. This "$1/f$" law is called the Zipf's law (Fig. 6.7). Zipf plotted the frequency of words from texts of several different languages, and found that on log-log coordinates it was a decreasing straight line with 45°.[1]

2. *Citation of scientific papers.* As we (not so elite) scientists painfully know, many papers get very few, if at all, citations, and a small fragment gets the majority.

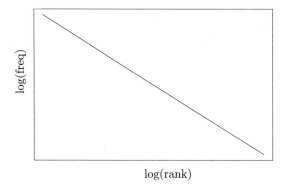

Fig. 6.7. The idealized Zipf's law.

---

[1] There is comprehensive website of papers on Zipf's law maintained by Wentian Li: http://www.nslij-genetics.org/wli/zipf/.

Redner's analysis [419] of the Physics Review papers and citations (353, 268 papers and 3, 110, 839 citations from July 1893 through June 2003):

- 11 publications with > 1000 citations
- 79 publications with > 500 citations
- 237 publications with > 300 citations
- 2,340 publications with > 100 citations
- 8,073 publications with > 50 citations
- 245,459 publications with < 10 citations
- 178,019 publications with < 5 citations
- 84,144 publications with 1 citation.

3. *Web hits.* The number of "hits" received by servers, say, during a single day: yahoo.com was reported to have the most traffic.

4. *Copies of books sold.* The copies of books sold shows a power law distribution. There are few bestsellers, and publishing houses complain that they are able to sell few copies of many titles. The Bible, Mao Zedong Quotations, and the Harry Potter series lead the list, (and hopefully Complexity Explained will be a moderate success, too).

5. *Telephone calls.* The number of calls received a single day by subscribers to AT&T. There was a customer who got 35846 calls.

6. *Magnitude of earthquakes.* It called Richter's law, and we will discuss in Sect. 9.3.2.

7. *Diameter of moon craters*

8. *Intensity of solar flares*

9. *Intensity of wars.* It is measured by relative battle deaths, i.e., the number of deaths divided by the total population of the countries.

10. *Wealth of richest people.* Wealth distribution is also far from being symmetric.

11. *Frequency of family names.* See Fig. 6.8. A mechanism of the generation of their distribution will be given in the next section.

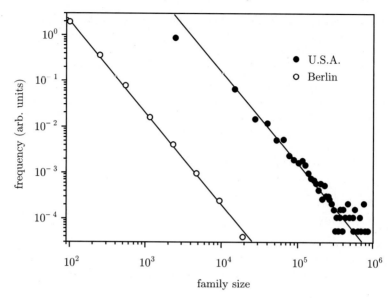

Fig. 6.8. The distribution of family names. From [579].

12. *Population of cities.* The size of a city seems to be inversely proportional to its rank order so that, the 100th largest city is a tenth the size of the 10th largest city. Size of the cities, and settlements are traditional examples of Zipf distribution.

13. *Income distribution.* Vilfredo Pareto (1848–1923) collected and analyzed statistical data of income distributions. Originally he investigated the allocation of wealth, which was characterized by the Pareto principle or the "80 − 20 rule" which says that 20% of the population owns 80% of the wealth.

6.3.2 Generation of Lognormal and Power Law Distributions

Here I will admit that I have a semi-serious hypothesis about the periodic recurrence of scientific fashions, that I like to mention during classes or lab meetings. The hypothesis states that scientific fashions return every 25 years, or so. The explanation is that we are able to learn something new during the ages 25–30 (basically after graduating from college), and we shall have sufficient power to propagate what we know around our age fifty.

The history of the power law distribution is a good example to justify some aspects of the hypothesis. The Pareto distribution was introduced in 1897 to characterize income distribution. Around 1925 Alfred Lotka analyzed the frequency distribution of scientific productivity, and George Udny Yule (1871-1951) gave an algorithm for the power law generation in biological context. Several years later Robert Gibrat (1904-1980) derived the rule of proportionate growth which may lead either to log-normal or power law distribution, depending on the details of the generating process. Zipf distribution was offered for word frequencies and city sizes in 1949, while Herbert Simon's algorithm [466] explains the formation of power law distribution based on similar concepts, what later became the celebrated principle of "preferential attachment". There was a (well, now) amusing duel between Herbert Simon and Benoit Mandelbrot [356] whether Simon's algorithm or Mandelbrot's suggestion to use an optimization principle is better. As we see now, they are somewhat complementary algorithms. In these years, namely 1957, William Schockley (1910-1989) the Nobel prize winner co-inventor of the transistor re-analyzed the model of scientific productivity and explained the log-normal distribution he found in an experiment with researchers of the Brookhaven National Laboratory. 25 years later, Montroll and Schlesinger ([358] went back to this paper:

"Shockley explained that, to publish a paper, one must (i) have the ability to select an appropriate problem for investigation, (ii) have competence to work on it, (iii) be capable of recognizing a worthwhile result, (iv) be able to choose an appropriate stopping point in the research and start to prepare the manuscript, (v) have the ability to present the results and conclusions adequately, (vi) be able to profit from the criticism of those who share an interest in the work, (vii) have the determination to complete and submit a manuscript for publication, and (viii) respond positively to referees' criticism..."

While skew distributions seem to be ubiquitous, there seems to be a consensus now about the non-existence of an individual mechanism to generate such distributions [356, 381].

Even more, there are several different questions to be answered: How did sociobiological mechanisms lead to the formation of asymmetric distributions from normal distribution? How are skew distributions generated from a degenerate distribution?

Reed and Hughes [422] suggested a relatively general mechanism of power law generation. In one sentence: "Randomly stopped random processes leading in average exponential growth generates power laws." This mechanism is based on two assumptions. First, there is stochastic exponential growth. Such kinds of growth processes can be generated by a *multiplicative* random process (instead of having an additive random term to the deterministic forcing function). Second, the time that the process is observed (e.g., "frozen", "killed") is chosen from an exponential distribution.

There are four classes of stochastic processes which generate exponential growth:

1. continuous time continuous state space (CCS),

2. discrete time continuous state space (DCS),

3. continuous time discrete state space (CDS),

4. discrete time discrete state space (DDS).

1. Geometric Brownian motion (GBM), is a continuous time continuous state space stochastic (CCS) process, where the logarithm of a random variable makes a Brownian motion as described by the *stochastic differential equation*:

$$dX_t = \mu X_t dt + \sigma X_t dW_t. \tag{6.8}$$

Here W_t is the special Wiener process, and dW_t has a normal distribution with mean 0 and variance dt. The conditional expectation is $E(X_t|X_0) = X_0 \exp(\mu t)$. GBM is able to generate a *double Pareto distribution*, where not only an *upper tail*, but also a *lower tail* exists. Somewhat more technically, the PDF of the double Pareto distribution is proportional to $x^{-\alpha-1}$ for $x > 1$, and to $x^{\beta-1}$ for $x < 1$ [420]. The distribution of total income in the USA shows this behavior, see Fig. 6.9.

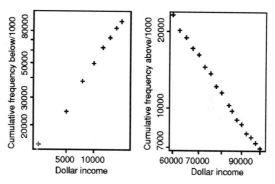

Fig. 6.9. Distribution of money income in the USA in 2000. Adapted from [422].

The Black–Scholes model (together with its versions) is maybe the most widely accepted financial model for option pricing [63]. An option is defined as "contract between two parties in which one party has the right but not the obligation to do something, usually to buy or sell some underlying asset". The model was exactly a GBM, and has been a big success.

2. A discrete time continuous state multiplicative process (DCS) is defined as
$$X_{n+1} = Z_n X_n, \qquad (6.9)$$
where the set of Z_ns are independent identically distributed random variables with mean μ. The conditional expectation is $E(X_n|X_0) = X_0 \mu^n$.

Depending on the nature of the multiplicative process there are three different outcomes. It is increasing, when $P(Z_n > 1) = 1$, and decreasing when $P(Z_n < 1) = 1$. It may be bidirectional when $P(Z_n > 1) > 0$ and $P(Z_n < 1) > 0$. Among others, the upper-tail behavior of the number of visitors to websites was explained by a multiplicative model combined with random killing [244, 356].

3. Homogeneous birth-and-death processes belong to the family of continuous time discrete state space processes(CDS).

Yule process

A population starts at time 0 with one individual. As time increases, individuals may give birth to a new individual, the chance of any particular individual giving birth during time $[t, t + dt]$ being λdt. This is called the linear birth process or Yule process.

From [9].

The long tail property of the distribution of the number of species per genus was considered by Yule in 1924 and the paper was rediscovered several times. Yule made two basic assumptions:

- The initial number of species in a genus is one. New species of the same genus are generated by a Yule process with a λ parameter.

- A new species of a novel genus appears at constant rate μ, and then it behaves as the previous assumption prescribes.

He used a pure birth process, and calculated the probability distribution of the number of species per genus. Extinction was neglected, assuming that it happens by cataclysmic events only. While the upper-tail distribution was explained by the randomly killed pure birth mechanism, a more realistic mechanism took into account the random individual extinctions. In this case the applied stochastic model contained a randomly killed birth and-death process [423].

4. A Galton–Watson branching process is a discrete time discrete state space process (DDS). It was originally applied to the extinction of British surnames.

Reed and Hughes [424] gave a model for the frequency distribution of family names. A Galton–Watson process described the growth in the number of persons with a given family name, and it was supplemented with a Poisson process to model new names created either by changing an already existing name or by immigration. Based on these assumptions the naive analysis gave the result that the probability of having m individuals with a given name for large m is

$$P[\text{exactly } m \text{ individuals have this name}] \approx m - 1 - k \quad \text{as } m \to \infty.$$

A more thorough analysis showed that the probability for large m is proportional to $Q(m)m^{-1-\kappa}$, where $Q(m)$ is a bounded, log-periodic function, and κ is determined by a the parameters of the immigration process and the offspring distribution. The log-periodic correction of the power law distribution was suggested by [206] reflecting scale hierarchies and may have a substantial role in predicting the behavior of complex systems. Mathematically it led to the extension of dimensions from fractals to complex numbers. We shall return to this topic in Sect. 9.3.6.

Other Mechanisms that Generate Power Laws

Newman [381] also gave a general overview about the different mechanisms of power law generation. We used here what he calls the "combination of exponentials" (where one exponential refers to growth and another to the killing time). Of course, exponential growth is related to autocatalysis or positive feedback. Several other mechanisms will be discussed in later chapters.

"Preferential attachment" was suggested to describe the generation of degree distribution (also called edge distribution) of evolving networks. It is a simple model with scale free behavior, and the edge distribution follows a power law. Such behavior has been found in many networks such as airport networks, scientific collaboration networks, and movie actor networks. The model is very simple, and as the Reader certainly knows, it became extremely popular [5, 43].

The model is extremely simple. One starts with a few connected nodes and then adds nodes one at a time. Each time a node is added, m links are made to existing nodes with a probability proportional to the number of edges k, already attached to each node. Thus, nodes which already have many links are more likely to be linked to the new nodes. The path length of these networks grows logarithmically. There are a number of variations on this model. For example, in non-linear models the probability of forming a link with another node is a non-linear function of k.

Power law behavior could also be generated at special "critical points". Two possible mechanisms, self-organized criticality and intermittent criticality will be reviewed in Sect. 9.3.1.

After studying the most important statistical distributions and the algorithms which generate them we discuss a difficult conceptual question. We often intuitively feel that complex structures are neither purely deterministic nor random. In the next chapter such kinds of complex structures are analyzed.

7

Simple and Complex Structures: Between Order and Randomness

7.1 Complexity and Randomness

A large family of complexity measures is related to the concept of randomness. Within this family there are two classes. For the first class complexity increases monotonically with randomness, and for the second one there is some maximum at some intermediate value (Fig. 7.1). Interestingly, these measures reflect different aspects of what we call complexity.

Algorithmic Complexity and Structures

The most well-known example for the first class is the Kolmogorov algorithmic complexity. The complexity of an arbitrary string of characters is the number of bits in the shortest program computing it on a standard universal computer. It says, if there is a data set (encoded, say, as a series of zeros and ones), and there is no other way to explain the generation of this data set, that it is *algorithmically complex*. But if the data set has some pattern, then one may write a computer program to generate this data set, and the program is shorter than the data set itself. The program "1. Write zero! 2. Write one! Go to 1!" can generate the infinitely repeated "01" string.

The algorithmic complexity of a string s denoted by $K(s)$ is defined by the length of its minimal description $d(s)$: $K(s) := |d(s)|$. $K(s)$ is maximal if the string is random.

Complexity: Between Order and Randomness

According to the second approach, the two extreme cases, namely perfectly ordered and perfectly random systems have very low structural complexity.

A system, which is the mixture of order and randomness, is structurally more complex [245, 217].

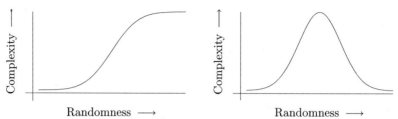

Fig. 7.1. Classification of various complexity measures based on their relationship to randomness. *Left*: first type. More random structures have more complexity. *Right*: second type. Purely ordered and purely random structures have small complexity. Systems neither totally ordered nor totally random have larger complexity.

> Unfortunately, the term *complexity* has been used without qualification by so many authors, both scientific and non-scientific, that the term has been almost stripped of its meaning. Given this state of affairs, it is even more imperative to state clearly why one is defining a measure of complexity and what it is intended to capture.
>
> Measures of statistical complexity: Why? From [173].

Complexity of Patterns

Grassberger [217] analyzed the complexity of patterns generated by simple algorithms. Specifically, a one-dimensional cellular automata created different patterns, as Fig. 7.2 shows.

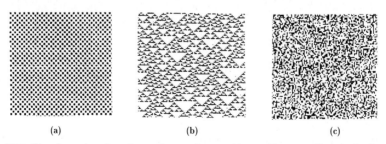

Fig. 7.2. Purely ordered and purely random patterns have small complexity. Patterns neither totally ordered nor totally random have larger complexity. Adapted from [217].

Based on this analysis it seems to be intuitively easy to accept that there are such kinds of (statistical) complexity measures which have low values for both random and highly ordered systems, and show maximum when a system shows an intermediate level of order.

Ordered and Random Structures

Crystal structures of solids show regularity and order, while certain solid, or solid-like substances, called as "amorphous solids" don't have well defined structures with long-range order. Glass is a typical example of such kinds of random structures, see Fig. 7.3.

Fig. 7.3. Crystal and amorphous substance. From http://www.geo.uw.edu.pl/ ZASOBY/ANDRZEJ_KOZLOWSKI/e_kr1.htm.

7.2 Structural Complexity

Chemical molecules are formed by atoms, and molecular structures have been subject of chemical research for many years. Intuitively it was plausible that there are "simpler" and "more complex" molecules.

One of the most important aspects of chemistry is to characterize the structure of molecules and connect structural properties to function. Chemical molecules are composed of atoms connected by chemical bonds, and the structure of the molecules can be represented by *graphs*. Molecules are obviously three-dimensional structures and a two-dimensional representation preserves the connectivity but not the full geometrical relationships. For a review of measures of network complexity motivated mostly by chemical applications see [66]. The structural analysis of chemical molecules leads us to introduce some basic concepts of graph theory.

7.2.1 Structures and Graphs

Some Basic Concepts of Graph Theory

A network or graph G is defined as a set V of nodes (occasionally also called vertices, points) and a set E of edges (links, ties). An edge (i,j) is a line which connects the i and j nodes. An edge may or may not be directed. Two (i and j) nodes are *adjacent* if they are connected by an edge (i,j). A directed (i,j) edge is said to point from node i to node j. In this section we will deal with undirected graphs, unless told otherwise.

The adjacency relation, say, *among atoms in a molecule* may be characterized by an adjacency matrix, where an element $a_{ij} = 1$, if a pair of nodes i, j are connected to each other, and $a_{ij} = 0$, if there is no adjacency. The *degree* (d_i) of a node (i) is the total number of edges originating and terminating from and to it (of course, the distinction exists for directed graphs only). The total degree (or adjacency) of a graph G is simply the sum of the node degrees which is the same as the total number of edges multiplied by 2: $A(G) = \sum_{i \in V} d_i = 2|E|$. A graph can be characterized by the average node degree, $\langle a_i \rangle := A(G)/|V|$.

A path in the graph is a sequence of adjacent vertices. If the first and last vertex is the same in the sequence the path is called a loop or cycle. The graph is connected when there is at least one path between any pair of vertices in it. If a graph has no cycles is called acyclic. A tree is a connected acyclic graph. A forest is a graph which is a set of trees, i.e., it is acyclic but not necessarily connected. A path graph with n nodes, $P(n)$, is a special tree in which the degree of any vertex is less then three. A star-graph, $S(n)$, is a graph containing one central vertex and $n - 1$ branches of length one edge; so this is a special tree in which all but one nodes have degree 1. In a complete (or full) graph, $K(n)$, any two vertices are connected by an edge. A path, a star and a full graph are illustrated in Fig. 7.4.

A *subgraph* is a graph generated from a graph by eliminating at least one edge or a node with its incident edges. A graph might have *components*, i.e., connected subgraphs or nodes that are not connected to each other.

Graph Distances and Their Role in the Structure – Function Relationship

The distance d_{ij} between nodes i and j is the shortest path distance from i to j. The Wiener index introduced by Harold (and not Norbert) Wiener in

Fig. 7.4. A path, a star and a full graph, with 20, 20 and 10 vertices, respectively.

1947 is denoted by $W(G)$ and defined as the sum of the distances of all pairs of nodes in the graph.

$$W(G) = \frac{1}{2} \sum_{i,j} d_{ij}. \tag{7.1}$$

Note that the Wiener index is not defined for unconnected graphs but of course this is not a big problem with chemical molecules, which is its main application area.

The average node distance $\langle d_i \rangle$ and average graph distance $\langle d \rangle$ (called as graph radius or average path length) are defined as

$$\langle d_i \rangle = \frac{W}{V}, \tag{7.2}$$

$$\langle d \rangle = \frac{W}{V(V-1)}. \tag{7.3}$$

Wiener index is the oldest topological index, and it was compared to physical-chemical properties of some organic compounds. Although formally it is meaningful for cyclic graphs, it was only considered for acyclic graphs, i.e., for molecules not containing loops. See Fig. 7.2.1 for an example.

Arthur Caley (1821-1895) used graphs to enumerate isomers of alkanes. (Isomer means that the arrangement of atoms is different while the number of the individual atoms is preserved.)

	1	2	3	4	$\sum_i d_{ij}$
1	0	1	2	2	5
2	1	0	1	1	3
3	2	1	0	2	5
4	2	1	2	0	5

$$W = \frac{1}{2} \sum_{i=1}^{N} \sum_{j=1}^{N} d_{ij} = 9$$

Fig. 7.5. Calculating the Wiener index for 2-methyl propane.

More specifically, the molecules called alkanes have the chemical formula C_nH_{2n+2} where C represents a carbon atom and H represents a hydrogen atom. The alkanes are acyclic saturated hydrocarbons. Each carbon atom C has four chemical bonds and each hydrogen atom H has one chemical bond. Because of these bonding properties of carbon and hydrogen, it is known that each n-carbon alkane is a tree which is an acyclic connected graph containing $3n + 1$ edges. Only the carbon skeleton is analyzed generally, so vertices refer to carbon atoms but not hydrogen atoms. The boiling point is a characteristic quantity for the alkanes and experimental data exists mostly on molecules with not larger than ten carbon atoms. Figure 7.6 shows a relationship between boiling points and Wiener index. The Wiener index proved to be useful for predicting the boiling point of larger molecules and the approach has been extended in many ways and became the basis of methods like quantitative structure-activity relationship (QSAR) and quantity structure-property Relationship (QSPR).

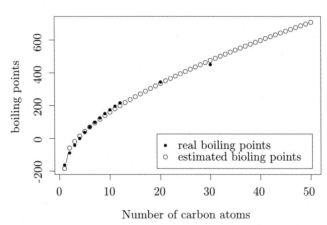

Fig. 7.6. The boiling point of alkanes versus the prediction based on their Wiener index. The prediction was calculated by nonlinear regression and has the following form: $126W^{0.2} - 185$.

Structural Complexity of Molecules: 3D Analysis

Chemical molecules can be considered as typed point systems. These systems consist of points with a different types for each point. The spatial position of each atom within a molecule is specified by numerical values in an appropriate coordinate system. Three spatial coordinates and one atom type coordinate define a four-dimensional space. Chemically it means that not only the existence of chemical bonds, but also their angles matter. Complexity measures were defined [571, 572] to express that larger diversity implies larger complexity. A molecule with the composition of CHClBrF has larger complexity than CH_4, since the atomic constituents of the first molecule is more diverse than that of the second. For molecules with the same atomic composition, but with different bond length, and angles, the more diverse is more complex: the C_5H_{12} (pentane) molecule exists in three different isometric forms, as Fig. 7.7 shows.

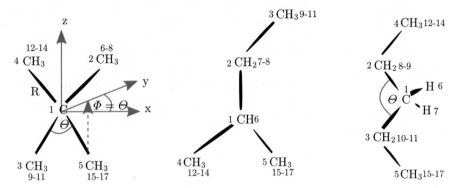

Fig. 7.7. The geometrical structure of the pentane isomers.

The absolute complexity of a typed point system compared to the maximum theoretical complexity of a system with the same number of elements is called its relative complexity. The relative complexity for neopentane $C(CH_3)_4$, and n-pentane $CH_3-CH_2-CH_2-CH_2-CH_3$ and isopentane $CH_3-CH_2=CH-(CH_3)_2$ is 9/58, 13/58 and 21/58. Isopentane is thus the most complex of the three noncyclic pentanes according to this measures.

Clustering Coefficient

A quantity which was known as transitivity in the social network literature gained much attention recently under the name clustering coefficient. Clustering coefficient is the probability that two nodes which have a common neighbor are connected by an edge. Mathematically the (local) clustering coefficient of node i is defined as [380]:

$$c_i := \frac{\text{number of triangles connected to } i}{\text{number of triples connected to } i}, \qquad (7.4)$$

the clustering coefficient of the network is the average clustering coefficient for its nodes.

7.2.2 Complexity of Graphs

Complexity of Graphs: Combination of Adjacency and Distance

The combination of adjacency (i.e., a measure of connectivity), and distance gives the possibility to characterize graphs on a more specific way. The ratio of $A/D = \langle a_i \rangle / \langle d_i \rangle$ is a possible complexity measure.

With a fixed number of nodes, A/D has minimum value for path graphs. For such graphs the connectivity is low, and distances are long. However, on a complete graph $K(n)$ the A/D has maximal value, since it is maximally connected, and the distance is minimal, each node is separated from the others by a single edge. Star graphs have intermediate complexity. Similarly, graphs with one cycle also show intermediate A/D value.

$$\frac{A}{D(P(n))} = \frac{6}{n(n+1)}, \tag{7.5}$$

$$\frac{A}{D(K(n))} = 1, \tag{7.6}$$

$$\frac{A}{D(S(n))} = \frac{n}{n-1}. \tag{7.7}$$

Complexity of Graphs: Diversity Measure

Huberman and Hogg [245] defined a complexity measure for interacting systems which can be described as tree graphs or forests. The complexity of such systems is based on their branching diversity. The definition of tree diversity is recursive. Let us have a tree T with b subtrees: $T_1, \ldots, T_b$. We want to count the number of interactions on each level. Let us assume that among the b subtrees k are nonisomorphic; the diversity of T is the product of the diversities of these subtrees multiplied by the number of possible interactions between them, a type of interaction is a non-empty subset of the k nonisomorphic subtrees as we count n-ary interactions:

$$D(T) = (2^k - 1) \prod_{j=1}^{k} D(T_{i_j}). \tag{7.8}$$

If a tree has no subtrees, i.e., if it consists of a single root vertex, then the product is empty and $D(T) = 1$. The complexity of a tree is defined as

$$C(T) = D(T) - 1. \tag{7.9}$$

This makes a regular tree (which has the same number of branches at each level) have a complexity of zero. See Fig. 7.8 for some examples.

7.2 Structural Complexity

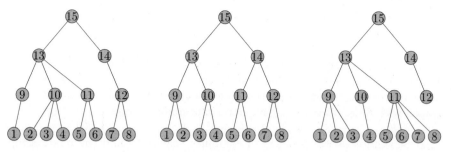

Fig. 7.8. Complexity of three trees with the same number of nodes and depth. The complexity of the left tree can be calculated as follows: $D(T_1) = \cdots = D(T_8) = 1$, $D(T_9) = \cdots = D(T_{12})$ as these contain only isomorphic subtrees, $D(T_{13}) = (2^3 - 1) \cdot 1 \cdot 1 \cdot 1 = 7$, $D(T_{14}) = 1$, $D(T_{15}) = (2^3 - 1) \cdot 7 \cdot 1 = 20$, $C(T_{15}) = 20$. The complexity of the next tree is zero, as it is regular, and the right tree also has complexity 20.

Complexity Measure for Graphs Containing Loops: Cyclomatic Number

Another graph-theoretical complexity measure is the cyclomatic number of the graph. The cyclomatic number $n(G)$ of graph G with n nodes, m edges and p connected components is

$$n(G) = m - n + p. \tag{7.10}$$

In a strongly connected graph, the cyclomatic number is equal to the maximum number of linearly independent circles. Since inserting a new edge in G increases $n(G)$ by unity, it can be applied as a complexity measure. This notion has been applied to measure *static software complexity*.

Cyclomatic complexity.

The cyclomatic complexity of a section of source code is the count of the number of linearly independent paths through the source code. For instance, if the source code contained no decision points such as IF statements or FOR loops, the complexity would be 1, since there is only a single path through the code. If the code had a single IF statement there would be two paths through the code, one path where the IF statement is evaluated as TRUE and one path where the IF statement is evaluated as FALSE. Cyclomatic complexity is normally calculated by creating a graph of the source code with each line of source code being a node on the graph and arrows between the nodes showing the execution pathways...

From http://en.wikipedia.org/wiki/Cyclomatic_complexity

Complexity Measure for Graphs: Density

Graph density is quite a simple complexity measure: it is the ratio of the number of edges and the number of possible edges in the graph. The larger the number of edges the larger the graph's density. Analysis and synthesis of graphs with different density have important applications.

It is a fundamental problem of computer science to find the relationship between three graph properties: the number of vertices, the (maximum or average) degree of the vertices and the diameter of the graph. (The diameter is the longest shortest path between any pair of nodes). E.g., if the maximum degree (d) and the diameter (k) is given then the maximum number of nodes (N) is requested. The "Moore bound" is a simple upper bound for this case is given as

$$N_{(d,k)} = \frac{d(d-1)^k - 2}{d-2}, \quad \text{if } d \geq 2. \tag{7.11}$$

The alternative problems are constructing graphs with (1) minimum degree for given N and d and (2) minimum diameter for given N and d. Two of these problems aim to construct graphs with minimal complexity (i.e., density), in the third problem the graph complexity is fixed by N and d and an optimal graph is requested with the given complexity [10].

Let's tell a little bit more about the construction principles. The Moore bound (7.11) is a theoretical upper bound for the number of vertices, but no algorithm is known to actually construct such a graph. Various graph constructions are suggested to approximate this limit and also the limits of the alternatives problems since these are (of course) strongly related. One class of graphs are extended chordal rings [289]. An extended chordal ring with N nodes can be given by a matrix (W) with p columns and $d-2$ rows, where d is the degree of every vertex and p is called the period of the graph and $d \geq 2$, $1 \leq p \leq N/2$ and $N \bmod p \equiv 0$ hold. The vertices of this graph are denoted by integer numbers between 0 and $N-1$ and vertices x, y are connected by an edge if and only if either

$$\begin{aligned} x - y &\equiv 1 \quad \bmod N, \text{ or} \\ x - y &\equiv -1 \quad \bmod N, \text{ or} \end{aligned} \tag{7.12}$$

$x \bmod y \equiv i$ and $y \equiv x \oplus_N w_{ij}$ for some $j = 1, 2, \ldots, d-2$. \quad (7.13)

Here $\oplus_N$ is addition modulo N. This definition means that we start from an ordinary ring graph with the desired number of vertices (N) and then add $d-2$ new edges for edge vertex in the following way: first we divide the vertices into p categories. If a given vertex x is in category i then we add edges from

x to vertices $x \oplus_N w_{i1}, x \oplus_N w_{i2}, \ldots, x \oplus_N w_{i,d-2}$. For small values of d and N all possible values of p can be calculated quickly and then the one with the shortest diameter can be selected. See Fig. 7.9 for an example.

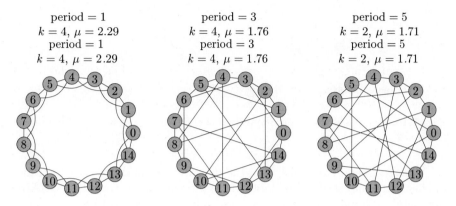

Fig. 7.9. Minimizing the diameter. Here we have $N = 15$ and $d = 4$, and we try all possible periods: $p = 1$, $p = 3$ and $p = 5$. The shortest diameter can be obtained with the $p = 5$ choice. μ is the average shortest path of the network, sometimes this is optimized instead of the diameter. (The diameter is the worst case, μ is the average case.) Figure from [289].

Complexity Measures: Tree-Width

A classic graph complexity measure is the tree-width of a graph. The basic assumption of this measure is that tree graphs are simple, and if a graph is less tree-like then it is more complex. The formal definition is as follows: for a graph $G = (V, E)$ its tree decomposition is a tree graph $T = (W, F)$, where each vertex of T is a subset of V ($W_i \subset V$ if $W_i \in W$) and T satisfies:

1. The union of all W_i vertices of T equals to V.

2. If u and v are neighbors in G then there is a vertex in T containing both of them.

3. For every v vertex in V the vertices in W containing v form a sub-tree in T. (Or equivalently, if W_i and W_j both contain a vertex v, then all nodes W_z of the tree in the (unique) path between W_i and W_j contain v.)

The width of a tree decomposition is the size of the largest T vertex minus one. The tree decomposition is not unique, a graph may have many different decompositions. The width of a graph is defined as the minimum of the widths of all possible tree decompositions. See Fig. 7.10 for an example.

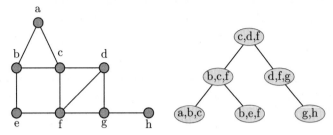

Fig. 7.10. One possible tree decomposition of a graph. The width of this decomposition is 2.

Yamaguchi and Aoki [573] used the tree-width to measure the complexity of chemical compounds. They obtained 9712 compounds from the LIGAND database [210] and calculated their tree-width. The tree-width of most compounds was between 1 and 3, and they found one single molecule with tree-width 4. The conclusion here is that the structure of chemical compounds in biological pathways is generally simple.

7.2.3 Fractal Structures

Loosely speaking fractal structures are generated by dynamic systems leading to chaos. By introducing a color code within the fractal sets related to the strange attractor of the chaotic process, figures with aesthetic value were generated; for a black-and-white version see e.g., Fig. 7.11, the Reader certainly has seen this on calendars.

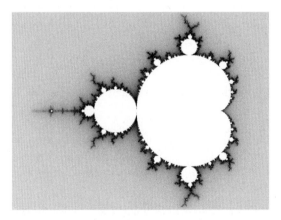

Fig. 7.11. A Mandelbrot set. It marks the set of points in the complex plane.

When the personal computers appeared, two German scientists, Heinz-Otto Peitgen and Peter Richter reacted rapidly [402], and published a book about the beauty of fractal structures, with many wonderful color pictures. While the pictures could be labeled as computer graphics, they were *not* the product of artist's intuition. They were generated by purely deterministic (and quite simple) algorithms.

The Mandelbrot set is the subset of the complex plane consisting of initial values c for which iterates of the function $z \mapsto z^2 + c$ do not tend to infinity. The boundary of this set is the Julia set.

More generally, the Julia set of a function (informally) consists of those points whose long-time behavior under repeated iterations of the function can change drastically under arbitrarily small perturbations.

The simplest visual representation of the Mandelbrot set is obtained if "good" points (initial values leading to a convergent series) are denoted. A finer version is when one calculates the number of steps to reach an absolute value of 2 (if this is reached once, the initial value is surely "bad") and uses the number of iterations to create a density plot. Julia sets are constructed in a similar way.

Fractal structures don't have integer, but fractional dimension. Points, lines, planes and bodies have zero, one, two and three dimensions, respectively. The self-similar structure shown in Fig. 1.6, the Koch curve has a dimension larger than one and smaller than two:

$$D = \frac{\log N}{\log r} = \frac{\log 4}{\log 3} = 1.26. \tag{7.14}$$

Here r is the measure of the linear size reduction, N is the D-dimensional measure of an object. It is length, area and volume in one, two and three dimensions.

Fractal Art? Jackson Pollock

Jackson Pollock (1912–1956), an "abstract expressionist", was known to develop painting technique based on random processes. A website about his work: http://www.artcyclopedia.com/artists/pollock_jackson.html.
".... He began painting with his canvases on the floor, and developed what was called his drip (or his preferred term, pour) technique. He used his brushes as implements for dripping paint, and the brush never touched the canvas..."
(from http://en.wikipedia.org/wiki/Jackson_Pollock).

His patterns (not only his technique) are now considered revolutionary. A characteristic artwork is shown in Fig. 7.12.

Fig. 7.12. Autumn Rhythm by Jackson Pollock.

There is recent scientific interest to see Pollock, who somehow anticipated fractal structures. Richard Taylor, a physicist and artist, spent a decade or so analyzing Pollock's artworks [508, 506, 507]. Pollock's dripped patterns were digitized, and analyzed by the standard methods of dimension analysis, and they were found to be fractals. The term "fractal expressionism" was coined to discriminate these artworks from computer generated "fractal art". As it turned out (25 years after Pollock) that many natural forms have fractal character, Taylor assumes that while his technique was seemingly random, natural rhythmicity also played a constructive role in creating his artworks. Pollock "must have adopted nature's rhythm when he painted..." [506]. More precisely, at the lower range of scale the fractal character is due to his technique, at a higher range due to his motions around the canvas (as the analysis of a film about Pollock during painting suggested).

David Mumford, former president of the International Mathematics Union speculated that "Zeitgeist" prescribed unintentional coincidence between discoveries in art and mathematics. He believes that it should not be totally accidental that the constructive role of randomness was found in mathematics and art in the same period. Randomness could be more effective than precise planning, as the artist Jackson Pollock and the mathematician Nicholas Metropolis (1915–1999) demonstrated. Metropolis discovered the Monte Carlo method to make stochastic simulation in the Manhattan project. *"Randomness is cool!"*, Mumford said.

Impossible Forms

In 1982 I read a short paper in a Hungarian art magazine about Tamás T. Farkas, a Hungarian artist, (see http://www.farkas-tamas.hu/). I was fascinated by his conscious effort to understand the organization principles behind impossible forms, and we published a paper [170] in the journal Leonardo.

The main intention of F. Farkas was to construct 'impossible' continuous forms by connecting the constituent elements in a logical manner. The term 'impossible' refers to the fact that most of these structures are impossible to realize in three dimensions. They may be visualized in a more-than-three-dimensional space, however, by certain mental procedures.

Many of his figures can be seen as continuous forms with no 'visible' elementary constituents. They have a 'holistic' character, i.e., any arbitrary point in the structure has the same importance as any other.

The continuous organization of lines in Fig. 7.13 emerges as an impossible eight branched form. In the center, a somewhat non-regular octagon can be seen. To interpret this figure one must realize that it is organized by eight different lateral views.

Fig. 7.13. Continuous form organized in eight directions. The eight-branched star is formed by continuously interacting triangles within a spatial oclagon. Eight 'impossible' triangles appear by visual analysis. The original artwork is colored. With permission of Tamás F. Farkas.

Science taught us that self-organizing systems can be characterized by the interaction of rigid deterministic and random effects. Can we use this concept also to artworks? I think we can relate self-organization to the graphics presented here: a family of figures could be generated by specifying an initial configuration and an algorithm to organize the form. Figure 7.14 exhibits an example of the cooperative organization of a 'chain of cells' in different planes. Another interesting aspect of the figure is the appearance of symmetry-breaking. "The balance between symmetry and symmetry breaking raises interesting questions not only in the natural sciences but also in aesthetics ... strict symmetry may be boring..." [227].

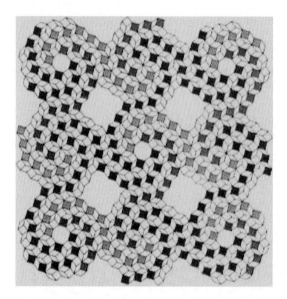

Fig. 7.14. Nine subsystems cooperate to form a complex system. The original artwork is colored. With permission of Tamás F. Farkas.

20 years later T. Farkas is still busy to interact with scientists to help visualize structural complexity by the perspective of a visual artist. In a recent paper written with György Darvas [122], who has a degree in physics and works on symmetry research, the confinement of quarks are visualized by artistic perspective, but based on scientific principles. Figure 7.15 illustrates the formation of a proton from three quarks, two so-called u quarks, and a d quark. As [122] writes:"The text*u* quarks are joined to their partners next to their to left-handed twists, and the text*d* quark next to its two right-handed twists...". Please note, that the figures does not pretend to be a physical model. The artists constructs new "impossible" forms, and a physicist may see some similarities to the organization of subatomic systems.

Fig. 7.15. Connection of quarks: from the perspective of visual art. With permission of Tamás F. Farkas.

7.3 Noise-Induced Ordering: An Elementary Mathematical Model

Self-organization is a vague concept in many respects, still a powerful notion of modern science. Specifically and counter-intuitively, noise proved to have beneficial (sometimes indispensable) role in constructing macroscopically ordered structures. A specific example showed how noise (actually in this example white noise) might play constructive role to enhance the functional dynamics of a system. So, from the world of art jump back to chemical kinetics. The role of external white noise in a mass action kinetic model was clearly demonstrated by Arnold et al. [23]. Let us consider the reaction network

$$ X \underset{k_2}{\overset{k_1}{\rightleftharpoons}} Y \qquad A + X + Y \overset{l_1}{\longrightarrow} 2Y + A \qquad B + X + Y \overset{l_2}{\longrightarrow} 2X + B. \qquad (7.15)$$

A and B are external, X and Y are internal components. Since the reaction is mass-conserving, the system can move under the constraint $C_1(t) + C_2(t) = N = $ const. for all t, where $C_1(t)$ and $C_2(t)$ are the quantities of X and Y at time t. Introducing $X(t) \equiv (1/N)C_1(t)$, and

$$ \alpha \equiv \frac{k_2}{k_1 + k_2} \quad \text{and} \quad \tilde{\beta} \equiv \frac{(l_2 B - l_1 A)/N}{k_1 + k_2} \qquad (7.16)$$

and making the special choice $k_1 = k_2$; the deterministic model is

$$ \frac{\mathrm{d}x(t)}{\mathrm{d}t} = f(x(t)) = h(x) + \bar{\beta} g(x) = \frac{1}{2} - x + \tilde{\beta} x(1-x). \qquad (7.17)$$

For the stationary case

$$-\tilde{\beta}x_s^2 + (\tilde{\beta} - 1)x_s + \frac{1}{2} = 0 \qquad (7.18)$$

holds. The x_s versus $\tilde{\beta}$ function is illustrated in Fig. 7.16.

It can be shown that $x_s \geq 0$ is stable, i.e., the deterministic model of the system can be characterized by one, stable stationary point.

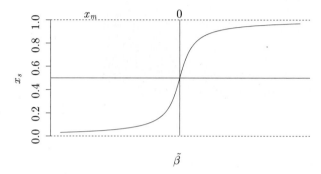

Fig. 7.16. No instability is displayed in the deterministic model.

Assuming that A and B are rapidly fluctuating quantities, the idealization of white noise can be adopted, and $\tilde{\beta}$ parameter is substituted by a β stationary stochastic process, in particular Gaussian white noise with expectation 0 and variance σ^2. For the associated stochastic process X_t the following Ito stochastic differential equation can be derived:

$$dX_t = \{\frac{1}{2} - X_t + \beta X_t(1 - X_t)\}dt + \sigma X_t(1 - x_t)dW_t. \qquad (7.19)$$

The fluctuation is state-dependent, i.e., the noise is multiplicative. Equation (7.19) can be associated with the Fokker-Planck equation:

$$\partial_t g(x,t) = -\partial_x \{\frac{1}{2} - x + \beta x(1-x)\} g(x,t) + \frac{1}{2}\sigma^2 \partial_{xx} x^2 (1-x)^2 g(x,t). \qquad (7.20)$$

Under the boundary condition, referring to g as a probability density function, the stationary probability density function is

$$g_s(s) = \mathcal{N}[\frac{1}{1(1-x)}] \exp\{\frac{2}{\sigma^2}(-\frac{1}{2x(1-x)} - \beta \ln[\frac{1-x}{x}])\}. \qquad (7.21)$$

The extrema x_m of $g_s(x)$ can be calculated from the relation

$$f(x_m) - \sigma^2 g(x_m)\frac{dg(x_m)}{dx} = 0. \qquad (7.22)$$

We get

$$\frac{1}{2} - x_m + \beta x_m(1 - x_m) - \sigma^2 x_m(1 - x_m)(1 - 2x_m) = 0. \tag{7.23}$$

This is a third-order equation for x_m, and it might occur that it has three positive roots. For the special case $\beta = 0$ (7.23) yields

$$x_m = \frac{1}{2} \quad \text{and} \quad x_\pm = \frac{1 \pm \sqrt{1 - 2/\sigma^2}}{2}. \tag{7.24}$$

For $\sigma = 2$ x_m is a triple root. The situations for $\sigma^2 < 2$ and $\sigma^2 > 2$ are given in Fig. 7.17.

Figure 7.17 illustrates that transition may occur which has no deterministic counterpart.

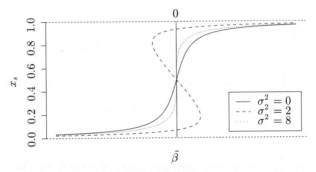

Fig. 7.17. Extrema of the probability density function as a function of β and for three values of the variance.

The interpretation is that noise has a constructive role in increasing complexity.

7.4 Networks Everywhere: Between Order and Randomness

7.4.1 Statistical Approach to Large Networks

The structural, functional or other relationship among the parts of a complex system can be well represented by a graph (or network). This representation is almost universal, it can be applied to such diverse systems as neural networks of the brain, food webs and ecosystems, electric power networks, systems of social connections, the global financial network, the world-wide web, etc. Interestingly enough the network representation is also universal in another sense: the graphs representing these very different systems are quite similar in some important respects. (So these systems are not that very different after all.) For example since the famous social psychological experiment of Stanley Milgram, it is known that from a certain point of view we live in a "small world". This simply means that in the majority of the systems taking only a small number of steps is enough to get anywhere, from anywhere. The friends of the friends of the friends of the friends of my friends is the total population of the planet. As it turns out, small world networks are somewhere halfway between randomness and deterministic structure.

Deterministic graphs were extensively studied by mathematicians interested in graph theory. The other extreme, *random graphs* were first studied by Erdős and Rényi in the late fifties [156, 157]. In one version of their random graph models one begins with N isolated nodes and connects every pair of nodes with a probability p. The edge distribution $P(k)$ equal to the probability of a node having k links is peaked around the mean value. Thus, every node is more or less the same as every other node, unlike many real social systems where entities have a large variation in their degree of connection with other entities. Also, random graphs are equally sparse everywhere, we cannot find any structures in them; they are random after all. It is not a big surprise that real networks are different: they have structure.

In the first approximation, they are quite clustered and far from being equally sparse: the friends of my friends are very often also friends of mine. So a minimal model of real networks should (1) show short characteristic path, just like random graphs do and (2) have high clustering. Watts and Strogatz constructed such a model [551, 550], and as usual the model was surprisingly simple: start with a regular lattice of vertices and rewire some edges randomly. The clustering of the lattice is high and it is not really affected by the random edges; these random shortcuts however make it possible to reach every node from every other node quickly in the couple of steps. Mathematically speaking, the probability that two adjacent vertices of a vertex are connected is quite high and independent of the total number of vertices (N); and the average shortest path length of the graph grows only as the logarithm of the number of vertices.

The small-world model aims to explain two properties found in real networks, the short average path length and the high clustering; it fails however to explain another universal network property: the extremely high variance of node degree. The degree of a node is simply the number of nodes connected to it and this is about the same for each vertex in an Erdős-Rényi type random graph: everybody is equal. The real world, however, is different: some vertices have very many connections, others many, few or none at all. Television reporters may have possibly tens of thousands of acquaintances, nerds only a couple of tens, with everybody else in between. If the Reader is familiar with Chapter 6 then this is not a very big surprise: real networks tend to have a power-law degree distribution. These networks are often called scale-free, as they have no characteristic scale for node degree: every kind of nodes are present, there is no such thing as 'typical node'.

The Barabási-Albert (BA) [5, 43] preferential attachment model offers a simple algorithm to generate scale free behavior. It was discussed in Sect. 6.3.2 in connection with generating edge distribution of networks described by power law.

An excellent annotated collection of papers on the structure and dynamics of networks was edited by three leading researchers of the fields (Mark Newman, László-Albert Barabási and Duncan Watts) [382].

7.4.2 Networks in Cell Biology

Network theory is contributing very much to understand the cells functional organization. Random Boolean networks were reviewed in Sect. 4.3.4. From retrospect, an obvious extension is to deviate from the assumption that each node (actually gene) is connected to the same number of other nodes.

Specifically, in *protein interaction* graphs (Fig. 7.18), the nodes are proteins, and two nodes are connected by a non-directed edge if the two proteins bind.

Protein interaction networks have a giant connected component and the distances on this component are close to the small-world limit given by random graphs. It turned out that the degree distribution of protein interaction networks (say, for yeast) is approximately scale-free (as it was found for the world wide web and the Internet by Laszlo Albert Barabasi and his coworkers and Faloutsos, Faloutsos and Faloutsos [169]), and these findings started a boom in network theory. The Reader of this book certainly knows [43].

A *metabolic network* is a directed and weighted tri-partite graph, whose three types of nodes are metabolites, reactions and enzymes, and two types of edges represent mass flow and catalytic regulation; see Fig. 7.19.

Fig. 7.18. Map of protein–protein interactions in yeast. Each point represents a different protein and each line indicates that the two proteins are capable of binding to one another. Only the largest cluster, which contains 78% of all proteins, is shown. In the original, colored version of the figure the color of a node signifies the phenotypic effect of removing the corresponding protein. (In the original figure color code was applied: red, lethal; green, non-lethal; orange, slow growth; yellow, unknown). Adapted from [257].

Mass flow edges connect reactants to reactions and reactions to products, and are marked by the stoichiometric coefficients of the metabolites. Enzymes catalyzing the reactions are represented as connected by regulatory edges to the nodes signifying the reaction.

Metabolic network representations indicate an approximately scale-free metabolite degree distribution.

The topological organization of cellular networks serves a lot of information about its function. Cellular networks are not communication networks designed for information processing and propagation. What important is the dynamics of the network prescribed by the equations of chemical kinetics. To each node a scalar $x_i(t)$ can be assigned, which represents its expression or activation level, and the edges denotes the influence (activation or inhibition) between nodes. The relationship between the topology of the network and qualitative dynamic behavior of the system resembles the problem of reaction kinetics related to the zero-deficiency theorem shown in Sect. 4.2.

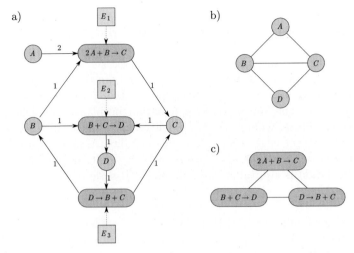

Fig. 7.19. Three possible representations of a reaction network with three enzyme-catalyzed reactions and four reactants. The most detailed picture, (**a**) includes three types of node: reactants (*circles*), reactions (*ellipses*) and enzymes (*squares*) and two types of edge corresponding to mass flow (*solid lines*) or catalysis (*dashed lines*). The edges are marked by stoichiometric coefficients of the reactants. (**b**) In the metabolite network all reactants that participate in the same reaction are connected, thus the network is composed of a set of completely connected subgraphs (*triangles* in this case). (**c**) In the reaction network two reactions are connected if they share a reactant. A similar graph can be constructed for the enzymes as well. Figure redrawn from [4].

7.4.3 Epidemics on Networks

In Sect. 4.4.2 we discussed some possibilities for modeling epidemics via dynamical systems. In the typical scenario, the total population was divided into a number of classes and then the equations defined how individuals can move from one class to another. In the SIR model there are three such classes: susceptibles, infectives and recovered people (or removed people if you like) and the equations are defined as

$$\dot{S} = -rSI, \tag{7.25}$$

$$\dot{I} = rSI - \gamma I, \tag{7.26}$$

$$\dot{R} = \gamma I. \tag{7.27}$$

These kinds of models are based on the fully mixed assumption: the probability that two individuals encounter is the same for everyone. This is however not a very realistic assumption for most diseases, e.g., in a disease attacking humans the encounters are usually constrained by the contact network, physical interaction is needed to transmit it. In mathematical terms this means that the SIR model should be extended to networks.

However simple this generalization might look, the resulting system is far more complicated to study than the simple fully mixed counterpart. One way to tackle it is to relate it to the bond percolation problem, a long studied area of physics. The simple bond percolation problem is as follows: given a network structure, what is the probability that a giant cluster is formed if we keep/remove the edges with uniform probability. A giant cluster is defined to be an infinite cluster in the infinite network limit. It can be shown that bond percolation is equivalent to a class of SIR-style problems on networks [216].

The major finding of the SIR model was the epidemic threshold, the fact if the ratio of r and γ is below a limit, then there is not outbreak in the system. It is thus a natural question whether there is such a threshold in the more realistic network-based model. The answer is not easy however, as it clearly depends on the structure of the network. One way to progress is to fix structural properties of the network and by this create a class of networks and then study this class.

The first finding was made for the class of networks with scale-free degree-distributions [400]. As we pointed out earlier, these networks can be commonly found among contact networks, see e.g., [317]. The sad result is that in these networks the epidemic threshold is zero, i.e., the disease always spreads. We need to mention the work by Callaway and his coworkers as well, they gave a formula for the epidemic threshold for networks with an arbitrary degree sequence (this class of networks is also called the configuration model) [89].

This result was later elaborated, and it turned out that if we also consider other network properties, like the correlation between the degrees of adjacent vertices [65] or network transitivity, [145] the threshold might reappear in SIR and also in SIS models. The epidemic threshold reappears if the network is embedded in a low-dimensional space, i.e., if we consider spatial coordinates as well [450, 549].

In addition to the references cited so far, Sect. VIII in [380] gives an excellent summary on epidemics spreading and other processes taking place on networks.[1]

---

[1] While deterministic models were mentioned here, stochastic models are also relevant. A recent stochastic model [421] for the spread of a sexually transmitted disease showed that while the number of infected people growths exponentially, the tree-graph of the infection tends to power law distribution.

7.4.4 Citation and Collaboration Networks in Science and Technology

Scientific Collaboration Networks

Erdős Number Project

I think, the majority of the Readers knows the story. Paul Erdős (1913–1996) was a legendary, very prolific mathematician, who spent his life by traveling from one mathematician to another, posed and solved problems, and traveled to another place. Erdős was the author of about 1500 papers, written with more than 500 mathematicians. For Erdős, mathematics was a social activity, and actually there is huge collaboration graph around him. The term Erdős-number was coined for the distance to Erdős in his collaboration network. Erdős himself has Erdős-number 0. His more than 500 coworkers have Erdős-number 1. Those, who did not collaborate with Erdős, but collaborated with Erdős's coworkers, have Erdős-number 2. There is an excellent website about the Erdős-number project, maintained by Jerry Grossman, a mathematician at Oakland University, Michigan: http://www.oakland.edu/enp/index.html.

The actual data of the Erdős-numbers on November 14th 2006 is seen in Table 7.1.

Table 7.1. The distribution of finite Erdős numbers

Erdős number 0 —	1 person
Erdős number 1 —	504 people
Erdős number 2 —	6593 people
Erdős number 3 —	33605 people
Erdős number 4 —	83642 people
Erdős number 5 —	87760 people
Erdős number 6 —	40014 people
Erdős number 7 —	11591 people
Erdős number 8 —	3146 people
Erdős number 9 —	819 people
Erdős number 10 —	244 people
Erdős number 11 —	68 people
Erdős number 12 —	23 people
Erdős number 13 —	5 people

The collaboration graphs of the mathematicians, of course is changing in time, and its evolution was analyzed by Grossman [223]. There are about 400.000 mathematicians in the databank of the Mathematical Reviews. Considering a collaborative graph, where the authors are the nodes, and edges exist if people (nodes) have common papers, we see that almost 70% of the

268,000 people have finite Erdős-number (an infinite Erdős-number means that there is no path from the author to Erdős), 50,000 mathematicians are who have papers with co-workers, and infinite Erdős-number, and still a significant number of mathematicians (84,000) published without collaboration. The collaboration graph was analyzed by graph-theoretical methods. The clustering coefficient was found to be 0.14 (i.e., high compared to random graphs). Average path length is eight, well, eight (and not six) degrees of separation. In any case, the collaboration graph has a "small world" character, with a large "Erdős component". Interestingly, those authors who belong to the "Erdős component", have 4.73 average degree (collaborators), while those, who collaborated, but don't belong to the "Erdős component", have 1.65 only (so they collaborated with one or two persons on the average).

Bridges to connect mathematicians to neurobiologists, economists and even to philosophers?

János (John) Szentágothai (JSz), one of the most distinguished neuroanatomist of the XX. century, has an Erdős number 2, since he has a common paper with Alfred Rényi published in 1956 [427] (actually about the probability of synaptic transmission in Clarke columns). It seems to be a plausible hypothesis that JSz is the bridge to connect the community of mathematicians to neurobiologists and even to philosophers. JSz has a common book [140] with two other scientific nobilities, Nobel prize winner neurophysiologist Sir John Eccles , and with Masao Ito. In this case, JSz should be a bridge between two separated communities. It is interesting to note, that JSz himself was thinking on the graph of the network of the cerebral cortex, in terms of what it is called today "small world". JSz hinted that the organization of the cortical network should be intermediate between random and regular structures [497, 500]. He estimated that "any neuron of the neocortex with any other over chains of not more than five neurons of average" ([22], p. 222). Eccles has a book with Karl Popper [412], so there is a direct math–neurobiology–philosophy chain.

Another non-mathematician with Erdős number 2 via Rényi is András Bródy, a Hungarian economist. They also published a paper in the same memorable year, in 1956 [426]. (It was about the problem of regulation of prices). So, another question is induced. Since most likely all people with number 1 are mathematicians, it would be interesting to know, how many non-mathematicians have Erdős number 2, and how any other scientific communities are involved in the collaboration graph?

The collaboration graph of the mathematicians is a good, and well-documented example of *social networks*. Co-authorship networks were analyzed in a number of papers. First, some studies are static in the sense that the only the actual structure of the networks are analyzed, but not their time evolution. Such studies were done by Newman [377, 378, 379] who analyzed networks from biomedical research, computer science and physics. He showed that these networks are small-worlds: their diameter is small and their clustering coefficient is large; they have a giant connected component which included most of the nodes.

Second, other studies focus on the time evolution of the structural properties. Newman [376] studied the Los Alamos E-print archive and the Medline database and found that the networks grow according to the preferential attachment principle. Similar conclusions were drawn by Jeong and his co-workers [258] by studying several networks, including the actor collaboration network and a neuroscience co-authorship network. Barabási and his colleagues [44] found that the average separation and the clustering coefficient decreases while the average degree and the relative size of the largest cluster increases in time in the coauthorship networks of mathematicians and neuroscientists.

Now we turn from scientific collaboration to analyze networks of patent citations.

Patent Citation Networks

Innovation plays a key role in economic development and the patent system is intended (and required by the United States Constitution) to promote *innovation*. The patent system promotes innovation by giving inventors the power to exclude others from using their inventions during the patent term. The power to exclude is a double-edged sword, however, because it benefits the original inventor, but imposes costs on later innovators seeking to build on past inventions. Thus, the proper design of the patent system is an important matter – and a matter of considerable current debate, e.g., [256, 106, 350]. Advances in computer technology and the availability of large patent databases have recently made it possible to study aspects of the patent system quantitatively. Since patents and the citations between them can be seen as a growing network, techniques from network theory [5, 380] can usefully be applied to analyze the patent citation networks .

In a cooperation with Katherine Strandburg, (who is a law professor with a PhD in physics), we elaborated in Budapest and Kalamazoo with Gábor Csárdi and Jan Tobochnik a network theoretical approach to the patent system, and set a kinetic model for the patent citation network growth [118]. (As usually, Gábor, who is graduate student at the moment of the writing of this sentence, and hopefully Dr. Csárdi, when you read it, made the lion share of the work under the supervision of three professors.)

Patentological Background

An application for a U.S. Patent is filed in the U.S. Patent and Trademark Office (USPTO). A patent examiner at the USPTO determines whether to grant a patent based on a number of criteria, most important of which for present purposes are the requirements of novelty and non-obviousness with respect to existing technology. Once a patent is issued by the USPTO, it is assigned a unique patent identification number. These numbers are sequential in the order in which the patents were granted.

Novelty and non-obviousness are evaluated by comparing the claimed invention to statutorily defined categories of "prior art", consisting in most cases primarily of prior patents. Patents are legally effective only for a limited term (currently 20 years from the date of application), but remain effective as "prior art" indefinitely. Inventors are required to provide citations to known references that are "material" to patentability, but are not required to search for relevant references (though they or their patent attorneys often do so). During consideration of the application, patent examiners search for additional relevant references.

Patent citations include potential prior art that was considered by the examiner. They thus reflect the judgment of patentees, their attorneys, and the USPTO patent examiners as to the prior patents that are most closely related to the invention claimed in an application. Patent citations thus provide, to some approximation, a "map" of the technical relationships between patents in the U.S. patent system. This "map" can be represented by a directed network, where the nodes are the patents and the directed edges the citations. Our research used a statistical physics approach inspired by studies of other complex networks to attempt to gain insight from this "map".

The analysis became possible by the existence of the extensive database of citations made available through the work of economists Hall, Jaffe, and Trajtenberg [231]. It is available on-line at http://www.nber.org/patents/. The database contains data from over 6 million patents granted between 13 July 1836 and 31 December 1999 but only reflects the citations made by patents after 1 January 1975: more than 2 million patents and over 16 million citations. Citations made by earlier patents are also available from the Patent Office, but not in an electronic format.

7.4 Citation and Collaboration Networks in Science and Technology

Modeling Framework: States and Kinetics

How to define a mathematical model framework for describing the temporal evolution of the patent network?

1. State variables. The first, obviously somewhat arbitrary step is to decide about the state variables. In our model, each patent is described by two variables:

   a) k, the number of citations it has received up to the current time step and

   b) l, the age of the patent, which is simply the difference between the current time step (as measured in patent numbers) and the patent number. Because a given patent may cite more than one other patent, several citations may be made in one time step.

   These two variables define what we call the "*attractiveness*" of a patent, $A(k, l)$, which determines the likelihood that the patent will be cited when the next citation is made.

2. Temporal change. The evolution of the network is modeled by a discrete time, discrete space stochastic dynamic system. Time is measured in patent number units, so that each "time step" represents the citations made by a single patent.

In every time step the probability that an older patent will be cited is proportional to the older patent's attractiveness multiplied by the number of citations made in that time step. We found that this simple model gives a very good approximation of the observed kinetics of the growth of the patent citation network.

More formally, the state of the system is described by $k_i(t)$ and $l_i(t)$, $(1 < i < N)$, where N is the patent number of the last patent studied and $k_i(t)$ and $l_i(t)$ are the in-degree and age, respectively, of patent i at the beginning of time step t. The attractiveness of any node with in-degree k and age l is denoted by $A(k, l)$. $A(k, l)$ is defined such that the probability that node i will be cited by a given citation e in time step t is given by

$$P[e \text{ cites node } i] = \frac{A(k_i(t), l_i(t))}{S(t)}, \qquad (7.28)$$

where $S(t)$ is the total attractiveness of the system at time step t:

$$S(t) = \sum_{j=1}^{t} A(k_j(t), l_j(t)). \qquad (7.29)$$

The total probability that node i will be cited in time step t is thus $E(t)A(k_i(t), l_i(t))/S(t)$, where $E(t)$ is the number of citations made by patent t. $A(k, l)$ and $S(t)$ are defined up to an arbitrary normalization parameter. To normalize, we arbitrarily define $A(0, 1) = 1$. With this normalization, $S(t)$ is the inverse probability that a "new" node, with $k = 0$ and $l = 1$, will be cited by a given citation during the next time step.

The $A(k, l)$ function determines the evolution of the network. It describes the average citation preferences of the citing patents (the inventors and patent examiners in reality). In this study, we measured and analyzed $A(k, l)$ for the United States patent system during the time period covered by our data. We find first that the parametrization by k and l consistently describes the average kinetics of the patent citation network. Of course, underlying patent citations are patentee and patent examiner evaluations of the significance of the cited patent and the technological relationship between the citing and cited patents that our probabilistic approach cannot capture. The way in which these "microscopic dynamics" are translated into the average behavior that we observe remains an open question.

Estimating the Attractiveness Function: An Algorithm

Let us assume that edges are added to the system one after another in a fixed order; if two edges are added in the same time step (i.e.,, by the same citing patent), their order is fixed arbitrarily for the measurement. Let e be an edge and let $c_e(k, l)$ be indicator random variables, one for each (e, k, l) triple, $(1 < e < E_{\text{tot}}, k \geq 0, l > 0)$, where E_{tot} is the total number of edges in the system. $c_e(k, l)$ is one if and only if edge e cites a (k, l) node (i.e.,, a node having in-degree k and age l) and zero otherwise. The probability that edge e cites a (k, l) node, i.e.,, that $c_e(k, l)$ is one, is thus given by

$$P[c_e(k, l) = 1] = \frac{N(t(e), k, l)A(k, l)}{S(t(e))}, \qquad (7.30)$$

where $t(e)$ is the time step during which edge e is added, $S(t(e))$ is the total attractiveness of the system right before adding edge e, and $N(t(e), k, l)$ is the number of (k, l) nodes in the network right before adding edge e. We thus have a formula for $A(k, l)$:

$$A(k, l) = \frac{P[c_e(k, l) = 1]S(t(e))}{N(t(e), k, l)}. \qquad (7.31)$$

In (7.31) it is easy to determine $N((t(e), k, l)$ for any (e, k, l), but $S(t(e))$ is unknown. Moreover, we have only a single experiment for $c_e(k, l)$ which is

7.4 Citation and Collaboration Networks in Science and Technology

not enough to approximate $P[c_e(k,l) = 1]$ properly. To proceed further, let us define a new set of random variables, each of which is a simple transformation of the corresponding $c_e(k,l)$ variable:

$$A_e(k,l) = \frac{c_e(k,l)S(t(e))}{N(t(e),k,l)}, \quad \text{if } N(t(e),k,l) > 0. \tag{7.32}$$

If $N(t(e),k,l) = 0$ then $A_e(k,l)$ is not defined. It is easy to see that the expected value of any $A_e(k,l)$ variable (if defined) is $A(k,l)$; thus we can approximate $A(k,l)$ by

$$\bar{A}(k,l) = \frac{1}{E(k,l)} \sum_{e=1}^{|E_{tot}|} \frac{\bar{c}_e(k,l)S(t(e))}{N(t(e),k,l)}. \tag{7.33}$$

Here $E(k,l)$ is the number of edges for which $N((t(e),k,l)) > 0$ for any $t(e)$, and $\bar{c}_e(k,l)$ is the realization of $c_e(k,l)$ in the network being studied.

To calculate this approximation for $A(k,l)$ we need to determine $S(t(e))$, which itself is defined in terms of $A(k,l)$. To determine $A(k,l)$ and $S(t(e))$ self-consistently, we use the following iterative approach:

1. First we assume that $S_0(t)$ is constant, and use (7.33) to compute $A_0(k,l)$, normalizing the values such that $A_0(0,1) = 1$.

2. Then we calculate $S_1(t)$ for each t based on $A_0(k,l)$ and use this to determine $A_1(k,l)$.

3. We repeat this procedure until the difference between $S_n(t)$ and $S_{n+1}(t)$ is smaller than a given small ϵ for all t.

Results

1. The $A(k,l)$ attractiveness function has two variables, so Fig. 7.20 and Fig. 7.21 show the one-dimensional projections of the $A(k,l)$ function. Interestingly, it seems to be a good approximation that $A(k,l)$ can be factorized as

$$A(k,l) = A_k(k) \cdot A_l(l). \tag{7.34}$$

A. Age-dependence

The measured $A_l(l)$ function for the patent citation network has two major features – a peak at approximately 200,000 patent numbers and a slowly decaying tail. Time is measured with patent numbers, and of course, 200,000 patent measures different time epoch on the time axis, since the rate of patenting is increasing rapidly. E.g., in 1998–1999, 200,000 patent numbers corresponded to about 15 months. We did not want to over-interpret the results, but it says that the maximal probability of the age of the cited patents is 15 months around 1999.[2]

The tail is best described by a power-law decay: $A_l(l) \sim l^{-\beta}$ with $\beta \approx 1.6$. The observation of this power law decay is an important result. It indicates that while typical citations are relatively short-term, there are a significant number of citations that occur after very long delays. Very old patents are cited, suggesting that the temporal reach of some innovations, which perhaps can be described roughly as "pioneer", is very long indeed. Moreover, because $A_l(l)$ is approximately independent of k – i.e.,, approximately the same power law decay is observed even for small k – the power law tail of $A_l(l)$ demonstrates that there is a significant possibility that patents that have gone virtually un-cited for long periods of time will re-emerge to collect citations. This slow power law decay of $A_l(l)$ thus suggests the unpredictability of innovative progress.

B. In-degree dependence

The measured $A_k(k)$ function increases monotonically with k, as Fig. 7.21 suggests. Higher in-degree always means higher attractiveness. Because the citation probability is proportional to the attractiveness, this means that the well-known preferential attachment, or the "rich get richer" effect is at work here – the more citations a patent has received, the more likely it is to receive another. The functional form of $A_k(k)$ is a power law over the entire range of k values. $A_k(k) \sim k^\alpha + a$, where $\alpha = 1.2014 \pm 0.0059$ and $a = 1.0235 \pm 0.0313$.

C. The attractiveness function: what does it tell us about the development?

The measurement procedure resulted in the form of the attractiveness functions, as:

$$A(k,l) = (k^\alpha + a) \cdot l^{-\beta}. \qquad (7.35)$$

The attractiveness functions is not only a variable, which characterizes the actual state of the system, but it has the character of a transition probability, which determines (of course, in stochastic sense) the development of the network. Since there are only very few networks, where the dynamics is well documented, patent citation networks seem to be very particular even from methodological point of view.

---

[2] Please note that the function is not symmetric, so the expected duration should not be the same.

7.4 Citation and Collaboration Networks in Science and Technology 233

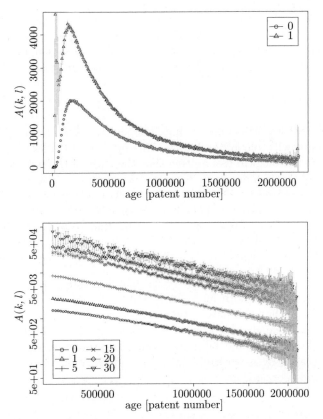

Fig. 7.20. The measured attractiveness $A(k,l)$ as a function of age l for various fixed values of in-degree, k. The *bottom figure* shows only the decreasing tail on log–log scales.

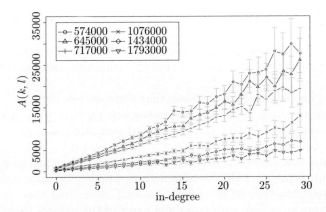

Fig. 7.21. The measured attractiveness $A(k,l)$ as a function of in-degree, k, for various fixed values of age, l.

We see that the preferential attachment rule works while the aging effects are also taken into account, and the fact that $\alpha > 1$ tends to the direction, which might loosely be called "stratification" – more and more nodes with very few citations and less and less nodes with many citations.

2. Rule change and its effects on the dynamics

Thus far, we have assumed a time-independent $A(k,l)$, which is reasonably consistent with our observations. This mathematical assumption materializes the patentological assumption that the rules of the patent system are unchanged, so the significant increase in the number of US patents granted (which is documented in [256, 230]) is due to acceleration of technological development. [256] suggested that there is a change in the level of rigorousness of the patent examinations however and we know that patent law also changed.

> Lerner and Jaffe describe the recent history of the patent system, in which patents were both more easily obtained and potently enforced. They believe that the patent review process of the U.S. Patent and Trademark office should be more rigorous and that the playing field between litigants should be more level. The protection for true innovators created by a workable patent system is vital to technological change and economic growth, they write.

To analyze the eventual time-dependence of the attractiveness function (i.e., the time dependence of the rule which prescribes the temporal evolution) we measured the parameters of the system as functions of time. To perform the fits, we averaged over a 500,000-patent sliding time window and calculated the parameters after every 100,000 patents. The measured α parameters are plotted in Fig. 7.22. There is a significant variation over time. The time dependence of the important β parameter was also explored, but no significant time dependence was observed to within the statistical errors.

The plot for the α parameter shows that there are two regimes. In the first regime, prior to about 1991, α is decreasing slightly with time, while in the second, starting around 1993, there is a significant increase.

We found that there has been a change in the underlying growth kinetics since approximately 1993. Since that time, preferential attachment in the patent system has become increasingly strong, indicating that patents are more and more stratified, with fewer and fewer of the patents receiving more and more of the citations. A few very important, perhaps "pioneer", patents seem to dominate the citations. This trend may be consistent with fears of

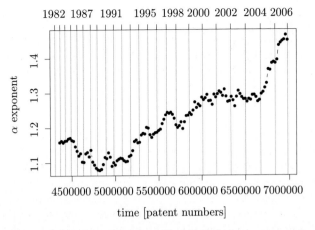

Fig. 7.22. The measured value of α as a function of time, measured as described in the text. The time in years is indicated by the *grey vertical lines*.

an increasing patent "thicket", in which more and more patents are issued on minor technical advances in any given area. These technically dense patents must be cited by patents that build upon or distinguish them directly, thus requiring that more citations be made, but few of them will be of sufficient significance to merit citation by any but the most closely related patents. These observations are consistent with recent suggestions that patent quality is decreasing as a result of insufficient standards of non-obviousness.

A Really Complex (Not Only) Network

The next chapter is about the brain, what we feel as the most complex device. It is a network of neurons, which are complex devices at their own right. The brain is much more, than a network, and we don't yet understand the role of intra- and interneuronal connections, which embodies our mind and I believe, our soul.

8

Complexity of the Brain: Structure, Function and Dynamics

8.1 Introductory Remarks

It is often said in colloquial sense that the brain is a prototype of complex systems. A few different notions of complexity may have more formally related to neural systems. First, structural complexity appears in the arborization of the nerve terminals at the single neuron level, and in the complexity of the graph structure at network level. Second, functional complexity is associated to the set of tasks being performed by the neural system. Third, dynamic complexity can be identified with the different attractors of dynamic processes, such as point attractors, closed curves related to periodic orbits, and strange attractors expressing the presence of chaotic behavior. In the book with Michael Arbib, and John Szentágothai [22] we tried to show that the understanding of the neural organization requires the integration of structural, functional and dynamic approaches. Structural studies investigate both the precise details of axonal and dendritic branching patterns of single neurons, and also global neural circuits of large brain regions. Functional approaches start from behavioral data and provide a (functional) decomposition of the system. Neurodynamic system theory offers a conceptual and mathematical framework for formulating both structure-driven bottom-up and function-driven top-down models.

In this section we integrate the basic experimental facts and computational models about neural organization behind the complexity of the brain. The newer results of brain theory have strong impacts on two fields of "applications". First, there is the question whether how the new results contribute to the better understanding of neurological and psychiatric disorders, and what kinds of new therapeutic strategies can be offered? The answer is in the better integration of scattered disciplines of studying brain and mind, from basic neuroscience via clinical and pharmacological methods to psychotherapy. Second, there is the recurring question about the transferable knowledge from neuroscience to computer engineering and related areas. How our knowledge about brain structure and function can contribute to design new types of intelligent artificial systems (maybe even with the ability to show and process emotions, too).

Computational models of these multi-level complex problems are the key methods of this integration.

8.2 Windows on the Brain

8.2.1 A Few Words About the Brain–Mind Problem

It might be true that the spectacular development in the neurosciences has widened rather than narrowed the gulf between whatever we know (or believe to know) about the structure and functions of neural elements and centers, and the global performance of the nervous system. This has been revealed in its higher functions studied by a number of disciplines, such as behavioral genetics, ethology, psychology (including the whole spectrum from psychophysics, through psycholinguistics, cognitive psychology, social psychology etc., to introspective psychology).

Consciousness, perception, mind, will, thoughts are notions that are believed to be in connection with brain processes. The question how mental activities are related to neural structures and processes has been a hot and much argued issue since the debates of ancient philosophers. Supposedly, connection between behavior and the "head" was recognized by Paleolithic people who occasionally tried to cure fellow cavemen by bilging the head.

At the end of this chapter, in Sect. 8.6.1 we shall briefly review the attempts to "solve" the brain-mind problem. from materialistic monism to interactionist dualism.

Cognitive neuroscience has the program to answer the philosophical question by scientific methods, and explain cognitive phenomena by neural mechanisms. So it is an explicit *methodological* reductionist approach (which is good). This section gives a very brief overview of experimental methods and the segment of knowledge a specific method provides for exploring structural and functional organization of the brain.

8.2.2 Experimental Methods: A Brief Review

Neuroanatomy

The first fundamental step substantiating modern neuroanatomy, the study of the structure of the nervous system was the discovery of Ramon y Cajal in the late 1800s who, using Camillo Golgi's silver staining methods, revealed the structure of nerve cells [88]. He discovered that these cells remain separate instead of merging into one another, which laid the foundation for the Neural Doctrine. Developments in histological techniques enabled neuroanatomists to trace nerve fibers and to discover projections from one part of the nervous system to another. Later, after the appearance of electron microscopy even finer parts, such as synapses or dendritic spines became examinable. Other anatomical methods include the localization of different chemicals – neurotransmitters, proteins, radioactively labeled deoxyglucose – in particular neurons.

Electrophysiology

Electrophysiologists make electrical recordings from the outside or inside of a single cell. Using patch clamp technique the characteristics of a single ion channel – a sort of molecular machinery in the cell membrane – can be recorded. On higher hierarchical levels, *populations of neurons* are examined using multielectrode techniques or optical imaging either in vivo or in vitro.

Brain Imaging

Different neurophysiological and brain imaging methods [87], developed to understand the function of the nervous system, might be categorized according to their temporal or spatial resolution, the temporal or spatial scale they are used on, their target structure in the hierarchy of brain structures or the quantity they measure. For example, electroencephalogram (EEG) and magnetoencephalogram (MEG) both offer the high temporal resolution required to measure the activity associated with brief sensory and cognitive events but their spatial resolution is relatively poor. The basic phenomena these techniques exploit is that nerve cells in the brain produce electrical impulses for communication which add up to generate macroscopically measurable quantities. EEG offers means to record, amplify and analyze the electrical field while in MEG measurements the magnetic field generated by electric currents flowing in the neural tissue are registered. While EEG and MEG are characterized by fine temporal but poor spatial resolution positron-emission tomography (PET) and magnetic resonance imaging (MRI) techniques represent the opposite tendency. These methods are based on the fact that increased activity of nerve cells increase the metabolism of these cells. Thus PET and MRI register signals related to metabolic processes and do not directly measure electrical events which are thought to be the basis of information representation in the brain. In a particular visual or auditory task for example, numerous different areas are activated in the time window specified by MRI's temporal resolution and consequently are seen concurrent on MRI images. On one hand the spatial resolution of EEG is poor but due to its high temporal resolution might reveal the temporal sequence of events, while on the other hand MRI can be used to study spatial sequences [87]. Novel experimental techniques use MRI or PET in various combinations with MEG or EEG completing each other's capabilities. fMRI became a popular, though somewhat controversial technique to find neural structures, ("f" stands for "functional"), which might be the anatomical substrates for certain neural functions. There is intensive ongoing research to find reliable correlates between blood flow and electrical neural activity.

Neuropsychological Studies

There are several famous studies, and three will be mentioned here very briefly.

Alexander Luria (1902–1977), a Soviet neuropsychologist studied a journalist, called Shereshevski, who apparently had a basically infinite memory. He was able to memorize long mathematical formulae, speeches, poems, even in foreign languages, etc. [323]. Shereshevski was diagnosed with *synaesthesia*, which is a neurological condition, when different senses are coupled. When he realized his ability, he performed as a mnemonist. His ability implied disorders in his everyday life, it was difficult to him to discriminate between events that happened minutes or years ago.

Phineas Gage suffered brain injury in consequence of an explosion in a railroad construction in 1848 in a small town in Vermont, and a 1 m long tamping iron passed through his brain. The analysis suggested that his frontal lobe was damaged, which might explain the (actually negative) changes in his personality. Antonio Damasio, a neurologist, reconstructed the famous story, and argued that mind and emotion are embodied in the brain, so Descartes, who stated the quasi-independence of the brain and mind is wrong [121].

Most likely the most well-known patient in cognitive neuropscyhology is HM. He suffered epilepsy probably in consequence of a bicycle accident, and later underwent surgery in which his hippocampus has been functionally removed. HM was studied, mostly by the now legendary neuropscyhologist Brenda Milner. HM (80 years old now, as I write this sentence), suffered from severe *anterograde amnesia*. It means that he was not able to learn new items, certainly not new facts or relationships between concepts. (He also had some moderate *retrograde amnesia*, he was able to recall memory traces he learned many years before the surgery, but did not remember events, which happened several days before his surgery). A retrospective analysis is given [111], and a story about HM's present state and his own view on his life was written in August 2006 [456].

Computational Methods

Using these and many other methods not to be mentioned here, neuroscientists have acquired great amount of information about the structure and function of the central nervous system [22]. Practically speaking, the huge extent of data was impossible to process by conventional ways and required new ideas that helped handling them. From the late 1950s the increasing numerical capacity of computers were invoked. On one hand, several techniques to numerically select significant data and analyze them were developed. *Neuroinformatics*, using the term in a narrow sense, offers quantitative methods for analyzing neuroscientific data. As it was said: "Neuroscience has far too diverse a set of foci for any single database to capture all that is needed by researchers. Instead, we have to understand how to build a federation of databases in which it is easy to link data between these databases to gain answers to complex problems."[20]. On the other hand, models serve for understanding the neural mechanism of phenomena found experimentally in the central nervous system. With computer techniques gathering ground and turning widespread, software packages to simulate the behavior of single neurons or even small networks of neurons became available, serving as a complementary technique for non-computer scientists as well. We shall discuss the details of neural models later in this chapter in Sect. 8.4.2. In a more broader sense neuroinformatics deals with the organization and process of neruoscientific data integrated with neural models.

8.3 Approaches and Organizational Principles

8.3.1 Levels

The brain is a prototype of hierarchical systems, as Fig. 8.1 shows.

More precisely, it is hierarchical dynamical system. At the molecular level the dynamic laws can be identified with chemical kinetics, at channel level with biophysical detailed equations for the membrane potential, and at synaptic

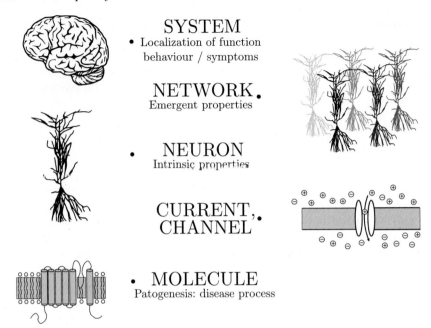

Fig. 8.1. The brain is a hierarchical dynamic system.

and network levels with learning rules to describe the dynamics of synaptic modifiability.

8.3.2 Bottom Up and top Down

In a *bottom up* brain theory, the emphasis tends to be on single neurons, plastic synapses, and neural networks built from these elements. Given a set of neurons, interconnected by excitatory and inhibitory synapses, the theory answers the question whether, and how, and with what involvement of self-organization or plasticity, the network is able to implement a given function. The general method is to build dynamical network models to describe spatiotemporal activity patterns and their change with synaptic modifiability to explain development, learning, and function.

Top down brain theory is essentially functional in nature, in that it starts with the isolation of some overall function, such as some pattern of behavior or linguistic performance or type of perception, and seeks to explain it by decomposing it into the interaction of a number of subsystems. What makes this exercise brain theory as distinct from cognitive psychology (or almost all of current connectionism) is that the choice of subsystems is biased in part by what we know about the function of different parts of the brain – whether obtained by analysis of the effects of brain lesions, imaging, or neurophysiology – so that there is some attempt to map the subsystems onto anatomical regions.

8.3.3 Organizational Principles

General Remarks

Recognizing basic *organizational principles* is the unique way to give an interpretation of the mechanisms underlying processes observed by all types of measurement techniques. Motoric response to different sensory information may vary on a broad spectrum, which means that interaction with the environment requires a sophisticated input-output relationship. Imaging techniques, neuropharmacological and lesion experiments revealed that there is a strong correspondence between the behavior and the activity of different brain areas, thus establishing a structure-function relationship. However, during behavioral and cognitive processes similar structure-function links can be unraveled using techniques operating on other hierarchical levels. Thus, we can draw the conclusion that similar hierarchical levels are present both in the structural and functional organization of the brain like the ones of the experimental techniques. Changes at the behavioral level (hearing, recalling memories, decision making) induce changes on all levels: on the system level (brain areas, e.g., auditory cortex, hippocampus, prefrontal cortex), on the network level (distributed computing, integrating devices), on the neural level (computational unit), on the synaptic level (information transfer) and on the molecular level.

A somewhat different question is how the external world is represented by neural activity patterns. Although it is beyond the current possibilities to give the formula for internal representation, basic governing principles can be sketched. One characteristic organizing principle of brain functions is the topographical representation of information. Furthermore, the self-organizing capability of the nervous system helps in the adaptation to the environment.

Topography

Neural information processing characteristically happens in a series of stages. For instance visual stimuli from the retina are passed to the visual cortex via the lateral geniculate nucleus and handed over to the parietal or temporal cortex. Although during this process sensory information may undergo different transformations, one thing remains unchanged: information coded by nearby neurons in the lower-level structure is coded by adjacent neurons in the higher-level structure as well. This principle is called *topographic order*. This topographic feature is well known also in the case of the somatosensory cortex too, where parts (i.e., cheeks, lips, fingers, etc.) of the body are mapped with relative size correlating with the number of receptors at the part in question.

There is a topographic organization in the auditory system, too. Vertebrate hair cell systems located in the inner ear structures are frequency selective and can be described by the place principle, i.e., each hair cell and neuron in the cochlea is tuned to respond to a specific frequency and that frequency can be determined based on its position in the cochlea. This theory proposes that the brain is able to tell which frequencies are being heard based on which neurons are firing.

SELF ORGANIZATION:
Development and Plasticity
of Ordered Structures

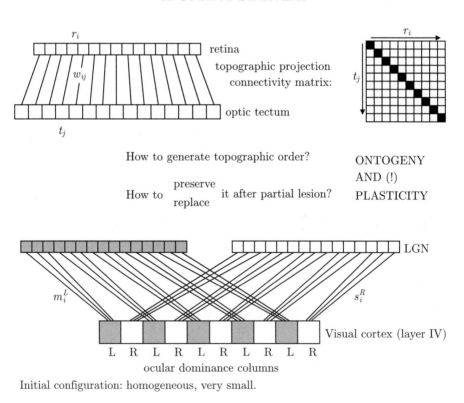

Fig. 8.2. Idealized explanation of the topographical organization.

8.3 Approaches and Organizational Principles 245

Models of the Development of Topographic Order

There are old debates about the formation of topographic maps in the nervous system (Fig. 8.2). In animals such as goldfish and frogs, the main visual center is the optic tectum. Retinotectal connections generate a topographic map which preserves visual information between subsequent layers of cells. Numerous mathematical models have been established to describe the mechanism of the formation of such ordered neural mappings.

The main issue concerns the role of the tectum in the formation of the mapping. There are two extreme hypotheses. The first assumes that chemical marker molecules mediate the target guidance between the pre- and postsynaptic neurons, the second states the correlation between pre- and postsynaptic activities by some Hebbian learning rules drives the development of connections.

Roger Sperry (1913–1994) asked whether the functional recovery after optic-nerve transection is established by specific affinities between matching retinal and tectal cells or by functional sorting out of largely random retinal regrowth. He made mismatch experiments (mostly with frogs and goldfish) by removing part of the retina, part of the tectum or parts of both. Early experimental results at first suggested that there was a strong tendency after a partial lesion for retinal ganglion cells to only grow back to the same tectal cells they contacted originally. According to Sperry's classical idea, reinforced by Attardi and Sperry [31] there is a specific point-to-point projection between retina and tectum. The chemoaffinity hypothesis suggests that order and orientation of the topographic map are uniquely derived from the interactions between pre- and postsynaptic cells. Models belonging to this family assume that growing axons are directed to the appropriate point by the aid of some "guiding substance". In the original version of the chemoaffinity hypothesis, each retinal cell was labeled with a unique chemical marker, and the complementary marker on the tectal neuron was recognized by some "molecular recognition" mechanism. Modified versions of the model used considerably fewer markers: two molecules in orthogonal gradients in the retina and complementary molecular gradients in the tectum. Newer experiments identified potential "cell adhesion molecules" to mediate synapse formation.

The second family of hypotheses and theories designed to explain neural development, learning, and conditioning are based on the modifiability of synapses by *correlated electrical activity of presynaptic and postsynaptic cells*. Model studies suggest that topographic mappings can be generated by activity-dependent self-organizing mechanisms described by some Hebbian learning algorithm (see more in Sect. 8.5.3).

We argued more than 20 years ago [160] by using a neural network model that some environmental noise is indispensable to generate globally ordered structures. As an excellent review [562] summarizes: "While many of the models could fit the available facts, the critical experimental findings concerning the nature of cell-to-cell recognition at the synaptic level remain unanswered."

As we learned from history of science, that earlier "either-or" questions will be answered with "both" (well, occasionally with "neither-nor"). There is a tendency to believe now that Sperry and Hebb is not oil and vinegar, but maybe vinaigrette [100]: "Artificially polarized statements might provoke debate within science or politics, but evolution is opportunistic."

Mapping of the Environment: Not Only Topographic

However, topography is not an universal organizing principle. While during navigation a map of the external world is represented in the rat brain, topography can not be found. Although topography in an internal representation of the environment seems to be natural, in the hippocampus, which is the possible locus of navigation, information storage is not accomplished this way: so called place cells coding a well defined portion of the environment by increased activity in this spot do not show topography. Place cells coding nearby locations are not necessarily neighbors, there are no synaptic connections showing their adjacent receptive fields. A possible explanation of this caveat is that while epithelial sensory input is derived from a closed surface (e.g., the retina in the case of visual stimuli, or the skin in the case of tactile stimuli), the spatial map has to account for an open environment, i.e., ever changing external maps have to be represented internally. Computational studies revealed that topographic connections are not necessary for the maintenance of spatial receptive fields if the input of place cells reflects the changes in the environment (increasing or decreasing the distance form a landmark) as a result of motion and not the environment itself (position of the landmark).

Self-Organization

The basic principle of neural functions is stated to be its self-organizing character [501]. The emergence of complexity by self-organizing mechanisms has been demonstrated both on ontogenetic and phylogenetic scales. Thinking in terms of dynamical concepts, ontogenetic development is associated with the temporal change of state due to interaction among the state variables (when these are considered to include the synaptic weights of our neural network), while phylogenetic evolution can be visualized as bifurcations in a parameter space which characterizes a whole range of evolutionary possibilities: continuous changes in the control parameters may lead to a discontinuous change in the state space. Ontogenetic development and phylogenetic evolution are closely related as dynamic processes [212].

According to embryological, anatomical and physiological studies the wiring of neural networks is the result of the interplay of purely deterministic (genetically regulated) and random (or highly complex) mechanisms. Fluctuations may operate as "organizing forces" in accordance with the theory of noise-induced transitions or stochastic resonance. Self-organizing developmental mechanisms (considered as "pattern formation by learning") are responsible for the formation and plastic behavior of ordered neural structures. Evolvability, the basis of self organization poses constraints on brain dynamics. Stable internal representation of the external world indicate the presence of attractors. Here, an attractor means one of the states of the system where the system settles after starting from a given initial condition. Self-organization needs these attractors to have a sufficient instability to be able to alter in order to adapt to the environment.

8.4 Single Cells

8.4.1 Single Cells: General Remarks

Neuroanatomists had long debates about the basic structure and building blocks of the brain. Though the "cell theory" (i.e., that the living material is composed of a discrete unit), was stated and partially proved in the 1830s, the general belief was that cell theory was not applicable to the brain. The *reticular hypothesis* assumed that neural tissue is rather continuous. Ramon y Cajal (1852–1934) adopted the staining technique of Camillo Golgi (1843–1926). While his technique was able to visualize only few randomly found cells, if a cell was stained, the very detailed branching patterns were visualized. Cajal strongly stated that according to his *neuron doctrine* the neurons are discrete entities, and they are the structural and functional units of the nervous system. While Cajal and Golgi shared the Nobel prize in Physiology or Medicine in 1906, their opinion was totally different: Golgi defended the reticular hypothesis. The appearance of the electron microscope proved (almost) that Cajal was right (see Fig. 8.3).

Although neurons divided by synapses are the discrete units of the nervous system, *gap junctions*, or electrical synapses, between several types of neurons provide a kind of continuous medium for signal processing. Electrical synapses

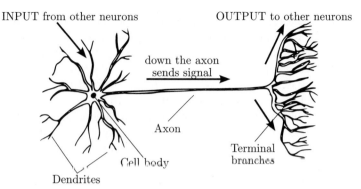

Fig. 8.3. Main anatomical parts of a typical neuron.

(which occur more often than it was believed) penetrating the cell membrane enable the electrical signal to be transferred to the postsynaptic cell without mediator chemical transmitters. A further means for non-discrete signaling is the use of non-synaptic communication via varicosities [539].

What Does a Single Cell Do?

Nerve cells are especially interesting from the physiological point of view not only because there is a membrane potential difference between the internal and external side of their membrane but because this difference is time-dependent unlike in any other cell of a living organism. For weak stimuli voltage changes decay rapidly to the resting potentials. If the stimuli are larger than a threshold, the trajectory of the membrane potential makes a rapid large increase followed by a decrease (i.e., emits an action potential or spike). This is a characteristic property of the excitable systems.

Neurons might be distinguished and categorized according to their general function: there are receptor or sensory neurons, interneurons and motor neurons. Sensory or afferent (carrying towards the brain) neurons are specialized to be sensitive to a particular physical stimulation such as light, sound, chemicals, temperature or pressure. Motor or efferent (carrying away from the brain) neurons receive impulses from other neurons and transmit this information to muscles or glands. Interneurons or intrinsic neurons form the largest group in the nervous system. They form connections between themselves and sensory neurons before transmitting the control to motor neurons.

Signal Generation, Temporal Patterns

Electrical impulses are considered to be generated at a specific site near the soma called the axon hillock. The resultant voltage at this site determines whether an action potential will or will not be generated and conducted along the axon to synapses that establish connection between neurons. When reaching the synapse at the end of the axon the action potential is likely to evoke neurotransmitter release, i.e., the emission of a certain chemical that transfers the signal of the presynaptic neuron to the postsynaptic cell, where it is converted back into an electrical signal to be summed with other similarly received signals from other sites in dendrites and the soma. Typical dynamical patterns, characteristic or regular spiking, fast spiking and bursting neurons can be seen in Fig. 8.4.

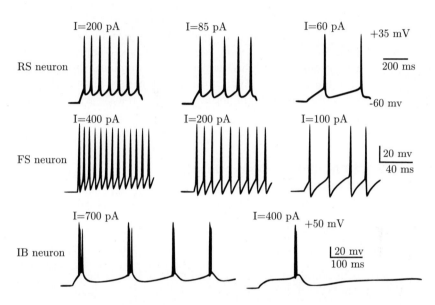

Fig. 8.4. Comparison of typical firing patterns, characteristic for regular spiking (RS), fast spiking (FS) and intrinsically bursting (IB) neurons with different excitatory currents. Modified from [251].

Single cell oscillations occur in consequence of the interplay of a few currents (e.g., the low threshold calcium current) and were demonstrated first mostly in invertebrates but later even in mammalian neurons (inferior olivary cells, Purkinje cells, thalamocortical cells and cortical pyramidal cells) [320].

Neurons may not only show simple oscillatory behavior, but also multirhythmicity. Multirhythmicity, a dynamical systems theoretical concept, means that the system can exhibit oscillations with more than one frequency. The (locally stable) attractors associated with the different periodic behaviors coexist, and are separated. In mathematical terms the initial conditions determine which rhythmicity is generated.

Neuromodulators may adjust the parameter values or can play the role of the "initial condition setter". Serotonin and dopamine serve as neuromodulators, and select among the multiple oscillatory modes of activity in e.g., the neuron R15 in the abdominal ganglion of Aplysia.

Action Potentials Versus Bursts: Unit of Information Transfer

Action potentials or spikes are regarded as elementary packages of information in the nervous system, although it is also suggested [319] that so called bursts might constitute the fundamental basis of neural information processing. Among others a burst seems to more reliable. If a postsynaptic cell does not respond to a presynaptic stimulus (which is an individual, single event) in consequence of a failure of the synaptic transmission, a burst (which is composed of a series of events) might induce synaptic release with a higher probability.

8.4.2 Single Cell Modeling: Deterministic and Stochastic Framework

Deterministic Framework: Compartmental Modeling Technique

The most successful model in theoretical and computational neuroscience is the Hodgkin–Huxley model. Sir Alan Lloyd Hodgkin (1914–1998) and Sir Andrew Fielding Huxley were carrying out experiments to determine how action potentials were generated in the squid's giant axon [237]. They constructed a mathematical model system accounting for the behavior of sodium and potassium membrane channels that was capable to reproduce the shape of an experimentally measured action potential:

$$\begin{aligned}
C\frac{dV(t)}{dt} =& \bar{g}_{\text{Na}} m^3\left(t, V(t)\right) h\left(t, V(t)\right) \left(E_{\text{Na}} - V(t)\right) + \\
& \bar{g}_{\text{K}} n^4\left(t, V(t)\right) h\left(t, V(t)\right) \left(E_{\text{K}} - V(t)\right) + \\
& \bar{g}_{\text{leak}}(E_{\text{leak}} - V(t)).
\end{aligned} \quad (8.1)$$

Here C is the capacitance of the membrane, $V(t)$ is the membrane potential, $dV(t)/dt$ is the change in membrane potential with respect to time, $\bar{g}_{\text{Na}}$ and $\bar{g}_{\text{K}}$ are the maximal conductance of sodium and potassium, E_{Na} and E_{K} reversal potentials for sodium and potassium, g_{leak} is the conductance for the so called leaky current and E_{leak} is the corresponding reversal potential, m, h and n are gating variables, the key variables of the model.

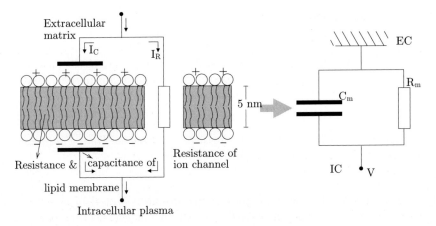

Fig. 8.5. The equivalent electrical circuit of a patch of axonal membrane.

This equation is based on the fundamental principles of electronic circuits (Fig. 8.5), i.e., Ohm's, Faraday's and Kirchoff's Laws. Current flowing through elements of this circuit is not the usual current of electrons, as it was thought for a while, but the current of some ions. Selective permeability of the membrane to a specific ion (chloride, sodium, potassium, etc.) is mostly determined by channels embedded in the membrane. A channel in this model is composed of gates that can be in an open or closed state, as described by the m, h and n variables. A channel is in the conducting state if all of its gates are opened. Gating variables are described by first order differential equations following the usual form of reaction kinetics:

$$\frac{dx(t)}{dt} = \alpha_x(V)(1 - x(t)) - \beta_x(V)x(t), \tag{8.2}$$

with x being one of the above three gating variables, α and β some membrane potential dependent rate functions. Equation (8.2) looks like as a chemical kinetic equation. The difference is the highly nonlinear voltage-dependence of the rate functions. These functions ensure a very fine-tuned mechanism for the generation and control of membrane potentials: they contribute very much to make the nervous system complicated and beautiful.

This model was formulated in the early 1950s and is still the standard of numerical simulation of neurons and serves as a framework for other excitable cells (as cardiac cells) as well. In 1963 Hodgkin and Huxley won the Nobel Prize for their discovery of the physical–chemical events underlying conduction of nerve impulses along neuronal axons.

Stochastic Framework

It has been already mentioned in Sect. 6.1.2 that the changes in the membrane potential can be considered approximately as a Brownian motion, and the paper [199] opened a new channel of research.

Membrane Potential as a Brownian Particle: Continuous Time Continuous State Space Stochastic Process (Diffusion Process)

The basic idea is that the dynamics of membrane potential of a neuron is supposed to make a random walk due to the net effects of excitatory and inhibitory synapses. In the original model Wiener process was assumed:

$$\mathrm{d}X(t) = \mu[X(t), t]\mathrm{d}t + \sigma[X(t), t]\mathrm{d}W(t), \tag{8.3}$$

$$T = \inf_{t \geq t_0} \{t : X(t) > S(t) | X(t_0) = x_0\}, \tag{8.4}$$

$$g[S(t), t | x_0, t_0] = \frac{\partial}{\partial t} P\{T \leq t\}. \tag{8.5}$$

The Italian mathematician, Luigi Ricciardi (from Naples) spent decades with stochastic single cell models [429]. He made the original model more realistic by substituting the Wiener process by Ornstein-Uhlenbeck process:

$$A_1(x) = -\frac{x}{\vartheta} + \mu, \tag{8.6}$$

$$A_2(x) = \sigma^2. \tag{8.7}$$

The introduction of non-stationary version of the OU model was necessary to describe certain effects, such as burst activity, when the firing frequency is subject to modulatory effects:

$$\begin{cases} A_1(x) &= -\frac{x}{\vartheta} + \alpha(t)e + \beta(t)i \\ A_2(x) &= \alpha_i(t)e^n + \beta(t)i^n \quad (n = 2, 3, \ldots). \end{cases} \tag{8.8}$$

As it often happens, Wiener process is less realistic, but it has solution in closed forms, the different versions of the OU process are more realistic, but computationally much more difficult.

First Passage Time Problems

When the voltage at a particular place on a neuron reaches a threshold, an action potential (nerve impulse) is produced. Many point processes in biology have similar origins as "first passage times"; that is, they occur when some underlying process first reaches a critical level or threshold.

Even for simple models of the underlying process (one-dimensional stochastic differential equations), very few analytical results are available for first passage times. Through simulation and heuristic approximation methods, several different types of behavior have been identified. The main current research activities are further development of approximation methods.

1. The *first-passage time of Markov processes to moving barriers* was solved by Tuckwell and Wan in 1984.

Let $\{X(t),\ t \geq 0\}$, $X(0) = x$ be a temporally homogeneous diffusion process with Ito stochastic differential

$$\mathrm{d}X = \alpha(X)\mathrm{d}t + \beta(X)\mathrm{d}W(t), \tag{8.9}$$

where $\{W(t),\ t \geq 0\}$ is a standard Wiener process.

The $T(x,y)$ random variable is the first time when X hits the Y moving barrier:

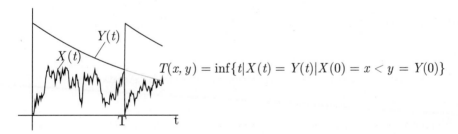

$$T(x,y) = \inf\{t | X(t) = Y(t) | X(0) = x < y = Y(0)\}$$

2. 20 years later the same authors determined the *time to first spike in stochastic Hodgkin-Huxley systems* [526].

- Hodgkin–Huxley equation with Gaussian white noise (with σ amplitude) approximating Poisson trains of excitatory and inhibitory post-synaptic potentials:

$$C_\mathrm{m} \frac{\mathrm{d}V(t)}{\mathrm{d}t} = I_\mathrm{membrane}(t, V) + \mu + \sigma \mathrm{d}W. \tag{8.10}$$

- Transition probability function $p(v, x, y, z, t; \hat{v}, \hat{x}, \hat{y}, \hat{z}, \hat{t})$ of the 4–dimension process (V, X, Y, Z) (V – membrane potential, X – potassium activation, Y – sodium activation, Z – sodium inactivation) satisfies the backward Kolmogorov equation:

$$-\frac{\partial p}{\partial \hat{t}} = \hat{L} p, \qquad (8.11)$$

where $\hat{v}, \hat{x}, \hat{y}, \hat{z}, \hat{t}$ are backward variables and $\hat{L}$ is a suitable differential operator for the noisy HH equations.

3. Firing time of a model neuron in the moving barrier scheme

The displacement $X(t)$ of a nerve cell's electrical potential from its resting value has been described by a diffusion approximation:

$$dX = (\alpha - X)dt + \beta dW(t), \qquad X(0) = x. \qquad (8.12)$$

- Constant threshold: not plausible → threshold declines after absolute refractory period.

- Time-dependent thresholds: e.g., oscillatory [561] or exponential [554] ⟶ moments of first-passage (firing) times can be calculated.

- The model can be extended with jump components:

$$dX = \alpha(X)dt + \beta(X)dW(t) + \gamma(X, u)\nu(dt \times du). \qquad (8.13)$$

Stochastic Resonance and Its Role in Neural Systems

Stochastic resonance (SR) is a mechanism, where noise plays a beneficial role in amplifying weak (often periodic) signals arriving to some nonlinear system. SR was found both experimentally and by model studies in various neurons and neural ensembles (such as in mechanoreceptor hair cells of the crayfish, mammalian cutaneous mechanoreceptors, temperature receptors, etc.

Noise can enhance the neuron's sensitivity. Consider a deterministic receptor neuron in the olfactory system devoid of random spontaneous activity; it fires an action potential when the receptor potential and consequently the odorant concentration exceeds a certain threshold. This means that any odorant concentration smaller than this threshold cannot be signaled to the brain. A "stochastic neuron", where the receptor potential presents a random (thermal) component added to the deterministic component that results from odorant stimulation, however, behaves differently. The presence of noise produces a finite probability for setting "signal+noise" to exceed the threshold value, and the cell activity may result in firing. Consequently, the neuron sensitivity at low concentration is greater than without noise. Specifically, mitral cells of the olfactory bulb may be able to show stochastic resonance, as it was demonstrated by simulations [52], as Fig. 8.6 shows.

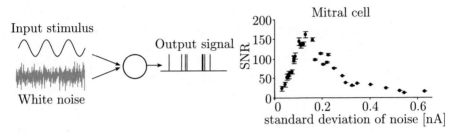

Fig. 8.6. The *left-hand side* explains that a weak, i.e., subthreshold phenomenon will be detected by the neuron after perturbed by a noise process. The *right-hand side* illustrates that the signal-to-noise ratio function shows a maximum for some intermediate value of noise intensity.

8.5 Structure, Dynamics, Function

8.5.1 Structural Aspects

The structural basis of neural functions is the network of neurons. Neurons are connected by excitatory and/or inhibitory synapses, and are responsible to generate different rhythms, associated to normal and pathological functions. They are the basis of pattern formation and pattern recognition, as well.

Hierarchy of Dynamics

Brain is prototype of hierarchical dynamical systems, as Fig. 8.1 showed. Neurons connected by synapses form networks to process, (in the general case time-dependent) inputs, as Fig. 8.7 shows.

Neuron

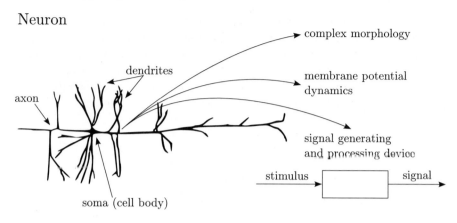

Synapse

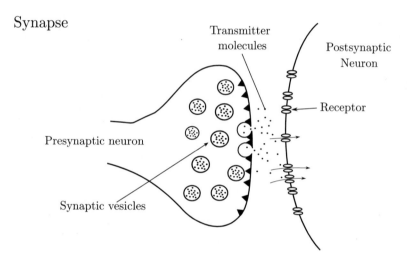

Network

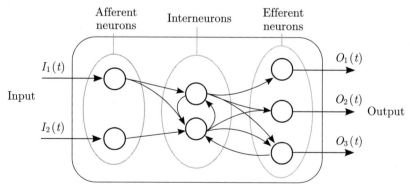

Fig. 8.7. Hierarchical organization of neural networks. A single neuron itself is a complex morphological and functional device. A synapse for mediating signal by neurochemical transmission also has a complex dynamics. A neural network process time-dependent input to output signals.

8.5 Structure, Dynamics, Function

A rather large subset of brain models takes the form of networks of intricately connected neurons in which each neuron is modeled as a single-compartment unit whose state is characterized by a single "membrane potential"; anatomical, biophysical and neurochemical details are neglected. Such neural network models are considered, at certain level of description, as two-level dynamic systems (assuming that the threshold of the neuron does not change with time; Fig. 8.8).

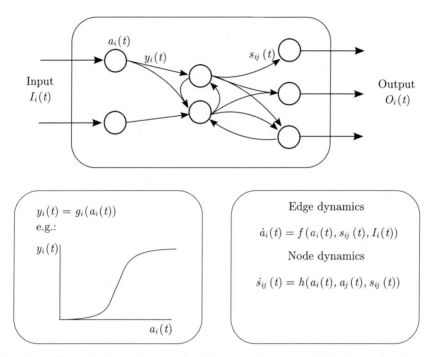

Fig. 8.8. Network dynamics is a double-dynamic system. Node dynamics is prescribed by activity dynamics, edge dynamics is defined by an appropriate learning rule.

Activity dynamics is often identified with the membrane potential equation, and the potential change (i.e., the form of f) is determined by the rate of presynaptic information transfer and the spontaneous activity decay. It is often assumed that the activity dynamics is given by the leaky integrator model of equation. Current learning theories generally assume that memory traces are somehow stored in the synaptic efficacies. The celebrated Hebb rule has been given as a simple local rule for explaining synaptic strengthening based on the conjunction between pre- and postsynaptic activity. Variations of learning rules will be discussed in Sect. 8.5.3.

Neurostatics

The structural analysis of neural networks developed very much, since (i) the scattered experimental data sets became more transparent by establishing integrated databases for connectivities, (ii) and by applying graph-theoretical methods.

One excellent database is CoCoMac (Collations of connectivity data on the macaque brain), which contains connectivity data of the macaque cerebral cortex [489, 290].

In our group in Budapest Fülöp Bazsó and his graduate student Tamás Nepusz in a collaboration with the neurobiologists László Négyessy and László Kocsis analyzed the visual and sensorimotor cortex of the macaque monkey by applying graph-theoretical methods. The actual neural connections of the cortex were transformed into a graph using the following principles:

1. Since the connectivity patterns of the individual neurons were unknown, neurons were grouped into *brain areas* based on anatomical considerations and the consensus between anatomists about the basic structure of the cortex. Every brain area became a single vertex in the graph to be studied.

2. Known connections between the areas were documented in the literature and in several databases (e.g., CoCoMac, PubMed). These connections were mapped to edges between the corresponding vertices of the graph.

The graph they analyzed consisted of 45 areas and 463 directed connections between them (Fig. 8.9). (In this case, directedness means that area A can connect to area B even if the opposite connection from area B to A does not exist). It turned out that the diameter of the graph is relatively small (5) compared to the number of areas involved, suggesting a small-world structure.

A shortcoming of this approach is that not all connections between areas are known, and known connections can also differ in strength: some areas are "wired" together more densely than others, but the actual strength is hard to quantify. Non-existing connections in the graph model does not necessarily mean that the two areas do not connect to each other: some of them do not, others might do, but the two cases are hard to distinguish due to methodological difficulties: checking the existence of some connections are hard and/or expensive to check. No dynamics was assumed, all the inferences were drawn from examining the statical structure of the cortex through this graph representation.

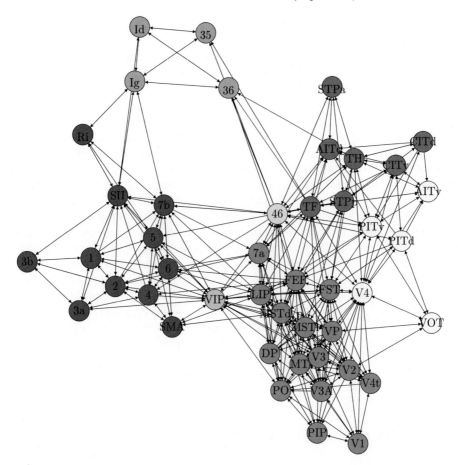

Fig. 8.9. The areas of the visual and somatosensory cortex of the macaque monkey and the known connections between them - represented as a graph. Colors (nor shown in this black-and-white version) represent clusters obtained by the Markov clustering algorithm [530].

Bazsó, Nepusz and Négyessy showed that graph-theoretical methods are able to discover the main structural division of the areas [373]. They weighted the edges of the graph by their betweennesses (the betweenness of an edge is the number of shortest paths in the graphs which include the edge being considered), and applied well-known graph clustering methods to identify the major structural parts. One of these methods which was particularly successful was the Markov Clustering Method or MCL in short [530]. The MCL algorithm was able to separate the visual and sensorimotor areas almost exactly, and it correctly identified the dorsal and ventral parts of the visual cortex. It also subdivided the sensorimotor cortex into two smaller parts. The method pointed out two areas (namely area 46 and area VIP (ventral intraparietal area) which might play a central role in the integration of the visual and tactile stimuli. This was in concordance with the prior assumptions of neuroanatomists, and was supported further by the calculated centrality scores of these areas.

Graph theory also enables us to study possible paths of information processing in the brain. Signals spread through neural connections, and the graph model maps neural connections between areas to directed edges. A directed path between area A and B in the graph corresponds to a possible route of information processing. With some simplification, we can say that the shortest paths between area A and B (where the actual length of a path is measured by the number of edges in the path) are of particular importance: to achieve the fastest processing speed, signals should follow these paths instead of the larger ones. By studying the shortest paths between the primary visual area $V1$ and the remaining areas, one can draw inferences on the most probable ways of how visual information gets to the sensorimotor cortex and how tactile information gets to the visual areas. This latter case is particularly interesting since it could explain how the primary visual area is able to respond to tactile information – a phenomenon observed in people suffering in early blindness. It turned out that all the shortest paths from the primary somatosensory cortex (area $3a$, 1 and 2) to $V1$ pass through VIP and then reach $V1$ via $MT, V3$ and PO. This observation provides further confirmation to the assumed central role of VIP in multimodal integration.

Columnar Organization

Experimental facts from anatomy, physiology, embryology and psychophysics gave evidence of highly ordered structure composed of *building blocks* of repetitive structures in the vertebrate nervous system.

The use of building blocks according to the modular architectonic principle [498] is rather common in the nervous system. Modular architecture is a basic feature of the spinal cord, the brain stem reticular formation, the hypothalamus, the subcortical relay nuclei, the cerebellar and especially the cerebral cortex. Columnar organization of the cerebral cortex was demonstrated by "physiological methods" first by Vernon Mountcastle in somatosensory cortex [365] and of Hubel and Wiesel [243] on visual cortex. "Physiological methods" here means that neuronal response properties remain relatively constant as one moves perpendicular to the surface of the cortex.

After the anatomical demonstration of the so-called cortico-cortical columns it was suggested by Szentágothai that the cerebral cortex might be considered on a large scale as a mosaic of vertical columns interconnected according to a pattern strictly specific to the species.

The functional significance of the cortical columns is somewhat nontrivial. The definition of the column is not clear. Some people use the expression for a narrow vertical chain of neurons ("minicolumns"), others use large-scale "hypercolumns". Szentágothai stated that the width of *cortical columns* is about $3 - 400 \mu$m. For an excellent balanced view about the scope and limits of the notion of neural columnar structures see [208].

8.5.2 Neural Rhythms

Neural rhythms with very different frequencies, from very slow oscillation related to deep sleep from high frequency oscillation occur in the brain. György Buzsáki wrote a beautiful book [85] about the topic. Here several functional and computational aspects of neural oscillators are reviewed.

Central Pattern Generators

A central pattern generator (CPG) is a network of neurons that produce rhythmic behavior in the absence of sensory input. Such phenomena as breathing, heartbeat, walking is related to CPGs. Relatively simple invertebrate systems, such as the crustacean stomatogastric ganglion, are capable of generating temporal patterns independently of peripheral reflex loops. The network structure and the transmitterology of this system has already been quite thoroughly uncovered. The fundamental temporal patterns can be explained by coupled oscillator models.

Phase Model of Two Coupled Oscillators

A mathematical technique to study the coupling effects is based on a class of oscillatory models, when the generating mechanism is neglected, and only the position of the oscillators are specified [116]. The actual state of a limit cycle oscillator can phenomenologically be represented by a single scalar variable: $\theta_i(t)$ specifies the position of the oscillator (i.e., its *phase*) around its limit cycle at time t, so it takes values between 0 and 2π radian in each cycle. The equation of motion is given by

$$\dot{\theta}_i(t) = \omega_i, \tag{8.14}$$

where ω_i is the frequency of the oscillation. The solution of (8.14) is

$$\theta_i(t) = (\omega_i t + \theta_i(0)) \mod 2\pi, \tag{8.15}$$

where $\theta_i(0)$ is the initial value of θ_i. The equations for two coupled oscillators are

$$\dot{\theta}_1(t) = \omega_1 + h_{12}(\theta_1, \theta_2),$$
$$\dot{\theta}_2(t) = \omega_2 + h_{21}(\theta_2, \theta_1). \tag{8.16}$$

Here h_{ij} is the coupling effect of the oscillator j to oscillator i. The coupling term is 2π periodic, since the rate of change depends on the phase of the oscillation only. The phase lag of oscillator 2 relative to oscillator 1 is defined as

$$\phi(t) = \theta_1(t) - \theta_2(t). \tag{8.17}$$

The combination of equations (8.16) and (8.17) leads to

$$\dot{\phi} = \dot{\theta}_1(t) - \dot{\theta}_2(t) = (\omega_1 - \omega_2) + (h_{12}(\theta_1, \theta_2) - h_{21}(\theta_2, \theta_1)). \tag{8.18}$$

Probably the simplest assumption for the coupling is that h_{ij} depends on the phase lag only, and zero phase lag implies zero coupling. This situations is called *diffusive coupling*. To make analytical calculations, often the coupling is set as $h_{ij} = a_{ij} \sin(\theta_j - \theta_i)$, we get

$$\dot{\phi} = (\omega_1 - \omega_2) - (a_{12} + a_{21}) \sin(\phi(t)). \tag{8.19}$$

A specific situation is when the time lag remains constant, i.e., $\dot{\phi} = 0$, which leads to

$$\phi = \arcsin(\frac{\omega_1 - \omega_2}{a_{12} + a_{21}}). \tag{8.20}$$

The value of the ratio of the frequency difference $\omega_1 - \omega_2$ to the net coupling strength $a_{12} + a_{21}$ determines the number of solutions. For small net coupling values the ratio absolute value is greater than unity.

Since the sin function takes values only between -1 and 1, there is no solution to (8.20) in this case, so the oscillators will drift with respect to each other. By increasing the net coupling a point can be reached, when the absolute value of the ratio is 1, and the system exhibits phase-locked motion. If the net coupling is positive (i.e., excitatory), then the faster oscillator leads the phase-locked motion by a phase between 0 and 90°. If the net coupling is negative (i.e., inhibitory), the slower oscillator leads by a phase between 90 and 180°.

Synchronizing a Chain of Oscillators

Ermentrout and Kopell [167, 287] studied a model

$$\dot{u}_k = F(u_k) + G^+(u_{k+1}, u_k) + G^-(u_{k-1}, u_k) \qquad (8.21)$$

with nearest neighbor coupling, where $F(u_k)$ specifies the free dynamics of the kth oscillator. To provide a general analytical study, some hypotheses are made on the coupling functions G^+ and G^-.

- Only phase, i.e., the angle between 0 and 2π which indicates progress around the limit cycle of each oscillator is relevant. This is true provided that the coupling is not so strong as to overcome the power of attraction of the limit cycle.

- The coupling terms are replaced by their averages over each cycle. This assumption works for the case of weak coupling only.

- Coupling is synaptic (chemical and not electric).

These assumptions reduce the equation to

$$\Phi_{k+1} = \omega_k + H_k^+(\Phi_{k-1} - \Phi_k) + H_k^-(\Phi_{k-1} - \Phi_k), \qquad (8.22)$$

where Φ_k is the phase of the kth oscillator, which has frequency ω_k. H^* are 2π-periodic functions. In consequence of the assumptions made, the coupling terms depend on the phase differences alone. A further assumption is the synaptic coupling hypothesis, $H^*(0) \neq 0$, to ensure that synchrony is not a solution even if the oscillators are identical. To solve the equations they work with the phase differences $\phi_k = \Phi_{k+1} - \Phi_k$ to get information about the spatial patterns of phase and frequency of the ensemble.

As Kopell remarks, biologically it is more plausible to regard the chain of oscillators as coupled to many neighbors, rather than to just the nearest neighbors. By using a multi-coupled structure it is possible to explain that sometimes frequency differences do not imply a phase lag. (In other words, phase delays may emerge from a different mechanism and not require frequency differences.)

In a further study, Kopell and Ermentrout [288] reduced the complex dynamics of neurons in the spinal cord of lamprey to a chain of oscillators with constant amplitude and a cooperative dynamics in the phases.

Networks of Limit Cycle Oscillators with Local and Long Range Coupling

We now consider synchronization in networks where individual oscillators are characterized by the phase variables Φ_i, oscillator i is driven by a random internal frequency ω_i taken from a distribution $g(\omega_i)$, and the interaction between oscillators i and k is homogeneous (i.e., we no longer have a chain but rather a fully connected network), and depends only on the phase difference $\Delta\Phi_{ik} = \Phi_i - \Phi_k$ of the two oscillators. The interaction law $f(\Delta\Phi_{ik})$ should be continuous and periodic. To enable the system to synchronize itself, $f(0) = 0$ must be chosen (contrary to the synaptic coupling hypothesis in the lamprey model), i.e., two synchronized oscillators ($\Delta\Phi_{ik} = 0$) do not drive each other but follow their internal driving force ω_i. Note the difference from the objective of Kopell and Ermentrout's lamprey model where spatial patterns, such as traveling waves, and not a synchronized state of an oscillator chain is required to model the network behavior.

Based on the now very popular Kuramoto model [302], Ermentrout [166] and others have studied the following equations,

$$\frac{d}{dt}\Phi_i = \omega_i - K \sum_{i,k \in N_i} \sin(\Phi_i - \Phi_k) + \xi_i, \tag{8.23}$$

which are a generic choice for a limit cycle system. (This is a somewhat imprecise notation, the equation should be understood as a stochastic differential equation.) Stochastic forces that act on the oscillator phases are described by the white noise term ξ. N_i is the set of all oscillators that are connected to oscillator i. Thus $N_i = \{i-1, i+1\}$ for a one-dimensional chain of limit cycle oscillators while $N_i = \{1, ..., N\}$ for a fully connected network. A fully connected oscillator network with dynamics (8.23) exhibits a strong tendency to synchronize phases if the noise and the width of the distribution $g(\omega_i)$ is not too large and the global coupling parameter K is sufficiently strong [452].

A special case of (8.23) is a network where all oscillators exhibit the same internal frequency ω. The coordinate transformation $\psi = \Phi_i - \omega_o t$ reduces (8.23) to

$$\frac{d\Phi_i}{dt} = -\frac{\partial H}{\partial \Phi_i} + \xi_i \tag{8.24}$$

with the energy-like function $H = -K \sum_{i,k \in N_i} \cos(\Phi_i - \Phi_k)$.

This system with local nearest neighbor coupling in two dimensions has been extensively studied in physics where it has been shown that phase synchronization is not a robust phenomenon. Any amount of noise, no matter how weak, will destroy a completely synchronized network state. Such a result indicates that the long range synchronization observed in the visual cortex cannot be explained with a local connection scheme but requires long range synapses. Indeed, long-range interconnections have been discovered in the first layer of the visual cortex of the cat [202, 203, 201, 124].

Gait Transitions in Locomotion

At a higher phylogenetic level, the locomotion of lower vertebrates, such as tadpole and lampreys, has been studied. Motor patterns underlying locomotions are produced by a spinal circuit. Not only the neural circuitry, but also the transmitters and the membrane properties are largely known [462]. Animal locomotion, even at more higher phylogenetic level, may be generated and controlled by a central pattern generator. However, the total motor pattern generator involved various feedback loops which use sensory input to adjust the motor pattern to current circumstances. Multi-legged animals exhibit several distinct patterns of leg movements; these basic patterns called gaits, e.g., walk, run, hop for humans; trot, canter, gallop for horses. Quadrupedal locomotion may be controlled by at least four coupled oscillators [105].

The general, model-independent properties of symmetrically coupled oscillators have been studied by group-theoretical methods. These oscillators have been identified with locomotor central pattern generators. Six symmetrically coupled oscillators have been considered as models of CPGs in insects, and the transitions between the gaits were modeled as symmetry-breaking bifurcations [103]. In particular, the relationship between the different network structures of the symmetrically coupled oscillators and the possible rhythmic patterns (associated to gaits) which they can generate were derived and listed. The symmetries of quadrupedal animal gaits have also been analyzed [104]. The analysis showed that minor modification in the network structure may imply a significant effect on the resulting gait. It is not easy to localize central pattern generators. By making symmetry analysis of animal gaits information about the possible network structure of CPGs can be obtained. Since the same network may produce different rhythmic patterns depending on the parameter values, the same locomotor central pattern generator may produce and control very different gaits.

There is a changing view on CPGs (see e.g., [273]). First, it seems to be evident, that though even single cells can produce in certain cases complex temporal behavior, the coordinated and controlled patterns are emergent network properties. In addition, the networks are not so rigid, or hard-wired: chemical modulators by modifying membrane properties may produce drastic changes in the behavior of the system, and even the network itself can be rearranged.

Coordination Dynamics

The practice and theory of coordination dynamics grew up from experiments and explanations of human bimanual coordination. There were experimental data of bimanual rhythmic finger tapping and finger and wrist abduction/adduction movement observed by Scott Kelso and others [273]. When speed is increased in an anti-phase tapping task (moving the hands anti-symmetrically), there was often an abrupt transition to an in-phase pattern There was a hysteresis, however, by subsequently decreasing the speed the system remained in the in-phase state. Starting the system from the in-phase state no transitions emerged.

The Haken–Kelso–Bunz Model

Conceptually and technically the theory came from synergetics, and the Haken–Kelso–Bunz model [229] was formulated. In its original form it described bifurcations in interlimb coordination between two components. This model considers moving limbs as coupled oscillators using the relative phase between the oscillators as the state variable. Rhythmic coordination is described in terms of the relative phase. An extended form of the model by taking into account non-identical components and additive noise is given by the equation (here I keep the original notation and usual physicist approach, e.g., [275], i.e., the "dot" should be understood as some (here not well-defined) stochastic differential):

$$\dot{\Phi} = \delta\omega - a\,\sin\Phi - 2b\sin 2\Phi + \sqrt{Q}\xi_t. \qquad (8.25)$$

Here $\delta\omega$ describes the heterogeneity of the components, i.e., the difference in the intrinsic frequencies of the two oscillators, so in the original model it was set as zero, a and b specifies the dynamics of the coupled system, and Q is the strength of the additive noise ξ_t.

The original HKB model ($\delta\omega = 0$ and $Q = 0$) was able to explain the fundamental experimental facts (Fig. 8.10).

I was never sure whether the HKB was a *constructed model* based on its mathematical properties, or it should be considered as a somewhat realistic model. I asked Viktor Jirsa, who was Hakens's student, and Kelso's close colleague. I can't do better than to copy his (slightly edited) mail:

"In the original 1985 paper there were two considerations on the basis of ϕ. First, ϕ must be 2π periodic, which means the $\dot{\phi} = -a\sin\phi - 2b\sin(2\phi)$ equation is 2π periodic also. This constrains the r.h.s to sin and cos functions.

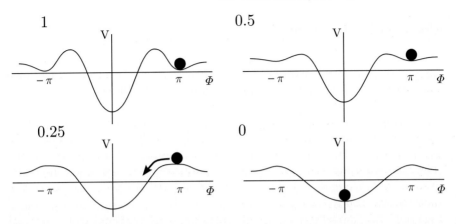

Fig. 8.10. Bifurcations in the Haken–Kelso–Bunz model of bimanual coordination from: http://dissertations.ub.rug.nl/FILES/faculties/ppsw/2004/m.h.g.verheul/c1.pdf.

Second, $\phi = 0$ is always a fixed point, hence these will be primarily sin functions or can be represented by sin functions. The next step is less rigorous and not discussed in the HKB paper: The r.h.s of the equation can now be decomposed into a Fourier series and can be truncated at the first order when you can get bistability. Such will be at $\sin 2\phi$. Though the truncation criterion is not fully rigorous, it holds extremely well, because there are not many stable states of ϕ in between zero and π. Hence it holds. Note that the equations on the oscillator level in terms of amplitude x are non-rigorous and ad hoc modeling.

In the attached paper [262] we present a more general derivation of a form which is more neural and has minimal assumption: two subnetworks exist which interact directly through a sigmoidal coupling (firing rate representation). These subnetworks capture the areas involved in motor and sensory coordination. Those activities can be mapped linearly (through a general convolution) onto the finger movement. The latter has been shown experimentally in Kelso's Nature paper [274]. Theoretically this has been spelled out in [261]. Then for periodic movements certain predictions are obtained for the brain activity which has been confirmed in MEG [262]. The assumptions are quite minimal that go into this work and are all neurally based."

While the coordination dynamics was established in the context of interlimb movement, gradually it has been extended to cortical structures, too [73]. In certain situations different cortical regions may have different intrinsic frequencies, called "broken symmetry". Such kinds of broken symmetry is a critical prerequisite for metastable dynamics, which they propose is a crucial feature of brain function.

Metastability and Local–Global Connectivity

The notion of *metastability* has been offered, as an important organizational principle of the brain in a few somewhat different contexts.

Friston [186, 187] used the terms complexity, dynamic instability and metastability synonymously, and suggested that brain dynamics is characterized by a series of transients. The Fingelkurts brothers[174] also emphasize the trade-off, better saying a sophisticated balance, between autonomous functioning of neural ensembles, and their coordination with others, as a key concept in their Operational Architectonics concept. Robert Kozma and Walter Freeman suggested the importance of the interplay between local and global interactions, They applied the physicists "percolation theory" to cortical networks, and demonstrated 'phase transition-like' phenomena in large-scale, random networks. They suggest that randomness not only can be tolerated in dynamic systems, but it plays a constructive role [183, 296]). I like the idea which Hans Lijenström [316] suggested by discussing the "stability-flexibility" dilemma. He argues that the architecture having extensive excitatory connections among neural populations connected mostly by inhibitory connections ensures robustness against both pruning (i.e., the decrease of connections), and external and internal fluctuations.

Sporns, Tononi and Edelman [515] emphasizes that *'functional segregation'* (i.e., autonomous operation of a functionally defined neuron population) is necessary to extract important features from sensory inputs quickly and reliably, while *'functional integration'* generates coherent perceptual and cognitive states allowing an organism to respond to objects and events.

"Metastability" or "relative coordination" may be an important property of cortical structures, which expresses that brain regions are general not completely synchronized and not completely independent. This is expressed in the *phase synchrony* versus *phase code* dichotomy.

Tsuda's chaotic itinerancy concept is also related to metastability. More precisely, the itinerant states in his model can be understood as an intermediate state between order and disorder [523]. It was also suggested [524] that chaotic itinerancy might be a mechanism for mediating the transition between synchronized and desynchronized states.

I think, at this moment the different approaches are not fully integrated, but something is in the air, and my expectation is that the analysis of the relationship among structure, function and dynamics will be emphasized much better.

Phase Synchrony Versus Phase Code

The finding of synchronized oscillation of multiunit activity in the visual cortex [219, 141] has generated much discussion. It seems to be remarkable from a functional point of view that rather remote columns oscillate in phase around 40Hz. It was suggested [115] that phase- and frequency-locked oscillations are strongly connected to the neurobiological basis of visual awareness. The theory was based on data from anesthetized animals, but later experiments [300, 369, 220] found similar results in awake animals. Synchronized cortical oscillations might be related to the neural basis of binding (coordinating, if you wish) different features (color, contour, orientation, velocity etc.) of the same object by forming coherently firing cells into assemblies [540] that code uniquely for the different stimuli.

Phase difference has a role in phase coding. A specific phase code will be mentioned in Sect. 8.5.2.

Normal and Pathological Cortical Rhythms

Table 8.1. Characterization of normal and pathological hippocampal rhythms based on their amplitude, duration and frequency. ENO: early network oscillation. SPW: irregular sharp wave. From [495].

Name	ENO	SPW	θ	β	γ	slow ripple	fast ripple
Amplitude (mV)	17–25	3.5	1	<1	<1	<1	<1
Duration (ms)	190–750	10–120				15–50	40–120	40–120
Frequency (Hz)	0.003–0.06	0.2–5	4–12	10–25	20–80	100–140	140–200

CPGs are not the only networks to generate rhythmic patterns in the central nervous system. From EEG studies it is known that there are some well defined rhythmic brain activities underlying behavioral phenomena (see Table 8.1 for a summary). For example it was observed that in the hippocampal formation learning and the composition of memory traces are accompanied by a series of specific oscillations. The so-called theta rhythm is characterized by a 1mV amplitude, 4–12 Hz frequency periodic electrical activity. It is observed while the animal is in the REM sleep period or during walking or while being engaged in sensory scanning or exploration. Theta activity often co-occurs with an other well-known rhythm, the gamma oscillation characterized by approximately 50mV amplitude and 40-100 Hz frequency. Both are population phenomena with correlated single cell activity. Theta activity is thought to be involved in time-locking cell activities, time-stamping for phase coding, increasing the signal to noise ratio in neural circuitries, regulating cellular learning. Gamma activity plays a role in the binding of perceived and recalled attributes of aspects and events. The two activities together could be enrolled in the formation of memory traces and in sequence recall. The mechanism underlying the generation of these brain activities and accordingly the relevant neural architecture is debated: theta activity might be delivered to the hippocampus from the septum by feed-forward inhibition or on the contrary might be generated intrahippocampally. Gamma activity is supposed to be generated locally in the hippocampus but could result from mutual excitation, mutual inhibition or probably from an excitatory-inhibitory loop. Recent computational studies suggest that a likely mechanism incorporating mutual inhibition and delays in the input of cells might simultaneously account for the generation of both the theta and the gamma rhythms in the hippocampus.

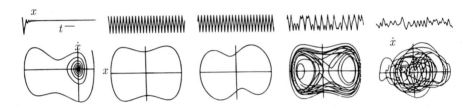

Fig. 8.11. Different modes of brain activity and their representation in the phase space: from stable fixed point to chaotic behavior. *Upper panel*: brain activity versus time. *Lower panel*: behavior of trajectories in the state space.

Epilepsy is a typical example of a dynamical disease. A dynamical disease occurs in an intact physiological system yet leads to abnormal dynamics. Epilepsy itself is characterized by the occurrence of seizures (i.e.,*ictal* activities). During epileptic seizures oscillatory activities emerge, which usually propagate through several distinct brain regions. The epileptic neural activities are generally displayed in the local field potential measured by local EEG. The epileptic activity occurs in a population of neurons when the membrane potentials of the neurons are "abnormally" synchronized. Both experiments and theoretical studies suggest the existence of a general synchronization mechanism in the hippocampal CA3 region. Synaptic inhibition regulates the spread of firing of pyramidal neurons. Inhibition may be reduced by applying drugs to block (mostly) $GABA_A$ receptors. If inhibition falls below a critical level, complete synchrony occurs. Rather arbitrarily, activity has been considered epileptic if more than 25% of the cells fire during 100ms. In vitro models of epilepsy offer a means to study the cellular mechanisms of the different types of epileptic phenomena by combined physiological and simulation methods. Several in vitro models of seizures have been developed, including electrical stimulation, low calcium, low magnesium and elevated potassium levels. Dynamical system theory offers a conceptual and mathematical framework to study epileptogenesis. Analytical studies based on bifurcation theory should clarify the possible operating modes of a given neural network (Fig. 8.11). The balance between excitation and inhibition is certainly an important control parameter, and its change may imply transition between the regimes. Epileptic activities may be considered as chaotic processes. There has been some hope that techniques of controlling chaos may offer new therapeutic and diagnostic tools for controlling epileptic activities. The question whether epileptic seizures can be predicted and controlled will be discussed in Sect. 9.3.4.

Chaos, however, may be controlled not only "from outside" the nervous system. The brain is not nearly as chaotic as it should be given its complex structure. One might notice that physiological parameters often avoid those ranges of values, which lead to chaos, and may think that it is controlled "from inside". A potential mechanism of shifting the parameter values from the region leading to chaos to a region leading to regular oscillation may be synaptic modification induced transitions.

Modeling Dynamics of Population of Neurons

In Sect. 8.4.2 some modeling aspects of nerve cells were introduced. The Reader, even without deep knowledge of computer science might be doubtful about the necessity of such kind of detailed description of single neurons in the simulation of large networks. Indeed, the computational time to account for every ion channel of every neuron would be enormous. Moreover in the study of large brain areas the information gained from every part of the complex, hierarchically organized neural tissue is superfluous. Still, to enable such studies neurons in models describing large brain areas are highly simplified. A certain approach is the one adopted by artificial neural network models, namely the use of oversimplified cell models quite far from biological reality. Using McCulloch–Pitts-like neurons [343] complex networks can be build and used for different pattern learning and recognition tasks. Large networks of simple elements are also used to study activity-propagation and different structure formation phenomena.

Another approach to describe both an average nerve cell and also the ensemble of nerve cells is a statistical one: statistical mesoscopic methods bind the local (microscopic) and global (macroscopic) scales by using anatomical and physiological data derived from the observation of single nerve cells and incorporating them as local parameters and variables into a general, global statistical description of neural fields. (Francesco Ventriglia, a neurocybernetican from Naples in a series of papers during 30+ years elaborated the kinetic theory of neural systems [535, 536, 537]. I believe that his fundamental works are not sufficiently appreciated.)

The adjective "local" means point-like here, and "global" may be interpreted as the whole. Obviously, local descriptions may have very different (and relative) meanings, even if we do not go below the single-cell level. (In any case, Ventriglia's model framework motivated us [222, 47, 52, 3, 161]. We tried to incorporate the statistical description of large neural populations with detailed single level description. I think, each research group has bedroom secrets. The story with our "population model" is ours, and I think I should not blab it out.) Viktor Jirsa [260] gave an excellent (almost: he also did not give credit to Ventriglia) review about the different families of neural network and neural field models, and mostly about the relationship of architecture and dynamics.

Dynamic Causal Modeling

While the electrophysiological results obtained by intracellular recordings have a proper corresponding theory based on compartmental modeling technique, data derived from the brain imaging devices lack coherent theoretical approaches. Though it became clear that the gap between conventional neural modeling techniques and brain imaging data should be narrowed, there is much to be done.

Dynamical causal modeling strategy offered by Karl Friston and his coworkers is a good first step into this direction [188, 488]. There are two underlying assumptions:

(i) There are n interacting areas, the state of each area is characterized by a scalar variable. The state of the system is modified by these inter-regional interactions, and by external input. The later has two effects, it modulates the connectivity, and has direct effects on regional activities. For the model framework and an illustrative example, see Fig. 8.12.

(ii) The neuronal state x should be converted into haemodynamic state y to be able to use the model for fMRI. Four auxiliary variables are assumed: the vasodilatory signal s, blood flow f, volume ν and deoxyhemoglobin content.

This is one possible model framework, and much work might be expected. I think the model framework could be used mostly for "inverse problems". Inverse problems start from data (actually fMRI time series), a dynamical model with unknown parameters is assumed, and the goal to get estimation for the parameters, actually for the effective connectivities.

Changes in the connectivities may be pathological, schizophrenia seems to be a "disconnection syndrome" [487] and may imply transition to pathological attractors, see Sect. 8.5.2.

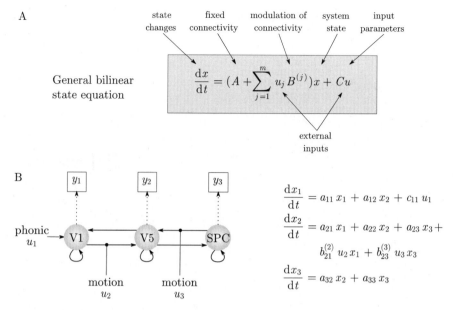

Fig. 8.12. This system consists of three areas V1, V5 and the superior parietal cortex (SPC). Their activity is represented by $(x_1, \ldots, x_3)$. *Black arrows* represent connections, *gray arrows* represent external inputs into the system and *thin dotted arrows* indicate the transformation from neural states into haemodynamic observations (*thin boxes*; see Fig. 8.13) for the haemodynamic forward model. Visual stimuli drive activity in V1 which is propagated to V5 and SPC through the connections between the areas. The V1 → V5 connection is allowed to change whenever the visual stimuli are moving, and the SPC → V5 connection is modulated whenever attention is directed to motion. The state equation for this particular example is shown on the *right*. From [488].

Neurological and Psychiatric Disorders as Dynamical Diseases

Neurological and psychiatric disorders can be interpreted as dynamical diseases. The concept emerged about ten years ago [481], and seems to be fruitful to explain a variety of disorders. For example, the emergence of seizures was explained by computational models to interpret transitions between the two states of a bistable system, namely between normal and epileptic activity [134]. The prediction and control of epileptic seizures became a hot, and controversial topic [321, 363]. It has become clear that dynamical models can predict seizure development and the administration of drugs could be designed accordingly providing novel therapeutic procedures for epileptic patients. Although, statistical analysis helps to predict the emergence of seizures, we still need to be cautious regarding its potential clinical applications [248]; see also Sect. 9.3.4.

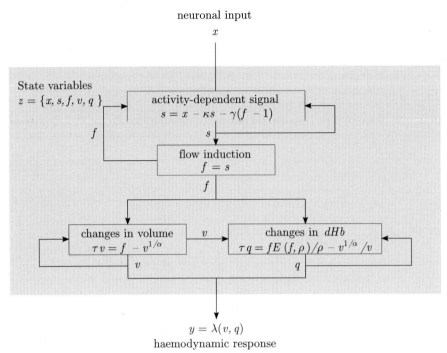

Fig. 8.13. A haemodynamic model for fMRI.

There are many other diseases where the dynamical characteristics seem to be relevant and the concept of "dynamical disease" can be applied. For example, ten years ago the question has been raised whether Parkinson's disease is a dynamical disease [138]. Along this line, more recently, a method was suggested for detecting preclinical tremor in Parkinson's disease [59]. It has been shown by using nonlinear dynamical theories and calculations [58] that patients with Parkinson's disease had an EEG series with higher complexity than normal persons during the performance of complicated motor tasks. The explanatory hypothesis for this increased complexity states that additional superfluous cortical networks are recruited due to impaired inhibition.

Depression is also thought to be a dynamical disease [366], and now it seems to be clear that there is a correspondence between clinical and electrophysiological dimensions [405], i.e., clinical remission and brain dynamics reorganization. Dynamical analysis of scalp EEG data is used to address the question whether schizophrenia originates from the reduction of functional or effective connectivities among brain regions [370] known as 'disconnection syndrome'. It needs more analysis to see how impairment of global (interregional) and local (intraregional) connections contribute to the emergence of schizophrenia. A functional computational model [72] suggests that schizophrenia might be the results of massive pruning of local connections in association cortex.

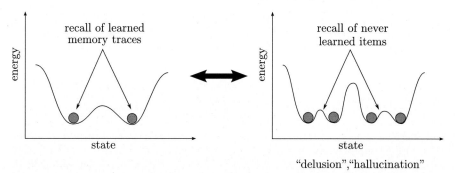

Fig. 8.14. Changes in the connectivities may imply changes in the attractor structure. Specifically, if the number of local minima is increased, in this pathological attractor structure never learned items could be "recalled". This scheme may be help to understand the dynamical framework of delusion.

Nonlinear dynamical methods gave new insights to study the neurodynamics in Alzheimer's disease [465, 259]. The analysis suggested that there is a reduced complexity and level of synchronization in the EEG patterns presumably due to impaired connectivity among different cortical regions. Dynamically evolving processes also can be captured in studies of migraines, i.e., the migraine aura lasts less than an hour, and precedes headache, and it is characterized by visual and other distortions. Hallucinatory patterns have been modeled by reaction-diffusion models leading to waves as has been observed in other excitable media (for a review see [481]).

Towards a Dynamical Neuropharmacology

Hippocampal theta oscillation was found to be in connection with mood and emotions and some associated illnesses such as stress or anxiety [184]. Modifying hippocampal theta oscillation by means of proper pharmacological agents the desired change in mood or emotions might be evoked.

The strategy is in accordance with the systems biological perspective sketched in Sect. 4.3. The implementation of the principle shown in Fig. 4.10 is visualized in Fig. 8.15.

In a cooperation with a physiology-oriented pharmacologist Mihaly Hajos (in the Pharmacia corporation in Kalamazoo, and after its acquisition by Pfizer, at Groton, Conn.) a set of model was developed to study the eventual physiological effects of drugs, which are supposed to be mood regulators [280, 282, 527, 528]. Typical experimental data are shown in Fig. 8.17.

276   8 Complexity of the Brain

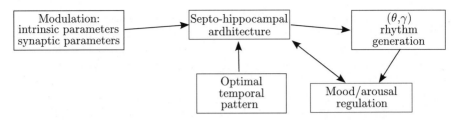

Fig. 8.15. Systems biological perspective to dynamical models of mood regulation. Septohippocampal system is supposed to be associated for anxiety disorders. Anxiety is correlated to larger theta rhythmic activity. Control of the system could be designed to shift the system from undesired state to a prescribed one.

A computational model was developed to realistically account for the generation of hippocampal theta oscillation in the CA1 region [226]. Five cell populations likely to play a key role in theta activity generation are described in the model extending the Hodgkin–Huxley formalism.

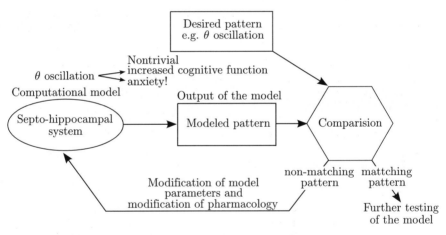

Fig. 8.16. Computational neuropharmacology – an idealized method for drug discovery. See text for a description. From [281].

Briefly, we suggest the following method (see Fig. 8.16 for a graphical explanation): First, connection between electrical brain activity and behavior or mood is identified [184, 574], yielding the so called *desired pattern*. In the presented case this pattern consists of firing rate and timing of firings of all modeled cell populations as well as the local field potential measured in the hippocampal CA1 region. Indeed, the identification of such a pattern is a highly nontrivial task. From a behavioral point of view, a certain pattern might correspond to multiple cognitive functions or moods. In the case of theta rhythms, it is known that on one hand anxiolytic drugs reduce hippocampal theta frequency [344, 221] by impairing the subcortical control of hippocampal theta activity, while on the other hand hippocampal theta rhythm is essential for the animal in different memory and cognitive tasks [397, 278]. The power of theta and alpha oscillations in humans reflect cognitive and memory performance [284] in a nontrivial way. Thus, to achieve good results, the desired pattern has to be detailed, containing all available information.

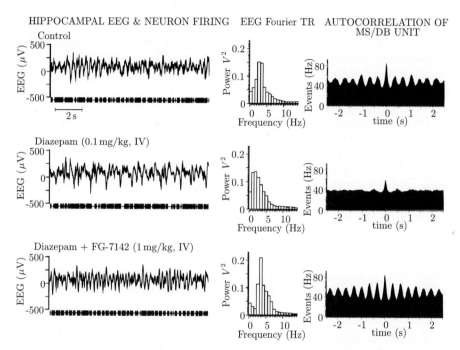

Fig. 8.17. Pharmacological control of hippocampal theta rhythm. Theta activity can be depressed and induced by the hippocampal CA1 by GABA allosteric modulators. The positive allosteric modulator diazepam, which increase chloride conductance, suppresses theta activity. Negative allosteric modulator FG-7142, which reduces transmission of GABA-A synapses, reverses the effect of diazepam.

Second, a *mathematical model* is constructed, which – when its output is interpreted – gives a *modeled pattern* comparable with the desired pattern. Measures to carry out this *comparison* have to be set up in such a way that an automatic parameter-space search can be performed based on the quantitative and qualitative results of the comparison.

Last, when the modeled pattern sufficiently matches the desired pattern, model parameters has to be read out and interpreted. In the presented case of hippocampal theta oscillation modulation by $GABA_A$ allosteric modulators, the important parameters were the synaptic strength between different cell populations. As a result of this modeling the most sensitive parameters to modulate theta rhythms were found.

Multiscale Modeling

We offered a new combined physiological/computational approach to drug discovery [17] by finding optimal temporal patterns. Using modeling results, the drug screening phase of the drug discovery process can be made more effective by narrowing the test set of possible target drugs. Besides this benefit, more selective drugs can be designed if the modeling method is able to identify some specific sites of drug action.

Preliminary works have been done to set a multi-scale model by integrating compartmental neural technique and detailed kinetic description of pharmacological modulation of transmitter - receptor interaction is offered as a method to test the electrophysiological and behavioral effects of putative drugs. Even more, an inverse method is suggested as a method for controlling a neural system to realize a prescribed temporal pattern. The general plan is illustrated in Fig. 8.18.

Our working hypothesis is that

- for given putative anxiolytic drugs we can test their effects for the EEG curve by,
- setting the kinetic scheme and a set of rate constants, (which of course may be different for different recombinant receptors),
- simulating the drug-modulated GABA – receptor interaction,
- calculating the generated postsynaptic current,
- integrating the result into the network model of the septohippocampal system,
- simulating the emergent global electrical activity,
- evaluating the result and to decide whether the drug tested has a desirable effect.

For the results of some preliminary modeling and simulations see [162].

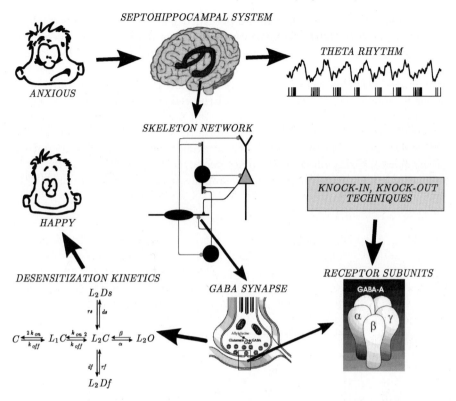

Fig. 8.18. Integration of detailed kinetic model and network models to test drug effects on system physiology and mood.

The general method we offer to test and design drugs is new, and now we have the conceptual framework to make significant progress. I believe that dynamical systems theory and computational neuroscience integrated with the well established, conventional molecular and electrophysiological methods, will offer a broad, innovative prospective in drug discovery and in the search of novel targets and strategies for neurological and psychiatric therapies.

Generation of the Double Neural Code

It is one of the central dogmas of neurobiology that information in the nervous system is embedded in spike trains. However, it is still a question which parameters of a spike train code for attributes of external stimuli. There are two fundamentally different ways of neural coding: *rate* and *temporal* coding. Rate coding implies that firing frequency, while temporal coding means that timing of spikes conveys information. So called *place cells* of the rodent hippocampus are delicate subjects for investigation of neural coding since both frequency (Fig. 8.19A,C) and timing of their spikes related to the local *theta field potential oscillation* (Fig. 8.19A,B) correlate with the position of the animal. Each place cell fires only when the animal is in a specific portion of the environment called the *place field* of the cell. Firing frequency increases until the middle and decreases in the second part of the place field so it is a unimodal function of the animal's position (rate coding). Timing of a spike can be defined as the phase of the theta cycle when the spike was emitted (*firing phase*). Firing phase in the place field decreases constantly so that it is a monotonic decreasing function of the animal's position (phase coding) [390].

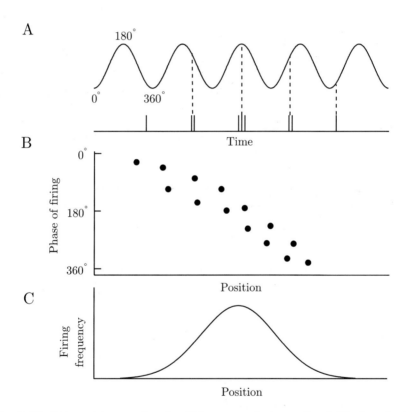

Fig. 8.19. Rate and temporal code of hippocampal place cells. *A*, Timing of spikes (vertical solid lines) is related to hippocampal theta oscillation. One cycle of theta is divided to 360°. *B*, Temporal or phase coding. Firing occurs in earlier and earlier phases relative to the theta oscillation so the firing phase is a monotonically decreasing function of position. *C*, Rate coding. Number of spikes fired in each cycle is increasing until about the middle and decreasing in the second part of the place field. This implies that firing frequency tuning curve of the place cell to be a unimodal function of position.

We proposed a mechanism to account for this *double code* and tested it by simulations first on an integrate-and-fire [309] and recently on a conductance-based model [247]. During exploration (when place cell activity can be observed) theta oscillation is present in the hippocampus, therefore we assumed that theta plays a key role in forming the firing pattern of place cells. First, we briefly describe our model on the generation of the doubly coding firing pattern of place cells [247] then discuss the multiple roles of theta in the suggested mechanism.

The dendrite of our model cell could generate membrane-potential oscillation in response to current injection. In response to tonic stimulation (constant injected current) the frequency of this oscillation was a continuously increasing function of the strength of stimulation (Fig. 8.20). In response to an oscillatory stimulation (sinusoid injected current) dendritic dynamics changed fundamentally. We kept the frequency and AC component (amplitude) of the sinusoid current at a fixed value and varied the DC component (average value of the current). Depending on the value of the DC component the dendrite was in different dynamical states: when the value of the DC component was relatively low the dendrite was forced to oscillate with the frequency of the stimulating current independently of the exact value of the DC component (Fig. 8.20, *regime I.*). However when the DC component exceeded a given value (Fig. 8.20, arrowhead) frequency of the dendritic membrane potential oscillation became dependent on DC component and the f-I curve became overlapping with that recorded in response to tonic stimulation (Fig. 8.20, *regime II*).

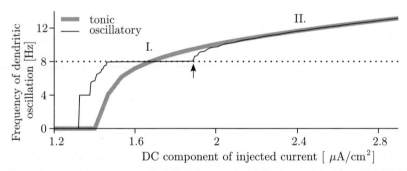

Fig. 8.20. Frequency of dendritic membrane potential oscillation as a function of the DC component of injected current (f-I curve). Above firing threshold tonic stimulation (*thick gray line*) resulted in increasing oscillation frequencies. Oscillatory stimulation (*thin black line*) resulted in frequency locking to the oscillatory input in a wide range of the DC component (*regime I.*): dendritic membrane potential oscillated with the same frequency (8Hz, *dotted line*) as the stimulating current. Above a given value of DC component (*arrowhead*) dendritic oscillation did not follow the oscillatory input current and its frequency curve quickly converged to the tonic depolarization-induced f-I curve (*regime II*).

In our two compartmental model both the soma and the dendrite received an oscillatory input of theta frequency. Input currents were in anti-phase and the DC component of dendritic input depended on the position of the animal (all the other parameters of the currents were kept constant). Outside the place field the value of the DC component was low so therefore the dendritic membrane potential oscillated with theta frequency, just in antiphase with somatic the one (Fig. 8.21). Firing probability was the sum of these two oscillations so it had a constant value. Firing threshold was defined to be above this value so the cell did not fire outside the place field. We assumed that the DC component of dendritic input is increased when the rat is inside the place field so that the dendrite starts to oscillate with a slightly higher frequency than theta. The sum of the two oscillations of different frequencies resulted in a composite oscillation whose local maxima occurred in earlier and earlier phases of the subsequent theta cycles. The cell fired in these phases which ensured the generation of the temporal code. The number of action potentials fired by the cell in each cycle depended on the value of somatic oscillation at the times of local maxima. As it was the unimodal function of the position of the animal peaking in the middle of the place field, rate code was also established.

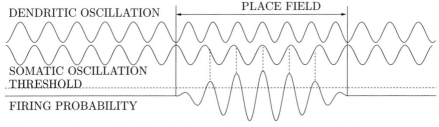

Fig. 8.21. Outside the place field somatic and dendritic oscillations were in antiphase with each other, so their sum, the firing probability of the cell had a constant value below the firing threshold. Inside the place field dendritic oscillation became faster due to the weaker theta modulation of its input therefore firing probability began to oscillate. When it exceeded the firing threshold the place cell could emit action potentials. In each cycle the peak of this oscillation occurred in earlier phase than in the previous one providing the phase or temporal code; the value of somatic membrane potential at the phases of local maxima increased until the middle and decreased in the second part of the place field which accounted for the generation of rate code.

Theta oscillation played essential role in our model in several ways:

- Theta as population cell activity. To define the firing phase of a cell there has to be some reference oscillation that the timing of the spike can be related to. This is the theta field potential oscillation in the hippocampus, which reflects the summed activity of cells populations.

- Theta rhythm as dendritic input. Each place cell only fires in a small portion of the environment [389]. In our model the cell fired only inside the place field because outside of it dendritic membrane potential oscillated with the same theta frequency (and in anti-phase) as the somatic one. Constant input can not generate an exactly theta frequency dendritic oscillation (its value had to be adjusted infinitely precisely) which is essential in our model in order not to have spikes outside the place field. However, in case of a theta-modulated periodic input DC component of the current may vary without changing the frequency of dendritic oscillation (Fig. 8.20, *regime I*). This way theta oscillation provides resistance for the mechanism against noisy inputs.

- Theta rhythm as somatic input. Rate code was based on that the cell fired at different somatic membrane potential values in the subsequent theta cycles (Fig. 8.21). Oscillation of membrane potential was necessary for this, which was provided by the theta modulation of somatic input.

8.5.3 Variations on the Hebbian Learning Rule: Different Roots

It had been known since the end of the nineteenth century that mature nerve cells cannot divide. Thus learning could not result from the proliferation of new neurons, therefore the locus of learning *must be the connections between cells*. Such kinds of phenomena are related to the neural basis of classical conditioning.

The Canadian psychologist Donald Hebb (1904–1985) marked a new era by introducing his learning rule and resulted in the sprouting of many new branches of theories and models on the mechanisms and algorithms of learning and related areas. He assumed that if two neurons connected by a synapse are active simultaneously, then there is a tendency to increase the strength of the connection between them.

Two characteristics of the original postulate [233] played key role in the development of post-Hebbian learning rules. First, in spite of being biologically motivated, it was a verbally described, *phenomenological* rule, without having view on detailed physiological mechanisms. Second, the idea seemed to be extremely convincing, therefore it became a widespread theoretical framework and a generally applied formal tool in the field of neural networks. Based on these two properties, the development of Hebb's idea followed two main directions. First, the postulate inspired an intensive and long lasting search for finding the *molecular* and *cellular* basis of the learning phenomena – which have been assumed to be Hebbian – thus this movement has been absorbed by neurobiology. Second, because of its computational usefulness, many variations evolved from the *biologically inspired* learning rules, and were applied to huge number of very different problems of artificial neural networks, without claiming any relation to biological foundation. Several families of rules sprouted from the original idea will be discussed. However, first we should note that what we qualify as Hebbian, post-Hebbian, and non-Hebbian learning rules may be somewhat subjective.

Classical Conditioning

The first attempt to model conditioning in terms of synaptic change was due to Hebb. Hebb's original intention was to connect the behavior of whole organisms to neural mechanisms by using concepts represented by cell assemblies. Specifically, classical conditioning involves the development of an association between two otherwise unrelated events over number of trails in which the events are temporally paired. Typically, the presentation of a neutral stimulus – one that does not naturally provoke behavior – is immediately followed with the presentation of unconditioned stimulus – an event that does not require training to produce a response-resulting in the eliciting of an unconditioned response.

Classical conditioning has been described by the Rescorla and Wagner's model [428]. They gave a formal model of conditioning which expresses that the capacity a conditional stimulus (CS) has to become associated with an unconditional stimulus (US) at any given time. The central idea of the Rescorla–Wagner model is that learning occurs if events violate expectations. More specifically, whenever the actual US level received on a trial differs from the level expected. The Rescorla–Wagner rule can be interpreted that the discrepancy between expected and actual values determines the measure of reinforcement. So, the rule and its many later modifications, are over the "unsupervised learning" paradigm. One drawback of the Rescorla–Wagner model is that it completely ignores the temporal sequence of information.

Development

The formation and refinement of neural circuits involve both the establishment of new, and the elimination of already existing connections. Specifically, the mechanism leading to synaptic elimination is called axonal or synaptic competition. Neuromuscular junctions and the visual system are the two best investigated examples, where synaptic competition plays an important role. A large variety of different generalized Hebbian learning rules applied for neural development was reviewed by [531].

The different mechanisms of competition elaborated in population biological context have been adopted in neural context.

In *consumptive competition*, in systems of consumers and resources (e.g., predators and preys, respectively), each individual consumer tries to avoid the others and hinders the others solely by consuming resources that they might otherwise have consumed; in other words, consumers hinder each other because they share the same resources. In neurobiology, competition is commonly associated with this dependence on shared resources.

In *interference competition*, instead of hindrance through dependence on shared resources, there is direct interference between individuals. This occurs, for example, if there are direct negative interactions e.g., aggressive or toxic interactions between individuals. In axonal competition, nerve terminals could seek to destroy each other by releasing proteases.

Long-Term Potentiation – Long-Term Depression

Long-term potentiation (LTP) was first discovered in the hippocampus and is very prominent there. LTP is an increase in synaptic strength that can be rapidly induced by brief periods of synaptic stimulation and which has been reported to last for hours in vitro, and for days and weeks in vivo [64].

The LTP (and later the LTD) after their discovery, have been regarded as the cellular physiological basis of Hebbian learning. Subsequently, the properties of the LTP and LTD became more clear, and the question arises, how LTP and LTD can be implemented in accordance with the phenomenological Hebb type learning. Formally, the question is how to specify the general functional F to serve as a learning rule with the known properties of LTP and LTD. Recognizing the existence of this gap between biological mechanisms and the long-used Hebbian learning rule, there have been many attempts to derive the corresponding phenomenological rule based on more or less detailed neurochemical mechanisms.

The time-course of LTP may be insufficient to sustain long-term memory, but there appear to be multiple LTP mechanisms, and one dependent on protein synthesis might serve long-term memory: inhibition of protein synthesis disrupts the maintenance of LTP, but leaves the induction of LTP relatively or totally intact. It is possible to relate properties and mechanisms of long-term synaptic plasticity in the mammalian brain to learning and memory.

Timing

Studies in cortical and hippocampal slices have shown that back-propagating action potentials may contribute to induce persistent synaptic potentiation or depression. The timing of presynaptic and postsynaptic action potentials play a decisive role in determining the sign of synaptic modification [329]. The temporal order of the synaptic input and the postsynaptic spike within a narrow temporal window determines whether LTP or LTD is elicited, according to a temporally asymmetric Hebbian learning rule.

Bi and Poo [60] showed that postsynaptic spiking that peaked within a time window of 20ms *after* synaptic activation resulted in LTP, while spiking within a window of 20ms *before* synaptic activation led to LTD. They suggested that a narrow and asymmetric window for the induction of synaptic modification should be taken into account.

The majority of the generalized Hebbian rules are based on statistical properties of presynaptic and postsynaptic activity (e.g., activity product, activity covariance etc.) without considering the detailed temporal structure of the spike patterns. Relative time spiking, however, has been taken into account even earlier (e.g., [494]).

Since changes in synaptic efficacy can depend on the precise timing relations of pre- and postsynaptic spikes, phenomenological 'temporal learning rules' generate opposite change in synaptic efficiency depending on whether the postsynaptic spike in advance of, or follow, the presynaptic spike. There is an attempt to show that differential Hebbian learning could be a proper framework to take into account the timing effects [433].

Algorithms of Learning Rules

The most general form of Hebb's rule to express the idea above, is that the synaptic weight from neuron i to neuron j changes according to:

$$\frac{\mathrm{d}}{\mathrm{d}t} w_{ij}(t) = F(a_i, a_j), \qquad (8.26)$$

where F is a functional, and a_j and a_i are presynaptic and postsynaptic activity functions (i.e., they may include activity levels over some period of time and not just the current activity values). To define specific learning rules, i.e., the form of F, a few points should be clarified.

1) What are the assumptions about the *locality* of the modifying signal? In many cases, the modification of a synapse between neurons i and j depends

on the state of these two cells alone, i.e., the mechanism is local. In this case teacher or external reinforcement signals are not explicitly involved: local synapses are the bases of the unsupervised learning.

2) How, if at all, do the presynaptic and postsynaptic cells *interact?* Consider first the potential answers for the "if at all" part of the question. The modification can be interactive, if both the pre- and postsynaptic cells are involved, and non-interactive, if either the pre- or postsynaptic cell alone influences the modification. The mechanism of the interaction may be conjunctional or correlational. In the first case, the co-occurrence of the pre- and postsynaptic activity is sufficient to cause synaptic change, while in the second case the covariance of the two activities has to be taken into account. (From a formal point of view, additive interactions – e.g., given with the function $F(a_i \pm a_j)$ – could have been defined, but they are considered as non-interactive rules. In other words, not only an entire rule, but even each term of it can be evaluated as interactive or non-interactive.)

3) What are the assumptions about the form of the *time-dependent* activity functions? In the simplest case, only the actual activity values are involved. In somewhat more complex situations, short-term averaged activity values determine the synaptic change. More generally, the *timing* of the activity values plays a role in the modification process.

The simplest Hebbian learning rule can be formalized as:

$$\frac{\mathrm{d}}{\mathrm{d}t} w_{ij}(t) = k\, a_i(t)\, a_j(t), \quad k > 0. \tag{8.27}$$

This rule expresses the conjunction among presynaptic and postsynaptic elements (using neurobiological terminology) or associative conditioning (in psychological terms), by a simple product of the actual states of presynaptic and postsynaptic elements, $a_j(t)$ and $a_i(t)$. A characteristic and unfortunate property of the simplest Hebbian rule (8.27) is that the synaptic strengths are ever increasing.

There are both brutal and sophisticated methods to eliminate the unpleasant property of ever-increasing weights which, unless compensated for, yields a network with saturated synaptic weights and thus no effective pattern discrimination. The qualification "brutal" was adopted for the situation when some external constraint (taking into account somehow the finiteness of resources) is applied to the internal mechanism. First, a predetermined upper bound can be given, such as the maximal value of the synaptic strength. Second, the so-called normalization procedure (which appeared already in [435]) gives a finite-sum constraint on all synaptic strengths, and can be interpreted as a competition of the presynaptic elements for postsynaptic resources (therefore it violates locality). Such rules may explain some aspects of neural development.

A weak form of the interactive rule (when correlational and not conjunctional interactions were assumed), namely

$$\frac{\mathrm{d}}{\mathrm{d}t} w_{ij}(t) = k\left(a_i(t) - [a_i(t)]\right)\left(a_j(t) - [a_j(t)]\right) \tag{8.28}$$

was offered by [435]. The activity variables fluctuate around a mean value $[x]$.

Depending on the sign of the correlation, the rule is capable of describing either synaptic enhancement or decrease. Covariance was suggested to induce associative LTD in the hippocampus [482].

Discussion: Over the Hebbian Paradigm

It is certainly not true that all learning rules could be interpreted in (even generalized) Hebbian sense. It is difficult, however, to draw the borderline between the Hebbian and "non-Hebbian" frameworks. One possible choice is to consider a learning rule Hebbian, if only two elements (one presynaptic and one postsynaptic) are involved. If we accept these limitations, we can determine what is labeled as non-Hebbian learning rule. Heterosynaptic plasticity and modifiability of synaptic triads and glomeruli – where *more than two cells* are explicitly involved in the modification process – could be understood also, as non-Hebbian. Such choice, however, would also exclude rules with the normalization procedure.

What is the relationship between the homosynaptic (or Hebbian activity-dependent), and heterosynaptic (or modulatory input-dependent) plasticity? It was suggested that Hebbian mechanisms are used primarily for learning and for forming short-term memory traces but they are not sufficient to recruit the events required to maintain a long-term memory [36]. In contrast, heterosynaptic plasticity commonly recruits long-term memory mechanisms that lead to transcription and to synaptic growth. When jointly recruited, homosynaptic mechanisms assure that learning is effectively established and heterosynaptic mechanisms ensure that memory is maintained.

In non-biological context, many types of supervised learning rules used in the artificial neural network community. The goal is to minimize the deviation of the actual response of the system from the prescribed one by modifying the synaptic strengths.

The spirit of the Hebbian idea survived more than a half century. (For a review of post-Hebbian algorithms, see [163].)

It will be interesting to see what kinds of phenomenological learning rules will be derived in the next several years starting from cellular level experimental and modeling studies of synaptic modifiability.

8.6 Complexity and Cybernetics: Towards a Unified Theory of Brain–Mind and Computer

8.6.1 Cybernetics Strikes Back

The Brain–Mind–Computer Trichotomy

The term "brain" is often associated with the notions of "mind" and of "computer". The brain–mind–computer problem has been treated within the framework of three separate dichotomies, see Sect. 11.2.4 in [22].

First, the brain – mind problem is related to the age-old philosophical debate among monists and dualists. Attempts to "solve" the brain-mind problem can be classified into two basic categories:

1) materialistic monism, leading in its ultimate consequences to some kind of reductionism; and

2) interactionist dualism, which is more or less some type of Neo-Cartesian philosophy.

The classification is, obviously, a crude oversimplification: a wide spectrum of monistic theories exist from Skinner's (1971) [468] radical behaviorism and Patricia Churchland's eliminative materialism [99] through Smart's (1981) [469] physicalism and Bickle's "ruthless reductionism" [61] to Bunge's [84] emergent materialism. Interactionist dualism has always been an influential viewpoint since Descartes defined the interaction between the spatially extended body and a non-corporeal mind. Though its modern version was elaborated by two intellectual heroes of the 20th century (Popper and Eccles 1977) [412] still it has been criticized or even ignored by the representatives of the "main stream" of the philosophy of mind, both as functionalists as well as by biologically-oriented thinkers. The philosophical tradition of hermeneutics, i.e., the "art of interpretation", which is a priori neither monist nor dualist, can be applied to the brain [523, 158, 165]. Even more is stated: on one side, the brain is an "object" of interpretation, on the other side, it is itself an interpreter: the brain is a *hermeneutic device*. This concept will be discussed later in this chapter. In similar vein, Arbib argues [18, 21] that our theories of the brain are metaphors, while the brain itself represents the world through schemas, which may themselves be viewed as metaphors.

Second, we have already mentioned that the brain – computer analogy got a new impetus by discovering the computational power of the extended

MCP networks. Arguments for the computer – brain disanalogy were listed by [107]. Digital computers are programmed from the outside; are structurally programmable; have low adaptability; and work by discrete dynamics; their physical implementation is irrelevant in principle; they exhibit sequential processing; and the information processing happens mostly at a network level. Brains are self-organizing devices; they are structurally non-programmable; they work by both discrete and continuous dynamics; their functions depend strongly on the physical (i.e., biological) substrate; the processing is parallel; and processing occurs for both network and intraneuronal information.

Inspiration from the brain leads away from the emphasis on a single universal machine towards a device composed of different structures, just as the brain may be divided into the cerebellum, hippocampus, motor cortex, and so on. Thus we can expect to contribute to neural computing as we come to chart the special power of each structure better. The brain may be considered as a metaphor for cooperative computation.

Third, the computational theory of mind holds that the computational metaphor is the final explanation of mental processes. "Connectionism" (Rumelhart and McClelland 1986) was an ambitious conceptual framework for a would-be general brain-mind-computer theory, but it is based on principles of "brain-style computation" that ignore many of the "real brain" data. The connectionist movement is thus directed more to the engineers of "near-future" generation computer systems and to cognitive psychologists. Connectionist models, in general, are dynamic systems (e.g., Farmer 1990) and in this respect offer useful concepts for brain theory. When the structure and function of the brain are studied by using theoretical methods, two concepts have to be emphasized: hierarchy and dynamics. The brain is considered a prototype of hierarchical structures, and can be studied at different levels, such as molecular, membrane, cellular, synaptic, network, and system level, as we saw in Sect. 8.3.1.

However, the study of networks relevant to brain theory is not limited to connectionism. The theory of networks with large numbers of nodes constituting very complex structures offers, at least in principle, a common model of natural and synthetic information processing systems. While research motivated by statistical physics adopts elementary units (e.g., neurons or computing elements) as nodes, the "neoconnectionist" school define networks with intermediate level computational-cognitive elements. Schemas [22] (Chap. 3) provide even more sophisticated building blocks for network models of "parallel distributed processing".

Many concepts of cybernetics returned in a somewhat different context. McCulloch's "embodied mind" notion [341], van Foerster's critiques on the controversial computational paradigm of AI and cognitive science, and the concept of circular causality all returned (many times without being credited) related to the concept of embodied cognition (say, [534, 431, 581]).

8.6.2 From Cognitive Science to Embodied Cognition

Cognitive Science

"Cognitive science is the interdisciplinary study of mind and intelligence, embracing philosophy, psychology, artificial intelligence, neuroscience, linguistics, and anthropology..." (from http://plato.stanford.edu/entries/cognitive-science/. Main stream cognitive science is based on two fundamental concepts: *representation* and *computation*. The central hypothesis of cognitive science assumes that the external world is represented in the mind by using different ways, such as logic, rules, concepts, images, and analogies. Mental representations are similar to data structures, and thinking is similar to computational operations on these data structures. A very good introductory book to cognitive science is Paul Thagard's Mind [512], which not only describe the strong and weak points of the different representational procedures, but answers the challenges of traditional cognitive science.

While the computational paradigm proved to be efficient, it certainly has its limits. Among others, the role of consciousness and emotions should be taken into account. Computation should be understood in a more general way, and the dynamic character of the mind (and brain) should be incorporated. Most importantly, the mind is not isolated, and its interaction with the whole body and its environment is important. These views led to the concept of *embodied cognition*.

Embodied Cognition

The central hypothesis of embodied cognitive science is that cognition emerges from the interaction of brain, the whole body, and of its environment. What does it mean to understand a phenomenon? A pragmatic answer is to synthesize the behavior from elements. Many scientists believe if they are able to build a mathematical model based on the knowledge of the mechanism to reproduce a phenomenon and predict some other phenomena by using the same model framework, they understand what is happening in their system. Alternatively, instead of building a mathematical model one may wish to construct a robot. Rodney Brooks at MIT is an emblematic figure with the goal of building humanoid robots [78]. Embodied cognitive science now seems to be an interface between neuroscience and robotics: the features of embodied cognitive systems should be built both into neural models, and robots, and the goal is to integrate sensory, cognitive and motor processes. (Or even more. Traditionally, emotions were neglected, as factors which reduce the cognitive performance. It is far from being true.)

One more step is the integration of realistic neural models to functioning robots. Our group in Budapest participates in a European project with the title "Integrating Cognition, Emotion and Autonomy":

Integrating Cognition, Emotion and Autonomy (ICEA)

European Integrated Project IST-027819

Integrating Cognition, Emotion and Autonomy - ICEA

ICEA is a four-year project on bio-inspired cognitive robotics and embodied cognition, bringing together cognitive scientists, neuroscientists, psychologists, computational modelers, roboticists and control engineers.

The primary aim of the project is to develop a cognitive systems architecture integrating cognitive, emotional and bioregulatory (self-maintenance) processes, based on the architecture and physiology of the mammalian brain. The twofold hypothesis behind this research is that:

- the emotional and bioregulatory mechanisms that come with the organismic embodiment of living cognitive systems also play a crucial role in the constitution of their high-level cognitive processes, and

- models of these mechanisms can be usefully integrated in artificial cognitive systems architectures, which will constitute a significant step towards truly autonomous robotic cognitive systems that reason and behave in accordance with energy and other self-preservation requirements.

from http://www2.his.se/icea/default.htm

While the whole field is developing rapidly, the important features of embodied cognitive systems can be identified, and will be discussed briefly in the next paragraphs.

Selective and Constructive Mechanisms of Development

The relationship between neural and mental development is debated in terms of selectionism, instructionalism and constructivism [478, 151, 418]. Selectionism [478] is based on (i) the generation of variabilities, (ii) the interactions of neural elements with environment, (iii) the selective amplification of certain neural elements. Instructionalism offers a unidirectional mechanism for learning, when information passes from the "teacher" to the "student". Constructivism emphasizes the importance of environmental driven development, and even learning is, at least partially, self-directed.

Selective stabilization hypothesis

Ontogeny and plasticity of neural structures, as well as their learning capability, should be considered as dynamic processes and generally they have to be explained by common or similar mechanisms. The postnatal ontogenetic development of the nervous system is determined by the interaction of the innate genetic program and environmental factors, since the information content of the genome does not seem to be sufficient to determine the precise formation of the topographic maps.

To find a compromise between the nativist (or preformist) and empiricist points of view the selective stabilization hypothesis was suggested by Jean-Pierre Changeux and his colleagues in the "Institut Pasteur" at Paris [96, 97, 98], as a third option. According to the former, the neuronal network is specified genetically, while the latter emphasizes (indeed overemphasizes) the role of the activity of the system to specify its own connectivities. The main conceptual advantage of the selective stabilization hypothesis is that it offers a gene-saving mechanism for specifying ordered neural structures.

In the last few years a new research field emerged to find the common foundations of autonomous mental development in robots in one side, and in animals and humans in the other [555]. The question is how the internal developmental programs (either genetically coded or artificially designed) interact with their environment by using their sensors and effectors to develop mental capabilities. At the very beginning of the book, in Sect. 1.1, one of the hero of structuralism and developmental psychology, Jean Piaget was mentioned. He described human cognitive development [406] by a constructivist developmental theory. Constructivism has roots in different disciplines, as philosophy, psychology and sociology. Its central idea is that human learning is constructed, that learners build new knowledge upon the foundation of previous learning. This view of learning sharply contrasts with one in which learning is the passive transmission of information from one individual to another, a view in which reception, not construction, is the key.

Nowadays, constructivism in the cognitive psychological movement can be understood as a reaction against the view that knowledge is the result of sensory perception only, and emphasizes the active role of the nervous system to construct an internal world.

Lev Vygotsky (1896–1934), a Russian psychologist and philosopher in the 1930s, is most often associated with the social constructivist theory. He emphasizes the influences of cultural and social contexts in learning and supports a discovery model of learning.

Quartz and Sejnowski [418] proclaimed a constructivist manifesto for the neural basis of cognitive development and the necessity of the interaction between cognitive and neural levels of description. They, however, didn't give historical analysis, so Piaget, Vigotsky and second-order cybernetics were not mentioned. Constructivism has roots, also in second-order cybernetics, we mentioned that Heinz von Foerster was a radical constructivist, who saw the limits of the computer metaphor. Tom Ziemke [581] reconstructed von Foerster's works from the perspective of modern embodied cognitive science, and pinpointed that he (i.e., HvF) had seen the necessity to integrate sensory, cognitive and motor functions, what people try to do now, well, fifty years after.

Dynamical Systems

Esther Thelen (1941–2004) and Elizabeth Bates (1947–2003), two leaders of the post-Piagetian developmental psychology, both of the died very early, adopted the concepts of dynamical systems theory for different child development [513]. Thelen was interested in the development of movements in infants, and Bates in language learning. Thelen was motivated to explain the emergence of movement patterns. She was influenced by what later became "coordination dynamics", and what we discussed in Sect. 8.5.2. She came to the conclusion that emergence of infant movements patterns is not determined by the nervous system alone, but body weights, elastic and inertial properties of muscles and environmental factors also influence the motor output. Bates, together with her colleagues used the connectionist model framework for simulating the dynamics of language development [151, 48]. As we know connectionist models describe temporal change of the states of activities of (formal) neurons and also of strengths of connectivities due to learning. The notions of dynamical systems theory, such as states, process, and attractor were offered as a (somewhat) alternative of the computational paradigm [514, 196, 548]. From the perspective of complex systems computation and dynamics are not complementary approaches [354, 117].

Action-Perception Loop and Internal Models

Why do we need cognition ultimately? We need it to guide actions. The cyclic nature of perception was suggested by Ulrich Neisser, a pioneer of cognitive psychology [374]. His analysis supports the view, that an organism is not a passive receiver but actively seeks information relevant to its actions. Thelen' works proved, that movements of neonates are based on *actions*, and not reactions or reflexes. Actions are determined by the goals and not by the trajectories of movement that form them. Perception does not give information on the present happening only, but tells also what is going to happen soon. Cognition has a role in extending the time-frame of prediction. Perception and action are inter-dependent. Actions need perception to guide them but actions also participate in the process of perception. Michael Arbib (McCulloch's student) in the second edition of his Metaphorical Brain [18] argued that real role of perception is to modify our *internal model of the world*, and showed how the integration of perceptual and motor schemas happen by adaptive neural networks. That book was published in 1989. Since than it became very clear that the challenge of the embodied cognition should be taken seriously. The book was written in this spirit, and I feel it now even more timely than ten-fifteen years ago. Internal models, mental models have fundamental role in thinking and reasoning, and it was revolutionary suggested by Kenneth Craik (1914 - 1945) [113].

> The idea that people rely on mental models can be traced back to Kenneth Craik's suggestion in 1943 that the mind constructs "small-scale models" of reality that it uses to anticipate events. Mental models can be constructed from perception, imagination, or the comprehension of discourse. They underlie visual images, but they can also be abstract, representing situations that cannot be visualised..." from http://www.tcd.ie/Psychology/Ruth_Byrne/mental_models/

Synthetic Modeling

Synthetic modeling is a model framework for simulating autonomous embodied systems by assembling from simple building blocks. This version of embodied cognitive models use realistic models of the neural system, and it is combined with some model (mathematical or physical) of the body. Gerry Edelman, the Nobel prize winner founder of the Neuroscience Institute at La Jolla and his coworkers have been working on both types of realization of this big project [11, 176].

Active Vision

Perception, even visual perception is active, as it was suggested both in context of machine and biological vision. Conceptually it goes back to the very influential psychologist James Jerome Gibson (1904-1979), who considered perceptual systems related to their action [200]. More technically, Ruzena Bajcsy [37] and Dana Ballard [41] solved traditional vision problems by relatively simple algorithm, if the active sensor gave only limited information.

To test their efficiency, abstract level biological(ly inspired) algorithms can be incorporated to robots, and/or virtual reality environment. Interactive mechanism select and process only that part of visual input which is relevant to the task to be solved. There is a rapidly increasing literature about this topic, two of them are mentioned here. The ontogenetic development of the receptive fields of evolutionary robots with active vision was modeled by Dario Floreano and his coworkers in Lausanne [177]. While the architecture of the neural network embedded in the robot was given, the synaptic weights were modified by a post-Hebbian algorithm. The evolved robots selects special features of the visual environment (e.g., edges to able to navigate). A recent extension of Ballard's work (together with his student Nathan Sprague, now my colleague at Kalamazoo College) developed a sufficiently complicated graphic model to test synthetic models of visuo-motor behavior [479].

8.6.3 The Brain as a Hermeneutic Device

Biological theories were always motivated either by the engineers' "device" or the philosophical approach. How, if at all, the "device approach" and the "philosophical approach" can be reconciled? I suggested (with the help of Ildiko Aradi) [158] that the brain is a hermeneutic device.

A Few Words on Hermeneutics

Hermeneutics is a branch of continental philosophy which treats the understanding and interpretation of texts. The *methodological hermeneutics* was established to interpret the texts. The *philosophical hermeneutics* emphasized the existential understanding instead of interpretation. The *critical hermeneutics* offers a methodologically self-reflective reconstruction of the social foundations of discourse and inter-subjective understanding. Finally, the *phenomenological hermeneutics* is an attempt to synthesize the various hermeneutic currents. For an excellent introduction for non-philosophers, see [326].

One of the most important concept of the hermeneutics is the hermeneutic circle. This notion means that definition or understanding of something employs attributes which already presuppose a definition or understanding of that thing. The method is in strong opposition of the classical methods of science, which does not allow such kinds of circular explanations. Hans-Georg Gadamer (1900-2002) writes [225]: "Understanding always implies a *preunderstanding* which is in turn prefigured, by the determinate tradition in which the interpreter lives and that shapes his prejudices."

The eventual role of the hermeneutics in (natural) sciences has been, and still is, very controversial. To his contribution to the Sokal hoax[1] the Nobel-prize winner physicist Steven Weinberg [553] wrote: "... A physicist friend of mine once said that in facing death, he drew some consolation from the reflection that he would never again have to look up the word "hermeneutic" in the dictionary.

Hermeneutics and the Brain

It is rather obvious that, despite the methodological success of the analytic sciences, the marvelous complexity of life and of the brain cannot be explained completely in terms of physics. Can science learn from philosophy, even philosophical aesthetics? It may and must be admitted that hermeneutics (and aesthetics) emphasizing the necessity of self-reflexive interpretation offers a fruitful approach to theoretical biology.

---

[1] The physicist Sokal submitted and published a paper in a journal about postmodern culture in 1996. The text was full with impossible pseudoscientific statements. Sokal wanted to demonstrate the lack of rigor of reviews in such kinds of journals. While I laughed on the absurd nonsense, the laugh did not come from my full heart. If an editor of a journal about cultural theory gets a manuscript from a professor of physics of a big university from New York, the bona fide way of thinking is this: "I like the ideology and social implications of the paper, the physics of the professional author is obviously good. So, I am ready to publish it."

In the Orwellian year Ichiro Tsuda (who suggested a mechanism based on chaotic itinerancy (3.6.5) for enhancing cognitive abilities), published a paper with the title " A hermeneutic process of the brain" in the Progress of Theoretical Physics, which is the journal of the Physical Society of Japan [521]. I was shocked by this paper. Professional physics journal don't really publish papers with philosophical connotations. Tsuda took the encouragement to submit a paper using philosophical terms about the brain to a physics journal, and what was more surprising, the editors accepted it.

In the same year the Hungarian translation of Gadamer's book (for the English translation: [225] was published. Gadamer states that the concepts of "truth" and "method" of the natural sciences cannot be applied - not even in principle - for the humanities. It can be asked, however, (and better if natural scientists do it) what the theory of interpretation, i.e., hermeneutics, offers for the natural sciences, and in particular for biology. Dennett analyzed [129] that not only texts, people and artifacts, but also biological organisms can be interpreted.

So what the device approach and the hermeneutic approach tells us about the brain? (For more details, see [158]). First, the brain is a physical structure which is controlled and also controls, learns and teaches, processes and creates information, recognizes and generates patterns, organizes its environment and is organized by it. Second, closed causal loops and self-referential systems implement the iterative nature of learning and interpretation offered by hermeneutics.

There seems to be a convergence between the "device approach" and the "philosophical approach" to the brain. Systems exhibiting "high" structural complexity *and* "high" dynamic complexity" (e.g., but not exclusively, chaos) may be candidates of being hermeneutic devices, since they are both object or subject of interpretation and interpreting agents. (It should be recalled, however, that even simple systems (such as the logistic difference equation) may lead to complex dynamics. So, the occurrence of chaos is not a sufficient condition for being a hermeneutic device.

Analyzing the neural and mental development it was stated [159] that (1) environmental influence is necessary for normal development; (2) selectionist mechanisms of neural development are based on the interaction between the innate genetic program and environmental factors; (3) mental development is involved not only in the representation but also in the creation of reality. (4) the human brain, which is a structurally and dynamically complex device, not only perceives but also creates reality: it is a hermeneutic device.

8.6.4 From Neurons to Soul and Back

To Kati, no matter what

The real success of the monistic paradigm would be to have a neural theory of mind, and whatever it means, soul. A controversial approach is still Sigmund Freud's psychoanalysis. While Freud was initially interested in the neural basis of psychic phenomena, psychoanalysis later developed its own language, methods, institutions, and even its isolated training methods. As is well-known, Popper claimed that psychoanalysis (as well as Marxism) is not a scientific discipline, since its statements cannot be falsified [411]. Psychoanalysis was perhaps the first formalized way of *psychotherapy*, a method based on asymmetric conversation between a trained psychotherapists and a client.

Freud had a background and interest in neurology, but in that period neuroscience was not sufficiently developed to display the underlying structural and other neural changes. He expected that decades later neurophysiology and neurochemistry will "...give the most surprising information and we cannot guess what answers it will return in a few dozen years of questions we have to put it ...".

But accepting that these methods did not exist, he developed a new terminology, ideas and methods which were revolutionary, but far from the strict scientific ones. He did not use the "brain language" but and abstract verbal description of the mind. Many of his followers, however, believed that it was nothing to do with the biology of mind.[2]

The intellectual atmosphere is slowly being changed, as the Nobel prize winner neurobiologist Eric Kandel started to build bridge between neurobiology and psychoanalysis [264]. Kandel, who was born in Vienna, wanted to be a psychiatrist and turned towards neuroscience to learn more about the brain. His reductionist approach has been very successful to explain the molecular mechanisms of learning.

Kandel specifies areas in which biology and psychoanalysis could interact. One issue is whether what kinds of structural changes can be established in the

---

[2] In Hungary, where the direct Freudian effect was strong, István Székács (1907–1999) who worked as a biochemist while psychoanalysis was forbidden in Hungary after the war, and later became a very influentially trainer of the new generation of psychiatrists, propagated the necessity of interaction of a general brain theory and psychoanalytical practice. The title of his Psychoanalysis and Natural Science [496] reflects well his attitude. Most likely he learned from János Szentágothai that general brain theories were in the air.

brain by psychotherapy, and how it may be an alternative of pharmacotherapy? "Until the last decade, the biological mechanisms of psychotherapeutic actions were thought not to be amenable to neurobiological investigation ..." [168]. It is no doubt that brain imaging methods are/will able to detect more systematically the structural and functional changes of the brain due psychotherapy. Recent results indicated that such disorders, as depression, obsessive-compulsive disorder, and anxiety treated by pharmacotherapy and psychotherapy result in changes in the same brain circuits. It is not to say that psychotherapy can substitute pharmacotherapy, but it worth to think on the interplay of the possible therapeutic strategies.

In a more general context, how can a brain be influenced from the outside? Mechanically (well, this is neurosurgery), electrically (i.e., by electroconvulsive therapy, also called as electroshock), chemically (neuropharmacology), and verbally (psychotherapy). In principle it should be a terrible difficult control-theoretical problem to find out how to combine the different strategies.

We don't really have neural based models of psychic phenomena. An extensively studied phenomena, namely *attachment* seems to be a good field to show the power of different approaches, from molecular to phenomenological. When you will read the remaining part of this chapter, you should remember that this is a field, which is traditionally very far from any type of model-based analysis.

Attachment Theory, Classical and Contemporary Cognitive Science

"Attachment is defined as an affectional tie that one person or animal forms between him/herself and another specific one (usually the parent) - a tie that binds them together in space and endures over time." ... (from Wikipedia).

Attachment theory was elaborated by the British psychiatrist John Bowlby (1907-1990) and his student Mary Ainsworth and it may be considered as a theory motivated by cybernetics. As opposed to the Freudian instinct and drive it assumes the emergence of adaptive controlled behavior due to the interaction of participants (actually an infant and her mother). For the origin of attachment theory, see e.g., [74].[3]

Infants show an evolved adaptive tendency to maintain proximity to an attachment figure (generally to their mother). Bowlby showed that this is a(n adaptive) process by which we develop a sense of safety in the world to people or things. If things go well, this leads to "safe attachment", if not, than to

---

[3] As a layman, I read first about the eventual role of implicit memory systems in attachment in Hungarian [403].

"unsafe attachment", i.e., detachment, insecurity, introversion, etc. The theory basically suggests a mechanism similar to the cyberneticians control processes.

Attachment can be conceived of as a force acting to minimize mother–infant distance in face of other forces acting on the mother–infant system. For example, in face of exploratory behavior, the strong tendency of the infants to explore their surroundings and thus to increase distance from mother. To study it under reproducible conditions, Mary Ainsworth developed a popular experimental paradigm called the "Strange Situation".

The Strange Situation experimental paradigm

This initially very controversial laboratory procedure for one-year-olds was originally designed to examine the balance of attachment and exploratory behaviors under conditions of low and high stress.

The Strange Situation is a 20-minute miniature drama with eight episodes. Mother and infant are introduced to a laboratory playroom, where they are later joined by an unfamiliar woman. While the stranger plays with the baby, the mother leaves briefly and then returns. A second separation ensues during which the baby is completely alone. Finally, the stranger and then the mother return. Based on [74].

There are three, characteristic behavioral patterns of the infants, labeled as avoidant, secure, and ambivalent, (according to newer classification, there is a fourth group, called disorganized, which might have a strong genetic component: a mutation related to a dopamine receptor [303]). The classical explanation reflects the perspective of main stream cognitive science, so as it was mentioned earlier, used mental representation. In a recent PhD dissertation Dean Petters in Birmingham presented a functional, high-level computational model [404] of attachment. The model is based on classical computer architectures, and assumed positive feedback loops, and was able to implement (i) the ability to switch goals, (ii) learn different predispositions to hold goals, (iii) learn from different actions/situations.

A set of algorithms with increasing complexity was offered. Some algorithms implement reactive architecture. Reactive architectures don't know what happened in the past, and what will happen in the future. The simplest is a reactive architecture with no capacity to learn, more complicated ones can learn by reinforcement. Deliberative architectures have memory and can make plans. Figure 8.22 illustrates a hybrid architecture to implement behavioral patterns described by attachment theory.

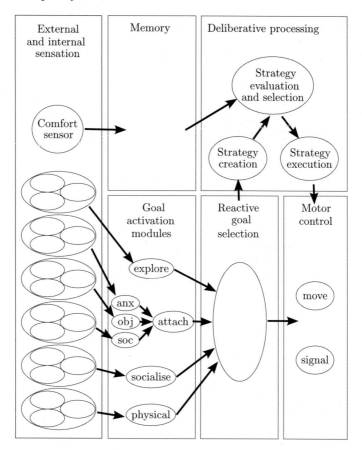

Fig. 8.22. A hybrid model. Deliberative mechanisms provide a secondary route to action, activated as a result of interrupts to reactive selection and arbitration and imposing different actions. The reactive goals are: exploration (explore); anxiety (anx); object wariness (obj); social wariness (soc); socialisation (socialise); and physical need (physical). The reactive goals of anxiety, object wariness, and social wariness are all combined in a single goal of attachment (attach) before they are considered for selection. The comfort sensor measures contact pleasure and represents this information symbolically. Adapted from [404].

This model, as it is, adopts the classical concepts of cognitive science from the perspective of computer science. Recently, however, it was suggested that attachment (and similar phenomena which traditionally belong to psychoanalysis) should be much better integrated into the contemporary cognitive science [180, 432]. The new question is whether "embodied cognitive science" or "cognitive neuroscience" might give a better framework.

Fonagy and Target [180] concluded:

"... The model of mind assumed by attachment theorists is consistent with the important discoveries of the first generation of cognitive psychology, but this approach has been supplanted by a number of recent developments collectively called "embodied cognition" or "enactive mind." Features of this new approach include (1) the increasing use of introspection as a research method; (2) a keen interest in the understanding of emotion as organizer as well as motivator of behavior; (3) rapidly advancing brain-imaging technology that has made cognitive neuropsychology into a brain as well as a mental science and led to increasingly functional cognitive accounts; and (4) a move away from reified laboratory studies and an interest in ecological approaches to cognition ...."

As we know, psychoanalysis, initially motivated by neuroscience, later was not influenced by it. It is probably timely to integrate psychotherapy/psychoanalysis with present research in neural plasticity, memory systems, learning mechanisms, emotions, attachment, neural, cognitive and emotional development. One important direction is related to the notion of *implicit memory* what people have without being consciously aware of having them. Psychotherapy can help to regulate personality by reorganizing attachment-related implicit memory systems [324, 189].

I believe that the complexity and dynamic organization of our cognitive and emotional structures cannot be understood without integrating traditional neuroscience, different directions of cognitive science with basic and clinical psychology including psychoanalysis and other psychotherapies. Computational models of these certainly complex phenomena seem to be the key tools of this integration.

We shall leave now the world of brain, but will use mental models to analyze decision making mechanisms in the next chapter.

9

From Models to Decision Making

9.1 Equation-Based Versus Agent-Based Model

9.1.1 Motivations

Agent-based modeling (ABM) has become a fashionable tool to simulate complex systems. Of course we know that dynamical systems theory offers a mathematical framework to model and simulate spatiotemporal phenomena. While there is not a strong difference between equation-based and model-based description, certainly not at the implementational level, agent-based model offers a somewhat different modeling philosophy and practice.

An agent is considered an autonomous computational unit, with the ability to bring decisions, or to put it another way, to make selections among possible strategies. Of course, if the definition is loose, we may interpret even a simple McCulloch-Pitts neuron, as an agent, who decides to fire, for a certain set of inputs, and remain at rest for others. Agents have time-dependent internal states, and their states change due to the interaction of other agents. Specific rules determine the interactions among agents. Agents may interact locally with their neighbors, or globally with all other agents. Interacting agents may be also the nodes of say, small world network, so an agent may interact both locally and partially globally.

Agents are supposed to show some elements of intelligence, or at least they are capable of showing some degree of autonomy, adaptation to environmental changes, communication with other agents, and occasionally goal-directed learning.

ABM has multiple intellectual roots. One is cellular automaton, which is a discrete time discrete state space dynamical system, where players are located on the grid points of the arena having a specified geometry. There is a rule for updating the dynamics of the state of each grid points, and the rule is the same for each point. Artificial life and artificial societies have been constructed based on these principles, as it will be reviewed soon.

Another intellectual source of agent-based modeling emerged as a reaction to the neoclassical theory of economics, which dominates the science of economics. As we learned from Brian Arthur (recently reinforced by the popular book [54]), the complex systems approach is not in accordance with the assumptions of the neoclassical theory. The comparison of the two approaches can be seen here:

- *Neoclassical economics*
  - Behavioral model for people:
  * Fully-informed
  * Rational
  * People interact only indirectly with one another (through markets)
  * Focus on equilibrium outcomes

- *Complexity approach*
  * People are adaptive
  * They interact directly with one another
  * Focus on dynamics
  * Methodology: equation vs. agent-based modeling?

Economy is a dynamic process. It is the emergent result of collective interaction of very different players (well, agents), from individual investors, to firms, consumers, banks etc, and ABM seems to be a very good method to simulate, how local interactions among players may lead to macroscopic behavioral patterns.

9.1.2 Artificial Life

From Cellular Automata to the Alife Movement

Cellular automata (CA) are also considered as intellectual roots of agent-based modeling. It is well-known that having been motivated by von Neumann model, John Conway *constructed* (this is the key word: "constructed": the algorithm does not pretend to implement a real mechanism, it is skilfully constructed to generate interesting phenomena) a set of rules, which was able to generate complex spatiotemporal patterns. The *game of life*, as it is called, also motivated the birth of the artificial life (Alife) movement. Alife models (written mostly by people belonging to the computer science community) grasp the *"possible"*, while biologists are interested mostly in the *"actual"*. I think, Alife models give insight to understanding real biological phenomena, and also "life, as it could be". However, the myth of simulation, which says "we understand what we can generate by computational algorithms" should not be taken literally. Alife certainly came from the tradition of cybernetics, and helped to popularize the notion that constructive bottom up models also contribute to understand biological dynamic organizations.

Artificial Chemistry

The Alife movement has strong branches. Artificial chemistry is a formal system. Its building blocks are the set of molecules, the set of reactions, and a rule, which describes the temporal evolution of the state. Formal chemical models may describe chemical events themselves, but may serve as a metalanguage as it was mentioned in Sect. 4.2.2. Artificial chemistry (just as formal chemical kinetics) can be used to model prebiotic evolution but also higher-level, even social systems [131].

Digital Evolution

Michael Conrad (1941–2000) was a pioneer of simulating evolution around 1970 studied a model of the evolution of an artificial open-ended ecosystem [108]. Individuals with given genotypes were defined, and these genotypes were interpreted as instructions, leading to phenotypes. The organisms competed in a simple one-dimensional world for the possession of resources which they use for self-repair and reproduction. Interactions between organisms for providing co-evolutionary selection pressure for increased complexity has been emphasized. 20 years later his evolutionary computer simulations became more realistic [109].

Somewhat in the spirit of Conrad (and a few others), an *artificial ecology* and *artificial evolution* was developed by the ecologist Thomas S. Ray by implementing the basic evolutionary and ecological processes in a computer. The system established is called *Tierra*. It was an abstract, not too biologically detailed model, but programs were able to mutate, replicate and recombine (also called crossover). Evolution was embedded in a computer, and programs evolved by the rules of evolution. Artificial evolution has two goals, first, to help in understanding real biological evolution, second, to provide evolutionary computational algorithms to solve hard optimization problem.

Tierra has been motivated by a few facts and requirements: (i) Life on Earth is a single example of evolution (sample size of one); it is possible (ii) to create and observe a new and independent instance of evolution, (iii) and broaden our perspective of evolution by increasing the sample size (to more than one) (iv) and to observe the properties of evolution in a (silicon-based) digital medium. According to this approach, real carbon-based biology is not the essence of an artificial evolutionary systems.

Of course, digital medium is nothing to do with our physics, Digital evolution is not constrained by the laws of our physics. Cyberspace is nothing to do with our 3D Euclidean space. The laws of conventional physics and chemistry of our organic medium is replaced by the rules of the program languages and operating systems used in the digital medium.

The idea behind the digital evolution is that we don't have any idea what might be the emergent result of it. Ray suggested a thought experiment. Let's assume we are all robots made of metal and our brains of silicon chips. We have no experience or knowledge of carbon based life. A robot brings several chemical substances, such as methane, ammonia, hydrogen, water, and some minerals to a scientific meeting. The robot asks: "Do you suppose we could build a computer from this stuff?" The answer is that some engineer-minded robot might see molecular computers, but neurons (not speaking about neural networks, and interacting brain regions) could be never imagined. As in the organic medium evolution established the unpredictable structures, digital evolution also should generate complex information processes.

Tierra motivated the construction of other platforms for the evolution of computer programs, e.g., *Avida*. It is en extension of Tierra at least in two respects. First, each program is stored in a different part of the memory, second the speed of the program running may be different for the individual programs, According to spirit of Alife, Avida does not *simulate* evolutionary processes, but "digital organisms in Avida evolve to survive in a complex computational environment and will adapt to perform entirely new traits in ways never expected by the researchers, some of which seem highly creative." (from the Avida website http://devolab.cse.msu.edu/, November 26th 2006). Programs able to evolve behave similarly in many respects that biological evolution. In the paper [310] it was shown how a complex function such as "equal" function evolved by simpler functions:

"Of course, digital organisms differ from organic life in their genetic constitution, metabolic activities and physical environments. However, digital organisms undergo the same processes of reproduction, mutation, inheritance and competition that allow evolution and adaptation by natural selection in organic forms."

Genetic Algorithm

Probably the best known concept related to digital evolution is the genetic algorithm (GA) formulated by John Holland in Ann Arbor, at the University of Michigan [239]. GA is one of the successful winner of the evolutionary computational algorithms (an excellent book about evolutionary computation is [178].

GA is generally used to solve optimization or search problems. There is a population of candidate solutions ("chromosomes") and the goal is to progress towards better solution by some algorithm. In the case of GA, this come from biological motivation, The solutions are encoded as strings, traditionally as binary strings of zeros and ones. A *fitness function*, which measures the quality of a solution, can be assigned the set of solutions.

A GA starts with the generation of a set of candidate solutions, chosen either randomly, or if there is some preliminary information, then around some estimated values. The next step is *selection*. A subset of the candidate solutions are selected controlled by the combination of (i) a fitness function based deterministic selection and (ii) some random effects. Now comes the next step: *reproduction*. The selected solutions are modified by *crossover* and *mutation*. Crossover is a procedure to establish "children" strings from the "parents" mainly by cutting and glue. The size of the population is fixed, so the worst solutions will be eliminated. A new generation of solutions (having the same size as the initial one) has generally a better fitness, than the previous. The procedure is iterated, and will stop when the fitness functions does not show an improvement.

GA and its different versions, and extensions are used now in problem solving. While the procedure is convergent, often it leads to local and not global optimum. A larger mutation rate (i.e., more randomness) helps to kick off the system from a local minimum.

Genetic Programming

GP is an extension of the genetic algorithm. While Holland adopted (typically binary) strings with fixed length, genetic programming uses trees. Hierarchical organizations better characterize the structure of computer programs, and proved to be much more flexible than string-based representations, as John Koza[1] one of the champions of genetic programming explains [292]. Of course, genetic operators, (mutation, crossover) should be defined for this geometry. An example, how crossover generates new programs is shown in Fig. 9.1.

---

[1] John Koza is also involved in the initiation to change the US electoral system and suggested a state-based plan for electing the president by national popular vote. There is a downloadable book about this issue; http://www.every-vote-equal.com/tableofcontents.htm; 27 November 2006.

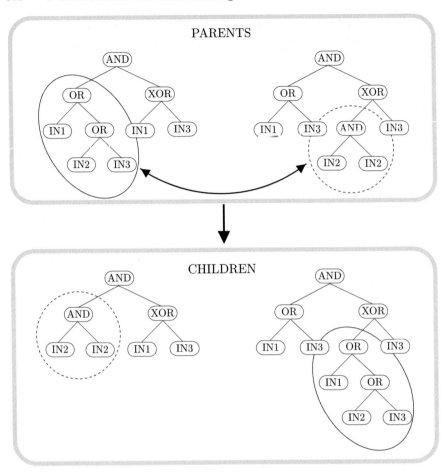

Fig. 9.1. Sub-tree crossover creating new programs from existing programs. Subtrees are selected randomly from two existing parse trees and are then swapped to produce child parse trees. http://www-users.york.ac.uk/~mal503/common/thesis/jpegimages/subtree_co.jpg.

GP is a method to detect and uncover "complexity". It helps to design and optimize computer programs. It also offers a technique for interpreting certain data sets. To be more precise, GP can solve certain "inverse problem" of (bio)chemical kinetics, i.e., starting from measured trajectories generates the network of chemical reactions. Similarly, it is a method for the automated synthesis of genetic networks, and also of analog electrical circuits. As it is stated, GP was able automatically produce computer programs that is competitive with human performance. The URL http://www.genetic-programming.com/humancompetitive.html (17 November 2006) gives a list of problems, which were solved by programs evolved by GP, where many of the solved problems were previously patented. GP produced human-competitive results, and seems to be an excellent method for novelty generation. For more details see [292, 293, 294, 295].

9.1.3 Artificial Societies

The spirit of the Alife movement was extended to artificial societies. Joshua Epstein and Robert Axtell published a book on "growing artificial societies", which still is a paradigmatic illustration of agent-based models of social phenomena. (Epstein was also the author of a book, as it as mentioned earlier on applying equation-based models of social dynamics ...). It could be considered as an outgrowth of the Schelling model.

At the Land of Sugarscape

The model is called Sugarscape [155], and it has several versions, with less difficult and more difficult agents – and thus society. Sugarscape is a simulated artificial world populated with artificial agents. The world consists of a square lattice with some sugar distribution. Sugar is the only source of food and is needed to stay alive Specifically, the sugar concentration had peaks in the northeast and southwest quadrants of the grid. The agents walk around on the lattice looking for sugars. The model is deterministic, only the initial conditions – i.e., the initial position of the agents – are random.

Agent-based models are theoretically completely equivalent to equation-based models, only the usual description is different: for the latter one generally it is a set of equations, for the former one it is generally a list of algorithms. While for Sugarscape the agent-based approach is obviously better suited let us here show an equivalent equation-based description.

This will be a discrete time discrete space deterministic model. Let N be the size of the $(N \times N)$ grid and M the total number of agents (dead or alive), these are parameters of the model. Other parameters are the maximum amount of sugar in the grid positions: $S^* = [s^*_{ij}]$, this is an $N \times N$ matrix of integer values between zero and four; the distance giving how far the agents can see on the lattice: e_i, $1 \leq i \leq M$, and the number of sugar units the agent digest in a time step: d_i, $1 \leq i \leq M$; these two parameters define the DNA of the heterogeneous agents.

The variables of the model are the actual amount of sugar at each grid position: the $S = [s_{ij}]$ matrix, vectors giving the actual coordinates of the agents: (x_i, y_i), and a vector giving the amount of reserves an agent has: w_i, basically her wealth in this model.

The variables are updated according to the following formulae:

$$s_{ij}(t+1) = \min\{ s_{ij}(t) + 1 - s_{ij}(t) \sum_{k}^{M} \delta(x_k(t), i)\delta(y_k(t), j) \, , \, s_{ij}^* \}$$
(9.1)

$$w_i(t+1) = \begin{cases} w_i(t) - d_i(t) + s_{x_i(t), y_i(t)}(t) & \text{if } w_i(t) \neq 0 \\ 0 & \text{otherwise} \end{cases}$$
(9.2)

$$(x_i(t+1), y_i(t+1)) = p_i(t+1)$$
(9.3)

where $p_i(t+1)$ is recursively defined as

$$p_1(t+1) = \operatorname{argmax}\{S_{ij}(t+1) | (i,j) \in P_1(t+1)\}$$
(9.4)

$$p_k(t+1) = \operatorname{argmax}\{S_{ij}(t+1) | (i,j) \in P_k(t+1) \setminus \bigcup_{l=1}^{k-1} \{p_l(t+1)\}\}$$
(9.5)

and the set of possible moves are defined as

$$P_k(t+1) = \{(x_n, y_k(t)) \in N \times N | \, |x_n - x_k(t)| \leq e_k\} \cup \\ \{(x_k(t), y_n) \in N \times N | \, |y_n - y_k(t)| \leq e_k\}.$$
(9.6)

Now compare this description to the normal algorithmic description of the Sugarscape world, as follows. A $N \times N$ grid is given, there is at most one agent in each grid point. In each time step first the agents eat up the sugar at their current position, then digest some amount of sugar given by a parameter which is different for different agents. If an agent has no sugar at all in her body after this, she starves to death and it is removed from the simulation. Then the amount of sugar increases by one at each grid position if it less than its maximum level. Then – still in the very same time step – the agents choose where to step next by looking around themselves. Each agent has a vision parameter, the number of steps it can see along each direction (but not diagonally). One by one in predefined order, the agents check the available cells within their sight and choose the one with the maximum sugar. Then the next time step follows.

While theoretically it is possible to analyze this system in the dynamical systems framework, this is obviously not done. Instead, some average measures are calculated from the state variables or their distribution is plotted and its time evolution is observed.

9.1 Equation-Based Versus Agent-Based Model

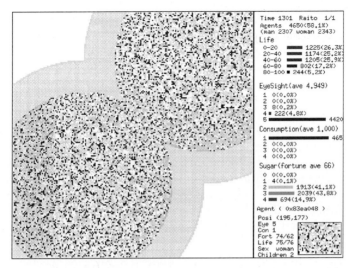

Fig. 9.2. One of the many implementations of the Sugarscape world. Various variables and parameter of the simulation can be seen on the *right panel*.

The Sugarscape simulations were traditionally run on a landscape with two sugar mountains, one in the northeast and one in the southwest with some deserts in the two other corners. The agents are placed randomly on the landscape and then one clicks on the "Run" button and waits for what would happen. See Fig. 9.2 for an example landscape.

This very basic model is able to explain how the skewed wealth distribution can be generated from totally symmetric and 'just' rules. (In fact it would be nice to see whether the skewed wealth distribution could be obtained by homogeneous agents as well.) The little differences in the agents skills and their starting position is enough to make a huge difference in the long run.

In the basic model agents only die if they starve to death and thus there is no need for reproduction. In the first extension of the model each agent is randomly tagged as female or male and in a more later version they also have age. Agents die at a given age. If an agent is old enough and has some sugar savings then it is considered fertile. If two fertile agents meet (i.e., they are on neighboring cells) then a new agent is created on a neighboring cell. The new agent will inherit also some of the wealth of the parents and also some random variation of their DNA (i.e., vision and digestion parameters). Introducing reproduction has a number of effects on the Sugarscape society. First, the average vision and metabolic efficiency starts to increase over time. Second, big oscillations started to occur in the size of Sugarscape population, unlike in the basic model. Third, the gap between rich and poor widened ever further.

The second extension added is trading. For trading another type of essential food is introduced: spice; agents needed both sugar and spice for survival. Also, the deserts are replaced by spice mountains in the landscape. From now on if two agents met they are able to trade spice for sugar and sugar for spice and the price is determined by the two interacting agents after some bargaining. The immediate result from the extended model was that trade is good. The average wealth of the society increased compared to the basic version. Another result was that the difference between the poor and the rich grew ever further. Finally, an important result was that even if the constructed supply and demand curves looked like examples from a microeconomics book, the Sugarscape society was never in equilibrium as predicted by traditional neoclassical economics. This is probably not as much as a surprise since in the Sugarscape world a number of phenomena, like horizontal inequality and incomplete information exist which are not allowed in traditional economics.

The last extension we mention here allowed Sugarscape agents to lend and borrow goods. This little addition had a huge impact: some agents started to borrow goods from their neighbors and lend it to others: banks appeared.

> For example, in the agent-based model Sugarscape (Epstein and Axtell 1996), each individual agent has simple local rules governing movement, sexual reproduction, trading behavior, combat, interaction with the environment, and the transmission of cultural attributes and diseases. These rules can all be "active" at once. When an initial population of such agents is released into an artificial environment in which, and with which, they interact, the resulting artificial society unavoidably links demography, economics, cultural adaptation, genetic evolution, combat, environmental effects, and epidemiology. Because the individual is multidimensional, so is the society.
>
> From: [154]: Epstein J (2006): Generative Social Science: Studies in Agent-Based Computational Modeling

How Did the Anasazi Native Americans Live or Could Have Lived?

Epstein now published Generative Social Science [154], a series of papers about the use of bottom-up social models. His intention is to show that complex social phenomena can be understood by local interactions among agents. A series of papers, very ambitiously, instead of studying virtual worlds, deals with the rise and fall of a real society, actually of the Anasazi native Americans. The historical fact is that the Anasazi abandoned Long House Valley, in about A.D. 1350 after they lived there for about 3000 years.

The Anasazi are a particularly good candidate for a simulation study since there are data available from various sources about the climate, crop production, and the actual place of the settlements in the valley. This makes it possible to compare the results quantitatively to the real events. The scientific question proposed whether it is possible that the Anasazi abandoned the place as a result of their normal, "internal" dynamics or there was some external force which made them chose this alternative. Note that as climate is part of the model, it is counted as internal here.

Let us briefly introduce the model. The unit of the modeling, the agent is one household, the smallest unit in archaeological records. Some attributes of the household are the model variables, these describe the state of the model: (1) the place of its residence, (2) its agricultural location, (3) the "age" of the household, everybody in a household assumed to have the same age, (4) the amount of corn the household owns. Most of the model parameters are set based on real data, these include: nutrition need per individual, the size of the household, the maximum length of grain storage, the parameters of the environment, the fertility of the household, etc. The model contained eight adjustable parameters: minimum and maximum death age, minimum and maximum age of the end of fertility, minimum and maximum fission probability, the average harvest and the harvest variance. (The actual death age, end of fertility and fission probability were chosen uniformly from the interval given with the parameters.) By the careful choice of these eight parameters the model was able to replicate the real world data very closely, except for one thing: the simulated Anasazi never abandoned the valley. This implies that as the model assumptions were *not* sufficient for explaining the phenomenon ("i.e.," the abandonment), there must have been some external cultural or environmental reason for it.

It is interesting to compare the Anasazi model to the Sugarscape world described earlier. In the Sugarscape model the (very minimal) assumptions were able to reproduce the real phenomena, so it shows that the assumptions are sufficient (but not necessarily required). Whereas in the Anasazi model the assumptions fail to reproduce one single aspect of the real events (and succeed to reproduce every thing else), this means that the assumptions are not sufficient for explaining this aspect.

ABM seems to be an appropriate methodological tool to test ideas about social mechanisms which may establish certain macroscopic patterns. It may help to test hypotheses about the spatiotemporal dynamics of these macroscopic patterns emerging due to interaction of agents characterized by their possible strategies and actual decisions.

9.1.4 Agent-Based Computational Economics

Agent-based computational economics (ACE) uses models of "virtual economic world" where different ("heterogeneous") agents participate. Interactions of agents specify the temporal development of an economy from a given initial condition. The agents are computational units, or, if you wish, basically softwares. These software modules implement rules, which might be a simple if-then instruction, but also could be a long set of instructions.

ACE certainly preserves the basic property of dynamical systems description, namely causality. It helps to understand the mechanism of pattern generation (empirical understanding), and to design "good economics" (normative understanding).

Decentralized market economies might be understood, much better as complex adaptive systems (occasionally labeled as CAS) than optimizing rational calculators. A CAS consists of large numbers of adaptive agents involved in local interactions. These microscopic interactions imply macroscopic patterns, which often show regularities, as shared market protocol or behavioral norms, which feedback to the microscopic interactions, as Tesfastion [511] states. He also classified the problems of ACE (of course we know, that every classification contains arbitrary elements): (i) Learning and the embodied mind; (ii) evolution of behavioral norms; (iii) bottom-up modeling of market processes; (iv) formation of economic networks (v) modeling of organizations, (vi) design of computational agents for automated markets; (vi) parallel experiments with real and computational agents (viii) building ACE computational laboratories.

A big enterprise of ACE was the establishment of the Santa-Fe Artificial Stock Market model.

The Santa-Fe Artificial Stock Market model

The goal was to set up an artificial stock market where players are inductive agents with bounded rationality, and to show that their interaction leads to real(istic) market dynamics [307].

There are n stock market traders, and two types of assets. The first is risk-free, pays a constant (obviously low) rate of return r.

First, the market dynamics should be specified. The state of the stock is characterized by a price p_t and yields a dividend d_t. The traders may act in each discrete time step. Actually either they trade (buy or sell), or don't, and the price is consequently updated by the equation

$$p_{t+1} = p_t + \beta(B_t - S_t). \tag{9.7}$$

Here B_t and S_t are the number of buyers and sellers in the actual time interval, and β is a parameter, which controls the "velocity" of the price change. The dynamics of the dividend is specified by a random walk (more precisely by an Ornstein - Uhlenbeck process, (studied in Sect. 6.1.3), $d_{t+1} = d_t +$ "Gaussian noise term". The fundamental value is a constructed quantity fv_t, $fv_t := d_t/r$. The agent knows the history of p_t, d_t and fv_t, and the goal is to maximize the money owned at the end of the simulation.

Agents are supposed to have inductive rationality, and evolutionary algorithms are adopted. The actions of the agents depend on rules that specify what to do for particular market conditions. The rules have variable strength, which are subject of change by some adaptive learning rule. In addition, new rules are formed by genetic algorithms, and the weakest rules are eliminated and are substituted by strongest rules (modified by random mutations).

The original model had two qualitatively different outcomes [396], one is equilibrium and homogeneous, the other is a nonequilibrium, heterogeneous. More specifically, under simple structural conditions (few agents, few rules per agent etc.) the system's state is in or around equilibrium. Prices are close to the fundamental values, so fluctuations are small. The agents behave by and large uniformly, based on the logic of the equilibrium theory (buy when the price is smaller than the fundamental value).

Nonequilibrium behavior has many faces. Occasionally there are large deviations from the fundamental values (identified as bubbles or crashes) due to collective interaction among agents. Agents will become heterogeneous, they may adopt highly different rules. Rules are subject to evolutionary changes. Wealth distribution shows a time arrow; initially the distribution is narrow, i.e., more or less everybody has the same wealth. However, the finding of some more advantageous rule by chance ("luck") may imply the emergence of much more wealth to some agents.

The Santa Fe Artificial Stock Market model was certainly a very influential attempt to make predictions on the market behavior in an environment when the agents have adaptive strategies. For the evaluation of the initial model and further artificial stock markets see [306].

9.2 Game Theory: Where We Are Now?

9.2.1 Classical game theory

As it was mentioned in Sect. 5.5, game theory was formulated basically by von Neumann and Morgenstern. It emerged as a mathematical theory of economics, and also proved to be an important tool for treating the problem of necessary cooperation to avoid (nuclear and other) catastrophes.

While the establishment of game theory was motivated by real games, the modern applications are related bringing decisions to social situations. Players have to choose among different strategies, and they take into account the possible strategies of all the other players. In *cooperative games* formations of coalitions are permitted, so finally the game is among different coalitions, and not among individual players. In the von Neumann and Morgenstern book [544] the term "coalition" appears in Chapter V, where zero-sum three-persons game were analyzed. A game is zero sum, when the total resource (award, penalty etc.) is fixed, and does not depend on the chosen strategies. In a non-zero sum game players interests are not necessarily in direct conflict.

Games, however, maybe *non-cooperatives*, too. In this case coalition formation is forbidden, as John Nash invented. The only non-cooperative game that von Neumann - Morgenstern treated was the zero-sum two persons game, which is necessarily non-cooperative. Nash excluded the possibility of coalition formation, and for these non-cooperative games he developed an equilibrium concept, which became later the Nash equilibrium. There is a situation, when each player choose a strategy, and nobody can better off by changing unilaterally her strategy. Nash showed that such equilibrium should exist for under very general conditions.

Let's see now how two or more players divide a cake among themselves!

Cake Division

The "cake division" problem is well-known. It is actually not a single problem, but a quite big set of different problems associated with the "proper" division of a set of goods (the cake) among n "players". In the simplest version of the problem the cake is homogeneous, in other versions it is heterogeneous, i.e., each player has her own preferences regarding to different pieces of the cake.

A division is called *proportional* if every player receives at least $1/n$ of the cake in her own measure. A division is called *envy-free* if no player would change its piece with another player. Note that if the cake is homogeneous then proportional and envy-free divisions are exactly the same and also note that an envy-free division is always proportional.

A proportional division however is not necessarily envy-free if the cake is heterogeneous for at least three players. It is possible that all three players consider their own piece to be at least $1/n$ of the whole but (say) one of them considers another piece to be even bigger.

Note that there is possible to come up with other definitions for the "fair" division, like Pareto-optimality.[2] Also note that the "preferences" of the players are usually expressed in terms of a probability measure on the cake and one needs to restrict the measure to ask meaningful questions, e.g., for k players we might need to assume that every subset of the cake can be divided into k equal parts according to the measures of any player, etc, see the cited references for details.

As Brams and Taylor [70] notes, there are in general four sets of problems associated with cake division. The first set includes existence theorems, i.e., determining the required and sufficient conditions for the existence of proportional or envy-free division. These theorems are usually not constructive, they only prove that such a division exists but don't help to find it [492, 493, 568].

The second set of problems is to give a *procedure* to obtain a desired, i.e., proportional or envy-free solution. As it happens many procedures given are not *algorithms* but so-called moving-knife solutions when typically a player or a referee continuously moves a knife over the cake and another player calls stop whenever she thinks the cake should be cut at the current position of the knife.

A two-player moving-knife solution proposed by Austin in 1982 [33] works as follows. First imagine that the cake is one-dimensional. (Some kind abstraction always necessary if want to formalize problems in the language of mathematics.) A referee moves a knife from the left edge of the cake (line) to the right edge until one player calls "stop" because she thinks the knife is at half-way point according to her measure. (Let this player be Player 1.) Now Player 1 places a second knife at the left edge of the cake and moves both knives together towards the right edge ensuring all the time that the piece between the two knives is $1/2$ in her measure. Player 2 calls "stop" whenever she thinks the piece between the two knives is $1/2$. As in the final situation, with the right knife at the right edge (and thus the left knife at the same place

---

[2] Pareto optimality means that no better solution exists for any player without making at least one player worse off.

where Player 1 called stop) Player 2's piece is larger than $1/2$ in her measure there will be a point when it is exactly $1/2$.

The third set is actually a subset of the second, but now we only allow strict *algorithms* (programmable on a Turing-machine if you like) and no moving-knife solutions. Some moving-knife solutions, like the previous example can be easily rephrased as an algorithms, others are not that easy.

The fourth set is the most interesting and relevant to us, so we will focus on this. This problem is to define a game in game theoretical sense in which each player has a non-losing strategy. In other words, we need to come up with a set of rules and a set of strategies for the players which ensure that they will receive at least $1/n$ of the cake in their own measure. (Or some other definition of "good" division for them.) The set of rules and the strategies are often called *protocols* in the literature [70]. Players in these games are rational and *risk-averse*, they don't choose a strategy which can result less than $1/n$ of the cake, even if this event is very unlikely.

The classic envy-free protocol for two players and heterogeneous cakes is the "cut-and-choose" protocol: the first player cuts the cake into two halves the second player chooses one of them. The strategy of the first player is to cut the cake into exactly two pieces according to her measure, this she will get at least $1/2$. The strategy of the second player is to choose the half which is at least as big as the other (according to her measure of course). While this protocol is envy-free, it is clear that the second player has some "advantage" as she can get a piece bigger than $1/2$, but this is impossible for the first player.

Let us first give a simple protocol k players and homogeneous cakes, ie. here each player has the same preferences. The protocol is given by mathematical induction.

1. If $k \leq 2$ then the problem is solved by the cut-and-choose protocol.

2. Divide the cake into $k-1$ pieces according to the protocol being described.

3. Each player divides her share into k pieces.

4. Player k chooses one from every other players' little pieces.

The strategy for dividing the cake is to divide equally and the strategy to choose is to choose the largest piece. Here is a short informal proof that the protocol is proportional. It is clearly proportional for two players. If the $k-1$ players divided the cake into pieces $a_1, a_2, \ldots, a_{k-1}$, $\sum a_i = 1$ then if player k follows her strategy she gets at least $a_1/k + a_2/k + \cdots a_{k-1}/k = 1/k$.

If a player divides equally after receiving her initial piece then she will keep holding an equal share.

For envy-free protocol for more players and heterogeneous cakes either, the inquiring Reader is advised to consult [70, 69] and references within.

We also didn't deal with the cases where the cake is not continuous but discrete, i.e., a set of items are distributed among players with heterogeneous preferences, please see [71] and [434].

Prisoner Dilemma

The most famous *non-zero* sum game is the Prisoner Dilemma, which analyzes the costs and benefits of two possible strategies, namely cooperation or defection. Here is basic example:

In this game the two players are partners in a crime who have been captured by the police. Each suspect is placed in a separate cell, and offered the opportunity to confess to the crime. The rows of the matrix correspond to strategies of the first player. The columns are strategies of the second player. The numbers in the matrix are the penalties: the first number is the penalty of the first player, the second is the penalty of the second player. Notice that the total penalty of both players is smaller if neither confesses so each receives 1. However, game theory predicts that this will not be the outcome of the game (therefore we have a dilemma). Each player having the intention to minimize her penalty, reasons as follows: if the other player does not confess, it is best for me to confess (0 instead of 1). If the other player does confess, it is also best for me to confess (2 instead of 3). So no matter what the other player will do, confession implies one year off the sentence.

The payoff matrix of the game describes the possible outcomes:

Player 1	Player 2 not confess	Player 2 confess
not confess	1,1	3,0
confess	0,3	2,2

Game theory predicts, that each player following her own self-interest will result in confessions (defection) by both players.

Trust in the other would minimize the total penalty. This is the basis of the dilemma. But can trust evolve? While the question is beyond the scope of classical game theory, it was investigated by evolutionary game theory. Specifically. first, a more sophisticated (i.e., actually more dynamic, or adaptive) version of the game, the *iterated prisoner's dilemma* was introduced by Robert Axelrod, a political scientist, in his book "The Evolution of Cooperation" [34]. The main point is that emergence of cooperation and trust based on *reciprocity*. This problem leads us to evolutionary game theory.

9.2.2 Evolutionary Game Theory

Evolutionary game theory elaborated by John Maynard Smith (1920-2004) and George Price (1922-1975). Price gave an interpretation of Fisher's fundamental theorem, basically by using an argument in Sect. 3.7.2. They introduced the concept of *evolutionary stable strategies* (ESS). If all members of a population adopts EES, there is not any mutant strategy, which can invade.

In Sect. 4.5 the equations behind evolutionary dynamics were discussed. These types of equations are the basis of evolutionary game dynamics. The interpretation of x_i in (4.43) is the frequency of strategy i, and the equations describe the evolution of strategies. As it was shown [394], these equations are basically equivalent to the Price equations.

Evolutionary Game Dynamics

In evolutionary game theory the payoff matrix is identified with fitness. If there are two players, A and B, there are three possibilities:(i) A meets A, both get a; (ii) B meets B, both get d; (iii) A meets B; A gets b and B gets c. Considering a population of A and B players, if x_a is the frequency of A and x_b is the frequency of B than the payoffs for A and B are as

$$f_A = ax_A + bx_B, \tag{9.8a}$$
$$f_B = cx_A + dx_B, \tag{9.8b}$$

Applying the frequency dependent selection, and using $x = x_A$ notation, we have

$$\dot{x}_i(t) = x(1-x)[a - b - c + d]. \tag{9.9}$$

Depending on the numerical values of the elements five qualitative outcomes are possible (the first two are symmetric) [387, 385]:

- $a > c$ and $b > d \longrightarrow$ A dominates B. The whole population tends to consist of A players.

- $a < c$ and $b < d \longrightarrow$ B dominates A. The whole population tends to consist of B players.

- $a > c$ and $b < d \longrightarrow$ A and B are bistable. The equilibrium point is unstable, and the initial condition determines whether the system converges to "pure A" or "pure B" state.

- $a < c$ and $b > d \longrightarrow$ coexistence of A and B. The stable equilibrium point is given as
$$x^* = \frac{d-b}{a-b-c+d}. \tag{9.10}$$

- $a = c$ and $b = d \longrightarrow$ A and B are neutral. Selection does not influence the composition of the population.

Examples for all cases can be found in the repeated prisoner's dilemma, where an interaction between two players consists of many rounds. "... Tit-for-tat (TFT) is a strategy which cooperates in the first round and then repeats whatever the other player did in the previous round. "Always defect" (AllD) is bistable with TFT if the average number of rounds is sufficiently high. "Always cooperate" (AllC) and TFT are neutral if there is no noise and can coexist in the presence of noise. AllC is dominated by AllD. ... [387].

In case of three strategies the situation is more complicated. Since rock smashes scissors, scissors cut paper, and paper covers rock, there is a non-transitive, circular relationship among the three concepts. Recently it was found that this strategy exists in the biological realm. One version of lizards live in three different forms (i.e., with orange, blue or yellow throats) with mating strategies which implies results characterized by circular relationship: "orange beats blue, blue beats yellow, and yellow beats orange" [12]. In a game with three possible strategies different long-term behaviors may emerge. It is possible that the three strategies coexist, or after increasing oscillation two strategies are subject of extinction.

Evolution of Cooperation

Is natural selection, i.e., a spontaneous mechanism sufficient to develop moral rules of cooperation from the interaction of self-interested players? Political scientist Rob Axelrod investigated this problem for many years. The starting points of his argument are that (i) biological evolution proved to be useful by adopting altruism; (ii) genetic algorithm used evolutionary principles successfully. Consequently, if the prisoner dilemma!iterated is played iteratively, strategies containing altruistic elements should perform better than purely selfish ones. The optimistic perspective was that evolutionary mechanisms may establish these altruistic phenotypes even from selfish ones.

It is well known that the "tit for tat" ("equivalent retaliation")[3] algorithm suggested by Anatol Rapoport (one of the founders of systems theory movement, as it was mentioned in Sect. 2.2.1) proved to be very efficient. The two steps of the algorithm are:
1. Cooperate!
2. Do what your opponents did in the previous step!

Long-term cooperation has evolutionary advantage to myopic selfishness. The more general rules behind evolution of cooperation is under investigation [325].

Evolution of fairness: the problem:

The Ultimatum Game is quickly catching up with the Prisoner's Dilemma as a prime showpiece of apparently irrational behavior. In the past two decades, it has inspired dozens of theoretical and experimental investigations. The rules of the game are surprisingly simple. Two players have to agree on how to split a sum of money. The proposer makes an offer. If the responder accepts, the deal goes ahead. If the responder rejects, neither player gets anything. In both cases, the game is over. Obviously, rational responders should accept even the smallest positive offer, since the alternative is getting nothing. Proposers, therefore, should be able to claim almost the entire sum. In a large number of human studies, however, conducted with different incentives in different countries, the majority of proposers offer 40 to 50% of the total sum, and about half of all responders reject offers below 30%.

The irrational human emphasis on a fair division suggests that players have preferences which do not depend solely on their own payoff, and that responders are ready to punish proposers offering only a small share by rejecting the deal (which costs less to themselves than to the proposers). But how do these preferences come about?...

From [388].

---

[3] The Hungarian language is not so economic, and it says that "Szemet szemért, fogat fogért", i.e., eye for eye, tooth for tooth) based on a principle taken from the Old Testament (Exodus 21).

Kin Selection

This algorithm is certainly motivated by biological principles. Cooperative hunting in group of animals required altruistic elements. *Kin selection* was suggested by William D. Hamilton (1936-2000) by using genetic arguments. Hamilton's rule says that:
$$r > C/B. \tag{9.11}$$

Here, C is the fitness cost to the altruist and B is the fitness benefit to the individual helped, and r is a measure of genetic relatedness. Hamilton's kin selection principle motivated Edwards Wilson to explain altruism, aggression, and other social behaviors in terms of biological evolution. His book on what he called *sociobiology* [563] dealt mostly with social animals (such as ants), and a single chapter with humans, provoked sharp debates. The opponents of sociobiology were headed by leading (admittedly leftist) evolutionary biologists, Richard Lewontin and Stephen Jay Gould (1941-2002) who attacked sociobiology for supporting biological determinism. Biological determinism may have as they argued, serious negative social consequences. Sociobiology has been replaced by *evolutionary psychology*, a less direct, more neutral theory to explain the evolution of human behavior and culture by mechanisms of natural selection [46].

Direct Reciprocity

Iterated prisoner's dilemma is a typical example of direct reciprocity. It is obvious that cooperation may emerge not only among relatives.

Cooperation, however, may exist also on more general form of mutual "reciprocity", and it might occur in the animal world between different species. An animal helps another and expects help back in the future ([518]): altruism might be mutual. The mechanism suggests that one player's cooperation increases the probability of the cooperation of the other later. Also, it turned out that while TFT proved to be twice the champion, it has some weakness as well; it is somewhat rigid. Sometimes defections are not intentional, and therefore a certain degree of forgiveness proved to be more advantageous than strict retaliation. A modified strategy, *Generous tit-for-tat* showed improved performance. An other algorithm *Win-stay, lose-shift* quickly corrects accidental mistakes and it exploits the other player if she chronically cooperates. It was suggested [325] that there is a simple relationship to characterize the situation when direct reciprocity can lead to evolution of cooperation: the probability w of the encounter between the same individuals should be larger than the cost-benefit ratio of the altruistic act:
$$w > C/B. \tag{9.12}$$

Reciprocal altruism is most likely very rare (or probably non-existent) in animal societies, cooperation is the result of selfishness or kin selection.

Indirect Reciprocity

Martin Nowak (roots in Vienna) and Karl Sigmund (Vienna) [386] offered a mathematical model to show that cooperation can emerge even if recipients have no chance to return the help to their helper. This is because helping improves reputation, which in turn makes one more likely to be helped. The indirect reciprocity is modeled as an asymmetric interaction between two randomly chosen players. The interaction is asymmetric, since one of them is the "donor", who can decide whether or not cooperate, and the other is a passive recipient. However, the result of the decision is not localized, it is observed by a subset of the population, who might propagate the information. Consequently, the decision to cooperate might increase one's reputation. Those people who are considered more helpful, have a better chance to receive help. The calculation of indirect reciprocity is certainly not easy. A cooperative donor would like to cooperate with a player, who is most likely a cooperator, and would not cooperate with a defector. The probability, q, of knowing someone's reputation should be larger than the cost-benefit ratio of the altruistic act:

$$q > C/B. \qquad (9.13)$$

Evolutionary game theory suggests that indirect reciprocity might be a mechanism for evolution of social norms.

Network Reciprocity

While in a well-mixed population pure natural selection would favor to defectors, interaction in real populations can better characterized by spatial structures or social networks, where the probability matrix of the binary interaction between individuals has some structure. Recently *evolutionary graph theory* was offered [315] to study the effect of the network structure on evolutionary dynamics. It was found, that for many network structures (including small-world and scale-free networks) a simple rule describe the condition of favoring cooperativity. The cost-benefit ratio should exceed the average number of neighbors, k:

$$B/C > k. \qquad (9.14)$$

Group Selection

The concept of *multilevel selection* offers that selection acts not only on individuals but on groups, too.

> A model for group selection
>
> A population is subdivided into groups. Cooperators help others in their own group. Defectors do not help. Individuals reproduce proportional to their payoff. Offspring are added to the same group. If a group reaches a certain size, it can split into two. In this case, another group becomes extinct in order to constrain the total population size. Note that only individuals reproduce, but selection emerges on two levels. There is competition between groups because some groups grow faster and split more often. In particular, pure cooperator groups grow faster than pure defector groups, whereas in any mixed group, defectors reproduce faster than cooperators. Therefore, selection on the lower level (within groups) favors defectors, whereas selection on the higher level (between groups) favors cooperators. This model is based on "group fecundity selection," which means that groups of cooperators have a higher rate of splitting in two. We can also imagine a model based on "group viability selection," where groups of cooperators are less likely to go extinct.
>
> From [325].

Under certain conditions, denoting by n the maximum group size, and by m the number of groups, group selection supports evolution of cooperation if

$$B/C > 1 + (n/m). \tag{9.15}$$

Constructive Evolution?

Evolutionary game theory suggests that selection and mutation, the fundamental principles of evolution can be supplemented with possible mechanisms for the evolution of cooperation. In each mechanism the benefit-to-cost ratio of the altruistic act should exceed some critical value. The appropriate ratio between competition and cooperation may be the driving force of constructive evolution.

9.3 Widening the Limits to Predictions: Earthquake, Eruptions Epileptics Seizures, and Stock Market Crashes

While the study of the eruptions of earthquakes, the onset of epileptic seizures, and the crashes of stock market traditionally are investigated by very different disciplines, and in departments, which differ very much in their scientific culture, complex system approach emphasizes the similarities and offers some common methods to predict the behavior of these systems, and/or understand the inherent limits of their predictability.

9.3.1 Scope and Limits of Predictability

Layman believe that extreme events both in nature and society, such as earthquakes, landslides, wildfires, stock market crashes, destruction of very tall tower buildings, engineering failures, outbreak of epidemics etc. are surprising phenomena, and their occurrence does not follow any rule. Of course, such kinds of extreme events (birth, death, marriages, career steps, etc) are rare, but they influence our everyday lives dramatically. Can we understand, assess, predict and control these events? (An extreme event should not be necessarily negative.) For a very recent excellent, edited book about extreme natural and social events is [6].

Complex systems theory offers different approaches to model these, generally large-scale (global) macroscopic events, which are generated from small-scale (local) microscopic interactions. There are two classes of theoretical approaches, which give quite different answers for the question of eventual predictability.

One possibility is to say, that big earthquakes are nothing else but small earthquakes, which don't stop. The consequence is that these critical events would inherently be unpredictable, since they don't have any precursors. This approach is called as *self-organized criticality* (SOC) and was championed by Per Bak [38].[4]

---

[4] Per Bak (1947–2003) formulated the problem this way: "How can the universe start with a few types of elementary particles at the big bang, and end up with life, history, economics, and literature? The question is screaming out to be answered but it is seldom even asked. Why did the big bang not form a simple gas of particles, or condense into on big crystal? We see complex phenomena around us often that we take them for granted without looking for further explanation. In fact, until recently very little scientific effort was devoted to understanding why nature is complex. I will argue that complex behavior in nature reflects the

Didier Sornette, originally a geophysicist, based on the analogy between natural and financial crisis, wrote an exciting, and therefore controversial bestseller [473] about the reasons, and the eventual predictability of the stock market crashes. According to his arguments, catastrophic events (or at least a class of them) result from accumulating amplifying cascades. Based on the hypothesis of this theory of *intermittent criticality*, many stock market crashes are generated by a slow building up of "subterranean forces", and their precursors may be detected. Were this hypothesis true, the predictability of these events may be imaginable.

First, a short comparative analysis of these phenomena is given.

9.3.2 Phenomenology

Earthquake Eruption

Richter Scale, Gutenberg–Richter Law and Others

We know from the media that the strength of the earthquakes are characterized by using the Richter magnitude scale. Accordingly, to the size of an earthquake a single number can be assigned. The magnitude (M) of an earthquake is proportional to the logarithm of the maximum amplitude of the earths motion. For example, an earthquake with magnitude 8, moves the ground 1000× more than a magnitude 5 earthquake.

There are many earthquakes, but only a few (well, still more than enough) make the headline news, as Fig. 9.3 shows.

The number of earthquakes of magnitude M is proportional to 10^{-M}:

$$\log N(M) = -bM, \quad b \sim 1. \tag{9.16}$$

$N(M)$ is the number of earthquakes of magnitude greater than M.

---

tendency of large systems with many components to evolve into a poised, 'critical state', way out of balance, where minor disturbances may lead to events, called avalanches, of all sizes. Most of the changes take place through catastrophic events rather than by following a smooth gradual path. The evolution to this very delicate state occurs without design from any outside agent. The state is established solely because of the dynamical interaction among individual elements of the system: the critical state is *self-organized*. Self-organized criticality is so far the only known general mechanism to generate complexity."

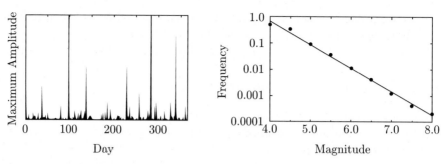

Fig. 9.3. (a) Magnitude of earthquakes on different days of 1995 in Southern California; (b) distribution of earthquakes with different magnitudes. From http://simscience.org/crackling/Advanced/Earthquakes/GutenbergRichter.html.

This statistical finding is called as the Gutenberg-Richter law. Earthquakes are not isolated events, after a mainshock in the same region there are a number of aftershocks with smaller magnitude. The temporal decay of these aftershocks is described with the equation

$$n(t) = \frac{K}{(c+t)^p}, \qquad (9.17)$$

where $n(t)$ is the number of earthquakes n measured in a certain time t, K is the decay rate, c is the "time offset" parameter; and the parameter p typically falls in the range $0.75 - 1.5$. For $p = 1$, it is called the *Omori's law*, stated by the Japanese seismologist Omori in 1894 (see Fig. 9.4).

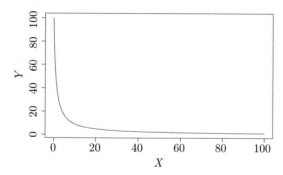

Fig. 9.4. Corrected Omori's law: The decay rate of the magnitude of earthquakes shows hyperbolic decrease.

There is another empirical law (*Bath's law*), which states that difference in the magnitude between a mainshock and its largest aftershock is 1.2, independent of the mainshock magnitude. The distribution of the magnitude of the aftershocks also follows the Gutenberg–Richter law.

Epileptic Seizures

Epilepsy, as a neurological disease, is characterized by abrupt, unprovoked recurrent seizures, exhibiting pathological electrical activity. There are several types of seizures in epilepsy. They can roughly be classified into *generalized* seizures and *partial* seizures. Generalized seizures arise from abnormal electrochemical activity affecting the whole brain at the same time; partial seizures arise from abnormal electrochemical activity affecting one part of the brain, but which may spread to other parts of the brain.

It was generally believed that the seizure is very really abrupt, and it begins just seconds earlier before its clinical onset. There are some indications, although the clinical relevance is still debated, that significant changes in the electrical activity of the brain could be detected by EEG (electroencephalogram), hours in advance of the onset of seizures. If we had a reliable precursor detection technique ([362, 248, 138]), antiepileptic drugs could be administered by some implantable devices to intervene in time to suppress the development of epileptic seizures.

To be able to make a real prediction, the EEG must show at least two different stages (see Fig. 9.5 [459]). First, there has to be a time period between the prediction and the earliest occurrence of the seizure, called seizure prediction horizon (SPH). This period is necessary for example, for drug administration. There should be another time interval, during which the predicted seizure occurs, called the seizure occurrence period (SOP). For obvious reasons, prediction methods should avoid false alarms: false predictions would lead to impairment due to possible side-effects of interventions.

Stock Market Crash: Some Comparative Analysis

Stock market crash is an abrupt fall of stock prices in a significant region of the stock market. Such kinds of crashes emerge due to the interaction of inherent economic factors and human actions.

There were a number of big crashes in the last few centuries. Kenneth Galbraith (1908-2006), the very influential economist wrote an easily readable small book with the title "A short history of financial euphoria" [191]. The

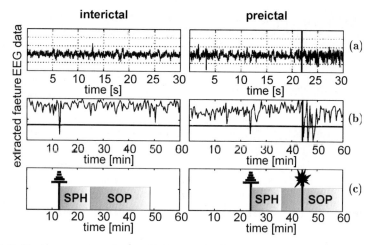

Fig. 9.5. Basic operation of a prediction method during an interictal and a preictal period. Seizure onset is marked by *vertical lines*. a Examples of EEG recordings and b exemplary time course of a feature extracted by a seizure prediction algorithm. The *solid, horizontal line* indicates the threshold for raising alarms. Alarm events and two consecutive time intervals characterizing a prediction, the seizure prediction horizon SPH and seizure occurrence period SOP are illustrated in (*c*). Note the different time scales for the EEG data and the feature time series. Adapted from [459].

book explains why and how the interplay of irrational expectations and specific economic factors generate boom and bust financial dynamics in each several decades.

Now historical crashes will be briefly reviewed. For more details see, e.g., "Stock Market Crash!Net' http://www.stock-market-crash.net/.

The Tulip Bulb Mania

The Tulip Bulb Mania emerged in Holland around 1635. Tulip bulbs first were bought for their aesthetic value, but as their prices increased, they became subject of buying and selling. People bought them to make a (large) profit. There was a month when its value increased twenty-fold. At a certain point the Dutch government attempted to control the mania. After its regulatory actions some informed speculator realized that the price could not become more inflated and started to sell bulbs. Other people soon noticed that the demand for tulips could not be maintained. Their attitude propagated very rapidly among people interested in the business, and soon panic, a social collective phenomenon, emerged. During six weeks there was a 90% reduction in its price (Fig. 9.6).

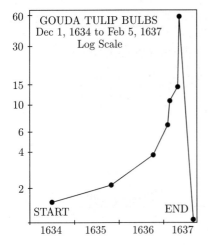

Fig. 9.6. Price dynamics of the rise and fall of the Tulip Bulb Mania. Adapted from http://www.stock-market-crash.net/tulip-mania.htm. ©Elliot Wave International.

The South Sea Bubble

A very bad financial *bubble*[5] occurred in England between 1711 and 1722 (Fig. 9.7).

The British government offered a deal to South Sea Company to finance a big state debt emerged during the War of the Spanish Succession. The South Sea Company traded with South America (excluding Brazil, as it was a Portuguese territory). After a rumor that South Sea Company had been granted full use of Latin American ports, "It became extremely fashionable to own South Sea Company shares". It turned out at a certain point for the leaders of the company that the actual commerce did not produce profit, and money was generated mostly from issuing stocks, so their shares became strongly overvalued. Soon after the owners started to sell shares, there was a panic among shareholders and the market crashed. Factors, such as *speculations*, *unrealistic expectations* and *corruption* contributed to the emergence of the bubble. Newton first saw the bubble, but later lost a lot.[6] Jonathan Swift also

---

[5] The term "bubble" was adopted in a poem of Jonathan Swift (1667–1745)
  The Nation too too late will find,
  Computing all their Cost and Trouble,
  Directors Promises but Wind,
  South Sea at best a mighty Bubble.
  from http://myweb.dal.ca/dmcneil/bubble/bubble-poem.html.

[6] "Sir Isaac Newton, scientist, master of the mint, and a certifiably rational man, fared less well. He sold his 7,000 shares of stock in April for a profit of 100 percent. But something induced him to reenter the market at the top, and

lost a large amount of money, so he was motivated to write a satire about the British society, which we know as Gulliver's Travels.

In a recent reanalyis of this bubble it was argued by Temin[7] and Voth [510] that "Hoare's Bank, a fledgling West End London banker, knew that a bubble was in progress and that it invested knowingly in the bubble; it was profitable to ride the bubble."

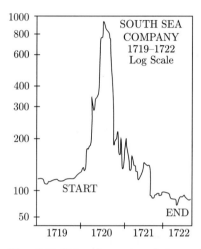

Fig. 9.7. Price dynamics of rise and fall of the South Sea Bubble. Adapted from http://www.stock-market-crash.net/southsea.htm ©Elliot Wave International.

Stock Market Crash of 1929

Since the stock market crash of 1929 is an emblematic event, it was extensively analyzed. Many investors adopted a high-gain or high loss strategy, called *leverage*, but the stock seemed to be very safe. These, so called *margin investors* knew that there was a bull market[8] from 1921, so believed that stock market always went up. When the bear market started after some intervention of the Federal Reserve (the central bank of the US, founded in 1913) panic emerged on Thursday, 24 October 1929. Margin investors became bankrupt.

---

he lost 20, 000. "I can calculate the movement of the stars, but *not* the madness of men." From http://moduleblog.nus.edu.sg/blogs/ec4394/archive/2006/10/30/1101.aspx (26 January 2007).

[7] This is the second citation for "Temin". The economist Peter Temin is the younger brother of the geneticist Howard Temin, who discovered the reverse transcriptase discussed in Sect. 2.1.2.

[8] The terms *bull market* and *bear market* are adopted for longer time periods when the prices are rising or falling, respectively.

Furthermore, since banks also invested their deposit in stock market, even the depositors' money was also lost (Fig. 9.8).

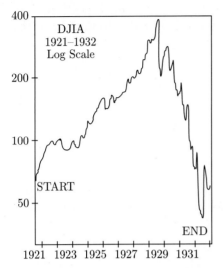

Fig. 9.8. Price dynamics of rise and fall of the stock market crash of 1929 Adapted from http://www.stock-market-crash.net/1929.htm ©Elliot Wave International.

Stock Market Crash of 1987

The Black Monday of 19 October 1987 is remembered as the largest (or second largest) one day stock market crash. While the bull market started in 1982, it accelerated in 1986. The growth of some companies by buying others was one driving force in the stock market. The appearance of personal computers was also considered as an excellent opportunity to increase profit. As it happened in other times, euphoria became the predominant attitude, and many new investors entered the market. Strong economic growth implied inflation, and the Federal Reserve's intervention by raising short term interest rates initiated instability. The big trading firms had *portfolio insurance*, which all of them wanted to use simultaneously. While the clients asked their brokers to sell, they were not able to act due to the large number of their clients. Together with the fall of the Dow Jones index, the national stock markets also collapsed. The central banks, however, managed to control the crisis. The reduction of short term interest rate by the Fed helped, and firms bought back their, now undervalued stocks (Fig. 9.9).

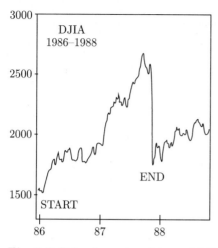

Fig. 9.9. Price dynamics of rise and fall of the stock market crash of 1987. Adapted from http://www.stock-market-crash.net/1987.htm ©Elliot Wave International.

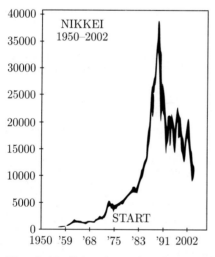

Fig. 9.10. Price dynamics of rise and fall of the Nikkei Bubble. Adapted from http://www.stock-market-crash.net/nikkei.htm ©Elliot Wave International.

The Nikkei Bubble

The bull market in Japan was driven first by the natural recovery from the effects of World War II. The ratio of people working in industry had strongly been increased, large corporations, who offered life-time employment, and received loyalty, had been formed. Japanese firms gained fame for "copying and improving Western products and selling them for much cheaper". In the next stage, the oil crisis combined with the development of high-tech Japanese car industry (as opposed to the low-tech US automotive technology) implied the permanent increase of the Japanese stock market. Finally, with the emergence of electronic giants (such as Hitachi and Sony) it was believed that Japan would dominate the whole microelectronics industry. Speculation started (also in the real estates market: the price of a home in the Tokyo area was much higher than even in Manhattan). To control the inflated economy interest rate was increased. The Nikkei index crashed from 40,000 to 10,000 during several months. As it happens, corruption came to light. The recovery period was extremely long: the Japanese stock market showed a fourteen year long bear market (Fig. 9.10). It is probably over: "Tokyo's Nikkei share average ended 2006 with a 6.9 percent gain for the year, marking its fourth straight year of growth and its longest bull run in nearly two decades", as the Reuter reports on December 29th, 2006, (see http://in.news.yahoo.com/061229/137/6anyi.html).

The Nasdaq Bubble

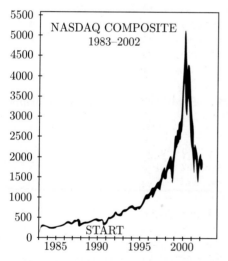

Fig. 9.11. Price dynamics of rise and fall of the Nasdaq Bubble. Adapted from http://www.stock-market-crash.net/nasdaq.htm ©Elliot Wave International.

There was a rapid recovery from the 1987 crisis (Fig. 9.11) The most effective driving force of this process was the propagation of personal computers to the everyday life, from all type of business applications to electronic entertainment. Hardwares, produced in different geographical regions, were very similar. What made the difference was the software. Stock prices of software houses showed a strong increase. Many small companies appeared in the stock market. Initial shareholders (often even employees of these companies), became rich very soon due to the increase of the values of these technological stocks. The Internet catalyzed the super-exponential increase of the stocks. The Nasdaq Composite Index increased during four years with from 600 to 5000 with a peak on March 2000 driven by the irrational expectations. Investors realized at certain point, that the abrupt enrichment of the "dotcom" companies was a bubble. The Nasdaq crashed to 800 for 2002. Today (Februar 2nd, 2007) its value is $2,475.88$ and showed 0.30 percentage increase for the day. (So, probably we are not close to the next bubble.)

History (Even Financial History) Repeats Itself

In all the six cases a very sharp (often, but not always super-exponential: please note that the vertical axes are in logarithmic scale) growth is followed by an abrupt fall. The early stage of the rise is economically justified, and it is related to the appearance of new products, generally related to technological changes. The initial rise is amplified, (more accurately) over-amplified by increased investment due to expectations for rapid and big profit. *Self-organized cooperativity* appears when many people invest synchronously. However, the velocity of the price increase due to a positive feedback mechanism cannot be sustained. The process becomes unstable. In an unstable situation small, local perturbations might have dramatic consequences, and say, a relatively small increase of the interest rate might initiate a crash.

Why there are relatively small number of crashes? In general cases people remember that price should have a peak sooner or later, and this expectation has a deterring effect for the potential buyers. In regular cases this feedback effect can stabilize the price. Stock market crash is the consequence of the impairment of the positive-negative regulatory loop.

9.3.3 Statistical Analysis of Extreme Events

It is not surprising to hear such kinds of questions:

- What is the probability of having big earthquake in California within a year?

- How large might
  - a possible stock market crash be tomorrow?
  - the lowest daily return (the minimum) be?
  - the highest daily return (the maximum) be over a given period?

Standard statistical procedures neglect data deviating"very much" from the others. (These data are called *outliers*). Extreme value analysis uses statistical methods to analyze the rarely occurring events. Very characteristically, extreme events occur in the tails of probability distributions as a function of the "size" of the events (such as energy, duration, etc.).

Gumbel Distribution and Other Extreme Value Distributions

"It seems that the rivers know the theory. It only remains to convince the engineers of the validity of this analysis." (Emil Gumbel)

Emil Gumbel (1891–1966) a famous pacifist, contributed very much to the establishment of statistical methods to describe extreme deviations from an "average" behavior. Extreme value analysis, a branch of mathematical statistics, estimates the probability of extreme floods, large insurance losses, market risk, freak waves, tsunamis, etc.

The Gumbel distribution is defined by the probability density function (PDF):

$$f(x; \mu, \sigma) = \exp(-(\exp(-x - \mu))/\sigma), \tag{9.18}$$

for

$$-\infty < x < \infty \tag{9.19}$$

It has a location (μ), and a scale (σ) parameter, but the Gumbel PDF has no shape parameter. The location is the value with the greatest observed frequency, while the scale characterizes the practical minimal and maximal values. This means that the Gumbel pdf has only one shape, which does not change. Figure 9.12 visualizes the parameter-dependence of the Gumbel distribution.

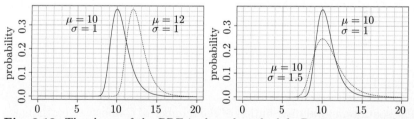

Fig. 9.12. The shape of the PDF is skewed to the left. By increasing μ the pdf is shifted to the right. As σ increases the PDF spreads out and becomes shallower.

While the Gumbel distribution shows a light-tail (exponential decay), other classes of the "extreme value distributions" behave different. The Fréchet distribution has heavy tail, and the Weilbull distribution is characterized by a bounded tail. Though the Gumbel distribution is not symmetric, still its two parameters could be well estimated.

The Gumbel distribution proved to be acceptable for describing the majority of Asian stock markets [119], but the Fréchet distribution fits modeling extreme rainfall better [291]. While the Gumbel distribution proved to be a very good method to estimate the probabilities of extreme events, such as floods, forest fires, financial losses, there are many new results (e.g., for its derivation random variables were assumed and so restricted to be independent and identically distributed), see [7].

Distributions with power-law tails (like the distributions of earthquakes and avalanches) have extreme value statistics described by Fréchet distributions. Distributions that have a strict upper or lower bound have extreme value distributions that are described by Weibull statistics.

Once again, in light of the recent re-emerged interest in power law distributions, what they are and how theory are used in analyzing extreme events?

Power Law Distributions

Long tail distributions have already been discussed in Sect. 6.3. Power law distribution is characterized by the distribution function

$$P(\xi > x) = ax^{-k}, \tag{9.20}$$

where k is the shape parameter.

In estimating the possible occurrence of extreme events it is important to know the probability that a particular sample will be larger than x:

$$P(x) = (x/x_{\min})^{-\alpha+1}, \tag{9.21}$$

while $\alpha > 1$.

The spatiotemporal distribution of the number of Californian earthquakes was analyzed [451], and showed power law distribution (Fig. 9.13).

According to the analysis of Mantegna and Stanley (presented in a highly cited paper on econophysics) [327] the S&P500 index initially shows lognormal distribution followed by a heavy tail power law distribution, as Fig. 9.14 shows.

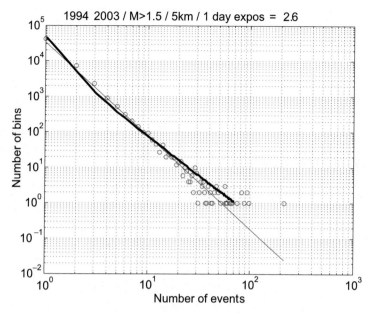

Fig. 9.13. Empirical probability density functions of the number r of earthquakes in the space–time [5×5 km^2 – 1 day], from California SCEC catalog. Adapted from [451].

Power laws appear to describe different financial fluctuations, such as fluctuations in stock price, trading volume, and the number of trades.

9.3.4 Towards Predicting Seizures

The analysis of EEG curves based on dynamical approach offered methods to separate the different qualitative regions before and during a seizure. The challenge is to define good measures.[9]

Measures based on the so-called recurrence points were offered by [314]. Three different measures were derived. One of them (called recurrence rate) measures the number of events, when a trajectory visits a predefined region of a state. While by visual inspection only two stages (inter-ictal and ictal periods) can be discriminated, the analysis helped to show the existence of a third stage, the pre-ictal period, which separated the two stages previously mentioned, as Fig. 9.15 shows.

---

[9] For the analysis of complex dynamics by recurrence plot analysis, see http://www.recurrence-plot.tk/, 16 February 2007).

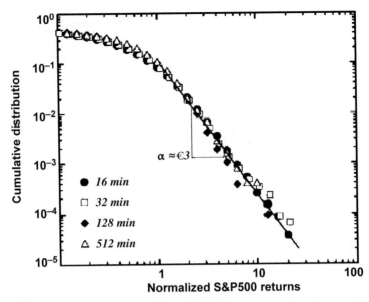

Fig. 9.14. log-normal for the initial part and power law heavy tail.

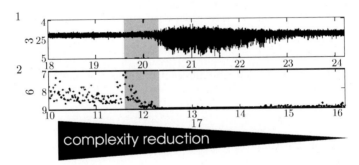

Fig. 9.15. The *upper figure* shows an EEG curve, while the bottom visualizes the temporal change of the recurrence rate. Seizure initiation is accompanied by lower signal complexity due to the higher synchrony.

In our own lab in Budapest, Zoltán Somogyvári analyzed the data came from Magda Szente physiological lab in Szeged [471]. The analysis of the slow dynamics of epileptic seizure led to the following results:

- epileptic seizures may begin much earlier than their clinical onset
- relative self-excitation is a control parameter to induce/suppress epilepsy.

9.3 Widening the Limits to Predictions

He also found that complexity reduction happens during the transition to the seizure, as Fig. 9.16 shows.

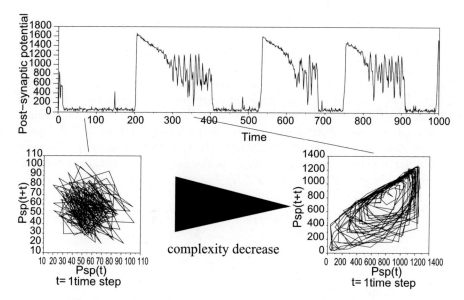

Fig. 9.16. There is a remarkable complexity reduction. *Upper panel*: postsynaptic potential in time. *Lower panel*: phase plane at the pre-epileptic and the seizure region. The *first plot* is shows some chaotic character, while the *second* is much more ordered.

We might have some sense about the general, formal mechanisms of the onset of critical states.

Common message to explain the emergence of epilepsy and emergence of earthquake:

Transition from pre-epileptic and preseismic states toward critical states:

- Gradually increased spatial and temporal correlations
- More dramatic change in energy release

better supports the "intermittent criticality" than the "self-organized criticality" hypothesis.

Based on the argument of [314].

9.3.5 Towards Predicting Market Crashes: Analysis of Price Peaks

Irrational expectations often initiate a super-exponential growth, which leads to instability and a crash. Are there stock-market behaviors controlled by general laws? Bertrand Roehner, an econophysicist from Paris, hopes to find the dynamic equations (i.e causal explanation) of price dynamics. He first starts with "modest goals", such as finding regular patterns behind speculations [436]. Specifically, the shape of price peaks is characterized by two symmetry parameters. An idealized curve is shown in Fig. 9.17. Three time points, i.e., the initial time, the time of the peak and the time of the stop of the fall, denoted by t_1, t_2 and t_3 respectively, should be determined. To each time point a price value, p_1, p_2 and p_3 is assigned. The duration of the bull market and bear market is measured by $T_{up} := t_2 - t_1$, and $T_{down} := t_3 - t_2$.

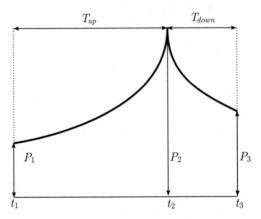

Fig. 9.17. Form of the price peaks. Based on Fig. 6.2 of [436].

The rising phase of curve leading to peaks is rarely slower than exponential, and the real important difference whether it is exponential (so it tends to infinite during infinite time), or super-exponential leading to finite-size singularities. The peak and bottom amplitude is defined as $A := p_2/p_1$, and $B := p_3/p_1$. Roehner suggests an empirical relationship to resilience: there seems to be correlation between A and B:

$$B = aA + b,$$

where the value of a is about $0.4-0.5$. From the positive sign of a implies that for larger peak amplitude, the bottom amplitude is also higher. With other words, large peak amplitude defends the system to fall to low.

As we have known since Newton, the best predictions can be given by dynamical models. So, let's discuss the different versions of dynamical models to catch the most important aspects of extreme events.

9.3.6 Dynamical Models of Extreme Events

Model of Self-Organized Criticality: Sandpile Model

Self-organized criticality suggests that the same effect may lead to small, but also to very large avalanches, so the outcome is not really predictable. A famous toy model was offered Bak, Tang, and Wiesenfeld 1987 [39].

Self-organized criticality: a generating algorithm

Each point $z(x, y)$ in the grid has a number n associated with it (say, average slope of the sand pile at that point).

For a randomly selected point:
$z(x, y) \to z(x, y) + 1$
if $z(x, y) >$ threshold, than
$z(x, y) = z(x, y) - 4$
$z(x \pm 1, y) \to z(x \pm 1, y) + 1$
$z(x, y \pm 1) \to z(x, y \pm 1) + 1$

If this redistribution results in n to be too big for any nearby grid points than start next iteration; else: stop and take randomly another point $z(x, y)$

Figure 9.18 illustrates the actual and simulated avalanche distributions. Avalanches show power law distribution, i.e., perturbation of a single point may result avalanche with very different size.

The model of self-organized criticality suggests that one mechanism towards catastrophic events are the occurrence of small events which don't stop, and by this way they are leading to unpredictability.

**"Explosions": Finite Time Singularities
and Intermittent Criticality**

Finite time singularity roughly speaking means that a dynamical variable gets an infinite value during finite time. This phenomenon is qualitatively differ-

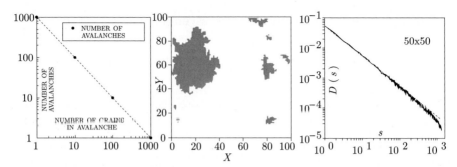

Fig. 9.18. Avalanches: self-organized criticality. *Left*: the number of avalanches and number of grains involved in an avalanche are plotted in log–log space. *Middle*: A region with avalanches While their sizes are different, each of them were triggered by the addition of a single grain. *Right*: the frequency distribution of avalanches with different size. From Introduction to Self-Organized Criticality & Earthquake; Winslow http://www.geo.lsa.umich.edu/~ruff/Geo105.W97/SOC/SOCeq.html.

ent from the exponential growth, when infinite value can be attained during infinite time only. Figure 1.3 showed both exponential and super-exponential growth. The simplest representative dynamical evolution equation leading to a finite-time singularity is:

$$\frac{\mathrm{d}x}{\mathrm{d}t} = x^m, \text{ its solution is: } x(t) = x(0)\left(\frac{t_c - t}{t_c}\right)^{-\frac{1}{m-1}}, \qquad (9.22)$$

with $m > 1$, while $m = 0$ leads to linear, and $m = 1$ to exponential growth, as Fig. 9.19 shows.

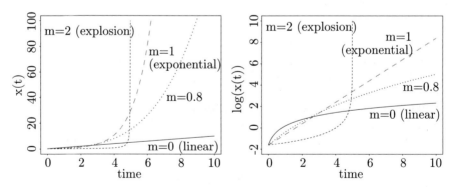

Fig. 9.19. Linear, exponential and super-exponential growth in linear and log-linear scales.

Such kinds of equation with $m > 1$ implements large, "higher-than-linear" (HTL) positive feedback, which seems to be a general mechanism behind finite time singularities. It ensures that the "instantaneous doubling time" tends to zero after a finite period.

General Mechanism for Finite Time Singularities

As cybernetics envisioned, the lack of the stabilizing effects of negative feedback mechanisms may lead to catastrophic consequences. If there are no mechanisms to compensate for the effects of HTL positive feedback, the processes lead to finite time singularities.

Positive feedback seems to be a general mechanism behind eruption of earthquakes, financial crisis (stock market crashes, hyperinflation), and epileptic seizures, as we review here briefly.

Positive Feedback Phenomena Leading to Earthquakes

The analysis of the role of positive feedback mechanisms in amplifying seismic activities led to the conclusion [454] that great earthquakes are critical singular events, and even simple models help to explain the generation of such kinds of singularities. Their statement is that, opposed to conclusions of the theory of self-organized criticality, the precursors of the dynamic process leading to singularities can be found.

What kinds of mechanisms may lead to accelerated seismic activity?

Models that have been proposed to describe accelerating seismicity before large earthquakes fall into two general classes: (i) related to mechanical feedback; (ii) based on the decay of the "stress shadow" from a prior event.

(i) The power law acceleration of regional seismicity is due to positive feedback in the failure process at constant large-scale stress. This feedback can be the result of stress transfer from damaged to intact regions, or it can result from the effect of damage in lowering the local elastic stiffness. Sammis and Sornette [454] showed that for both cases simple models lead to finite-time singularities.

(ii) The phenomenon of "stress shadow" inhibits seismic activity after a big earthquake. After the 1906 earthquake in San Francisco, the seismic activity dramatically reduced for about seventy years. The recovery of a stress shadow might be followed by accelerated seismicity resulted from the increasing stress as the shadow is eroded by tectonic loading. Finally this acceleration may lead to finite time singularities.

Positive Feedback Phenomena Leading to Financial Crisis

The onset of two types of financial crisis, stock market crash and hyperinflation is generated by the positive feedback between the actual and the expected growth rate.

Large stock market crashes, are the social analogues of big earthquakes (as all analogies, this is also has scope and limits). Sornette's view is that the stock market crash is not induced by single local events (such as a raise in the interest values, or other governmental regulations), but due to the unsustainable velocity of price increase. This speculative increase will take the system or more and more unstable situation. Finally, the market collapses due to any small disturbance.

While the conventional economic theory, which is based on the equilibrium between demand and supply, the complex systems approach suggests that partially due to our susceptibility to imitate each others behavior, there is a period when both demand and supply increase which finally lead to singularities. Equilibrium theory works well when the negative feedback effects expresses its stabilizing effect to the increase due to positive feedback. While in "normal situations" the activities of "buyers" and "sellers" neutralize each other, in "critical situations" there is a cooperative effects due to the imitative behavior ("everybody wants to buy since everybody else has already bought"). So, the positive feedback is HTL, and the increase is unsustainable. For the comparison of the "equilibrium" and the "intermittent criticality" approaches see Fig. 9.20.

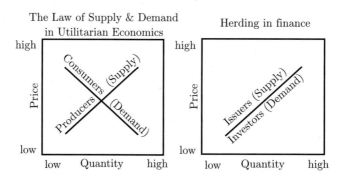

Fig. 9.20. Comparison of the equilibrium and the "intermittent criticality" approaches. The equilibrium between demand and supply can be stabilized due to the interplay between positive and negative feedback (*upper panel*). In case of the dominance of HTL positive feedback the process tends towards crisis, since the growth cannot be sustained.

Positive Feedback Phenomena Leading to Epilepsy

The stable dynamic operation of the brain is based on the balance of excitatory and inhibitory interactions. The impairment of the inhibitory synaptic transmission implies the onset of epileptic seizures. Epileptic activity occurs in a population of neurons when the membrane potentials of the neurons are "abnormally" synchronized. Both experiments and theoretical studies suggest the existence of a general synchronization mechanism in the hippocampal CA3 region. Synaptic inhibition regulates the spread of firing of pyramidal neurons. In experimental situations inhibition may be reduced by applying drugs to block (mostly) GABA-A receptors (Fig 9.21). If inhibition falls below a critical level, the degree of synchrony exceeds the threshold of normal patterns, and the system dynamics switches to epileptic pattern. Collective phenomena occurring in neural networks, such in case of disinhibiton-induced epilepsy have been studied successfully studied by combined physiological computational studies by Roger Traub and Richard Miles. Their book [517] is one of my favorite, and influenced me very much.

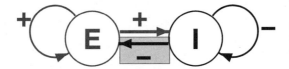

Fig. 9.21. An excitatory–inhibitory network supplemented with self-excitatory and self-inhibitory connections.

A Key to the Prediction?: Log-Periodic Corrections

Log-periodic correction of the dynamics towards crisis

On a refined scale it was found for different systems and variables (from acoustic energy to stock market price) that the "power law dynamics" is modulated with log-periodicity, as Figs. 9.22 and 9.23 show.

To describe also the log-periodic correction of the "power law", dynamics leading to finite-time singularities has been corrected.

In case of stock market crash, the simple price equation

$$\log[p(t)] = A + B(t_c - t)^\beta \tag{9.23}$$

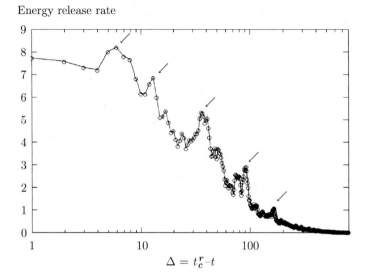

Fig. 9.22. Energy release rate of approaching rupture. Based on [472].

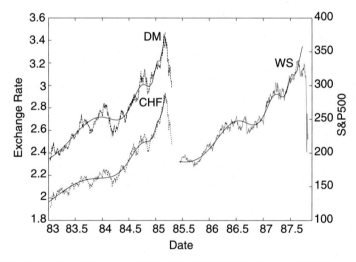

Fig. 9.23. Examples of log-periodicity: The S&P 500 index on Wall Street prior to the October 1987 Crash. The US dollar against the Deutschmark and Swiss Franc prior and around the collapse. The continuous line is a regression curve with a power law superimposed log-periodic oscillation. Based on [472].

9.3 Towards Predicting Market Crashes: Analysis of Price Peaks

has been refined by Sornette to as

$$\log[p(t)] = A + B(t_c - t)^\beta \to$$
$$\log[p(t)] = A + B(t_c - t)^\beta [1 + C \cos(\omega \log((t_c - t)/T))]. \tag{9.24}$$

From a formal point of view, log-periodicity reflects the fact that critical exponent or more generally dimensions can be "complex", (of course, here "complex" means such kinds of numbers which when squared give negative values).

Fractal dimensions have an important role, as we know, to characterize strange attractors related to chaotic dynamics. The notion of the "complex fractal" means a further generalization of the notion. The measured variable shows "power law" dynamics related to the real value of the exponents, and a log-periodic corrections due to the imaginary part of the exponent.

What did we learn?

The theory of complex systems suggests:

- extreme events:
  may be predicted
  their precursors can be detected

- there are methodological similarities to analyze and model different "critical events" occurring in physical, life and social phenomena

- unbalanced higher-than-linear positive feedback is the source of crisis

- there are initial results and many open problems.

10

How Many Cultures We Have?

10.1 Complexity as a Unifying Concept

10.1.1 Systems and Simulations

Complex systems theory, as we may understand it, preserves the best traditions of the general systems theory of von Bertalanffy, mentioned in Sect. 2.2.1. While he was a biologist, who worked on the basic principles of the life, he also explored the universal laws of organizations. It seems, however, that his main interest was improving the human condition by applying "systems thinking". Systems thinking is an important concept, and generally we feel that "complex problems" should be approached by different methods. Roughly speaking, a problem is "simple" when a single cause and a single effect can be identified. Probably the soft, and too abstract (better saying, empty) methods of systems thinking generated some revulsion, mostly among those who believe in the power of mathematical models of specific phenomena.

> Collective wisdom about specialists and generalists:
>
> The specialist knows everything about nothing, while the generalist knows nothing about everything.

Systems theory suggests that "the whole is more than the sum of its parts". This is a controversial statement, since we never know how it comes down in practice. Others may have similar feelings: " I usually hate this slogan but here it holds in a spectacular way", writes Karl Sigmund in his Customer Reviews,

(actually about [385]),
http://www.amazon.com/Evolutionary-Dynamics-Exploring-Equations-Life/
dp/0674023382, 30 April 2007).

While most likely the quest for *universal laws* was not a feasible project, complex systems might have *universal features* [35]. The price to get universal features is to give up the details. Physicists are traditionally very good in neglecting the details to see the big picture. (As a humble chemist by formal training, who has been working mostly on computational neuroscience, I agree with those who believe that not the devil but the angels live in the details.)[1]

Tamás Vicsek is a particularly successful Hungarian physicist, who designed simulation experiments of very different crowd behaviors, from bacterial colonies via flocking of birds to collective phenomena of humans; the latter e.g., in theater (many hands clapping) and in soccer stadium (Mexican wave - La Ola - , and panic) [49, 485, 338, 150, 209]. His recurring method is to show with surprisingly simple models how collective phenomena emerge by self-organized mechanisms. There is no need to external instruction to form global order, interaction with local neighbors is sufficient. "Surprisingly simple" means that he has sufficient courage to use such models, e.g., of waves in excitable medium (mentioned in Sect. 4.2.3) to describe waves in human population. Actually the Mexican wave model was motivated by the Wiener-Rosenblueth model of the cardiac wave propagation [503] (and it is just a coincidence that Arturo Rosenblueth (1900–1970) was a Mexican physiologist). However complex a human is, her state is characterized by a scalar variable with three possible values. She may be in an active, inactive or refractory state. There is a somewhat more detailed model, but the approach is the same. "Many scientists implicitly assume that we understand a particular phenomenon if we have a (computer) model that provides results that are consistent with observations and that makes correct predictions..." [538]. Model making is a combination of art and science. The most important question to be answered is not what we should put into the model, but *what to neglect*. Vicsek's success indicates a possible direction.

10.1.2 The Topics of the Book in Retrospective: Natural and Human Socioeconomic Systems

A large part of the book illustrates through examples how causal dynamical systems work.

---

[1] "Mert az angyal a részletekben lakik." "Petri György: Mosoly" ... "For the angel is in the detail." György Petri: Smile; Translated by Clive Wilmer and George Gömöri. (Thanks to Máté Lengyel.)

While *mechanical* systems are supposed to belong to the class of simple systems, even the mechanical clock, the symbol of a periodic machine Universe, (i) utilizes the concept of feedback and (ii) might benefit from a little chaos. Idealized (frictionless) mechanical systems show time reversibility, macroscopic processes have arrow of time. The first and second laws of *thermodynamics* reflects the constancy and change of nature.

Chemical kinetics uses both deterministic and stochastic models. Autocatalytic reactions are examples of positive feedback: the larger the concentration of a reagent, the larger the velocity of its own production. Such kinds of reactions are the ingredients of the generating mechanism of complex temporal and spatial patterns.

Reaction–diffusion systems, somewhat counter-intuitively, are the basis of biological pattern formation. While diffusion is a process which is driven by spatial gradients to eliminate inhomogeneities, the coupling of certain spatial processes (such as diffusion) and local processes (such as chemical reaction) may lead to the formation of spatial structures.

Systems biology adopts the perspective of a multilevel approach, and was developed recently as a reaction to the overwhelming success of molecular biology. It adopts different techniques from simulation of chemical reactions to analysis of biochemical and genetic networks. A challenge of the classical dynamic description is that cells are self-referential systems. It is far from being clear what is a really appropriate mathematical framework to describe the dynamics of such kinds of systems.

A higher level of *biocomplexity* is related to dynamics of *ecological networks*. A key question is whether how an ecological system preserves both its stability and diversity. Ecological communities seem to be connected together by weak relationships.

Epidemic models have also have large practical significance. Too much data have been accumulated to model the spread of such diseases, as HIV and SARS. The most important result of the classical epidemic model studies stated (in accordance to real data) that epidemics is a threshold phenomenon. The spread of an epidemic in human or computer networks is much more complicated, in certain situations the spread does not have any threshold.

One big family of *evolutionary dynamics* uses generalized versions of the replicator equation to describe selection and mutation. The units of replication may be very different, from molecules, via genes to behavioral strategies.

Neurodynamics has several different functional roles. First, it deals with the generation and control of normal and pathological brain rhythms. Epilepsy was mentioned, as an example, where unbalanced positive feedback leads to pathological behavior. It is also an example, where new methods of prediction may have important practical consequences. In relationship with another neurological disorder, anxiety, my own interest was reviewed: a new way of drug discovery can benefit from integrating the kinetic models of drug effects into the conventional network dynamics and to design drugs which are able to generate some "best" physiological temporal patterns.

Second, the emergence of complexity through self-organizing mechanisms has been studied on both the ontogenetic and the phylogenetic time scale. Self-organization phenomena are related to normal ontogenetic development and plastic behavior occurring at different hierarchical levels of the nervous system. An important question here is the balance between determinism and randomness in the nervous system. Mental development is seen not only as a simple "consequence" of neural development, since mind not only represents, but also creates reality.

Large sets of seismic data have been accumulated with the hope of being able to predict eruption of *earthquakes*. Unbalanced positive feedback was also mentioned, as a possible mechanism of amplifying smaller seismic activities.

The scope (and limits) of dynamical models of socioeconomic systems were illustrated by a number of examples.

Segregation dynamics is the popular, demonstrative example of the power of simulations in social sciences. By simulations the effects of different preference functions on the emerging spatial patterns can be analyzed. The core models of social epidemics and opinion dynamics were reviewed. Simple assumptions, such as interactions among three basic populations (infected, susceptibles and removed) reflect the characteristic features of behavioral patterns for the propagation of ideas.

The *war* and *love dynamics* were examined by toy models. There are different qualitative outcomes. Occasionally there are pure winners and losers. Under other conditions, everybody survives at a certain fixed level, or there is a periodic transition among different states (say, love and hate...).

Economic activities, such as *business cycles* often show oscillations, and fluctuations. Technically oscillations are deterministic phenomena generated by the interaction of variables, such as investment and saving. Probably the real patterns are not truly oscillatory, and chaotic processes (including their control) may be more relevant. Fluctuations are random phenomena. One of the key discoveries of the new discipline, econophysics, is that financial time series have much larger fluctuations, than a Gaussian distribution would predict. Stock market data, among others, should be subject of statistical analysis of extreme events.

Dynamics of illicit *drug market* seems to be chaotic, due to the interaction of positive and negative feedback loops among the addicts and sellers populations. Model-based control strategies may offer methods to shift the system to a predictable periodic state, or even a low level fixed equilibrium.

The availability of the on-line database of the US *patent citations* opened the possibility to explain the temporal development of the network of citations by a dynamical model. We assumed that the attractiveness of the patents (combined by their age and popularity) characterizes sufficiently well the state of the system. The development of the network was given by a simple probabilistic rule, and worked well.

Models of *evolution of cooperation* are now extended. Originally it was assumed that altruistic mechanisms work for gene-sharing communities, but it seems to be reasonable that different levels of reciprocal interactions are the driving force behind constructive evolution.

I am fully aware of not mentioning in this book many important issues, which are often related to complex systems. Global warming and terrorist networks have been hot topics, and complex systems research has already offered methods to analyze them. Also, I mention here that Yaneer Bar-Yam's Dynamics of Complex Systems [42] thematically obviously overlaps with the present book, but of course you may notice the ten years difference (and the different interest of the authors).

While complex systems approach has the power to unify different levels, methods, problems from physical via biological to social systems, there is no magic bullet. Complexity systems have a number of ingredients.

10.2 The Ingredients of Complex Systems

Paradoxes. We have seen two types of *paradoxes*, one in linguistic situation, as it was discussed in Sect. 1.2.3 and another related to multistable perception of figures mentioned in Sect. 2.2.3. Paradoxes are characterized by some deviation between the expected and the actual behaviors of a system, and it is generally a consequence of false assumptions. In a logical situation the observer's opinion oscillates between a "true' and a "false" value.

Circular and network causality. As opposed to simple systems, where causes and effects (actually most often a single cause and a single effect) can be separated, a system is certainly complex, if an effect feeds back into its cause. Biological cells, ecological networks, business relationships and other social structures are full with such kinds of feedback loops.

Organizational closures. As Robert Rosen suggested, a cell is internally controlled, and it main feature is its organizationally closure. Of course, it does not mean that it is not thermodynamically open. But the connection of these two types of open/closed properties, I think, was never addressed appropriately.

Finite-time singularity. Unbalanced positive feedback may lead to finite-time singularities if the self-amplification is larger than the threshold necessary to exponential increase. In such systems the characteristic variable tends to have infinite value during finite time. In chemical systems this phenomenon is identified with explosion. In other systems, such in case of stock prices, the superexponential increase cannot be continued for "ever" due to the unstable nature of this process, and is followed by a compensatory process (i.e., stock market crash).

Chaos and fractals. Chaos it is often, erroneously, identified with complexity. As we know, systems with low structural complexity (as the logistic map) also may lead to chaos. Chaos is certainly dynamically complex, and different measures of complexity, such as fractal dimension, have been defined. The extreme sensitivity to initial conditions are very important property of chaos, and lead to fractal attractors. It generated a big excitement when it turned out that chaos generated fractal structures that are rather extensively found in nature and society.

Emergence. Complex systems (understood now as population of homogeneous or heterogeneous units) may show collective phenomena, which cannot be predicted from the behavior of the constituents. Self-sustained oscillation (a technically not very complex phenomena) and chaos may emerge in consequence of interactions of specific variables in certain regions of parameters. Levels of hierarchical organizations, from subatomic to cosmic, are emergent products. Organization principles, which regulate the emergence, were suggested to exist from quantum physics, via chemical kinetic and life itself to, say, the evolution of social behavior, as cooperation, and the formation of urban segregation, panic, stock market crash, etc.

Self-organized complexity. Many events/phenomena are characterized by probability distributions with long tails, which follow the power law relationship. Phenomena with large fluctuations, as stock market crashes, or natural disasters are extremely rare events. Seemingly different phenomena might be generated by similar mechanisms. Power law distribution is a specific case of the property of power law scale-invariance, which assumes a relationship between two variables in the form of

$$y = ax^k, \tag{10.1}$$

where a and k are constants, the latter is called the scaling exponent. Statistical physics has a formalism to treat phase transitions. The characteristic variable of the transition behaves as (10.1), where $x := |T - T_c|$, T is the temperature, T_c is its critical value, and k is the critical exponent. One of the big successes of statistical physics is the demonstration of common critical exponents for various phenomena. So a large family of phenomena belong to the same universality class, independently from the details. Maybe angels live in the universal, too. *Self-organized criticality* and *intermittent criticality* are general mechanism suggested to explain the emergence of self-organized complexity in physics, the brain, finance, and in many other situations. While both approaches is based on the integration of different space- and timescales, mechanisms based on intermittent criticality may be predicted. Many specific examples are being studied now by these methods.

10.3 Complexity Explained: In Defense of (Bounded) Rationality

Complexity, Explanation, Rationality: What We Wanted to Achieve?

Complex systems theory has a double soul and double strategy. First, it suggests that there is an already existing methodology, the theory and practice of dynamic modeling, and we believe that many social phenomena, traditionally studied by descriptive-normative methods, can be attacked by them. Second, we clearly see the limits of our own methods, mostly for situations, when different space- and timescales are integrated, when networks of positive and negative loops are interconnected, when the observer is not independent from the observed phenomena, etc. Whatever is the situation, physical, biological and social complexity could and should be understood as an emergent product of the interplay among constituents, and it is difficult to imagine other possibilities, at least within the scientific framework. This view does not exclude the possibilities of having a feedback from the "whole" to the "parts".

From Natural Science to Humanities and Back

Whether we like it or not (actually *not*, we can't do too much just to see the reduction of the prestige of science, the somewhat increasing influence of pseudoscientific ideas. "Intelligent design" pretends to be science. Many of us feels that it is a strange and inadequate situation that science must defend itself. Of course, critics of science appeared in novels of giant writers, such as in Flaubert' Bouvard and Pecuchet, in Canetti's Auto-da-fé and Hesse's The Glass Bead Game [175, 90, 235]. (To the memory of Péter Balassa.)

Complex systems approach comes predominantly from natural sciences, but we are fully aware of the existence of limits of model-based scientific thinking. Still, scientists have the ambition to tell something about the possibility of understanding and controlling our uncertain world. Also we know, that only very few scientists are celebrities, and only a very very few of us has any influence on media and popular culture.

Science War or the Crisis of Modernity

While the science war was initiated by accusing social scientists to use "fashionable nonsense", and already Popper (a rare philosopher whom scientists like) claimed that the criteria of rationality in social sciences are much less solid, than in natural sciences. Well, fellows from social sciences/humanities pass judgments on scientists - these uneducated hicks - because of their lacking consideration of the crisis of modernity.

Among my coeval and elder colleagues there is a common understanding that, after all, the real stuff is Science: we measure and calculate, and the caravan keeps moving on, however slowly, accompanied simply by the unharmful barking of some philosophers.

The rationale of classical Science is : experiment-measure-calculate. The engineering version is plan-construct-control/command, while humanities rely on understanding, interpretation, participation.

However beautiful are constructions based on definitions, axioms, statements and proofs, they rarely reflect faithfully what is called reality. Mathematicians, the most respected rational fantasy players already left a big battle behind, and the rigorous, formalist program of mathematics could not be completed. Imre Lakatos (1922–1974)'s work on the philosophy of mathematics analyzed the way of progress in mathematics. After his early death it was written [570]:

"The thesis of 'Proofs and Refutations' is that the development of mathematics does not consist (as conventional philosophy of mathematics tells us it does) in the steady accumulation of eternal truths. Mathematics develops, according to Lakatos, in a much more dramatic and exciting way - by a process of conjecture, followed by attempts to 'prove' the conjecture (i.e., to reduce it to other conjectures) followed by criticism via attempts to produce counter-examples both to the conjectured theorem and to the various steps in the proof."

The alternative of classical rationality is not at all irrationality. As Herbert Simone suggested, the concept of "bounded rationality" is more appropriate to describe our behavior (fortunately) than perfect rationality. A new type of rationality, what we might call "resigning rationality", asks what we can do in situations, when we have difficulties with concepts of objective external observer and of objective reality. Can we accept the existence of "tacit knowledge" of Michael Polanyi [410], or the possibility of self-reflection?

Model and Truth

By thinking in and with models we consciously give up a language, which implies that we might be the only ones, who are able to know the only truth. We don't say that "these and these are the facts, and only the facts matter". A more appropriate language says: "By assuming this and that, and adopting the set of these rules, we may imply this and that". This is not a shame. We may admit that we are not the owners of the final and infallible truth.

The Age of Fallibility

What can we do after accepting the limits if classical rationality? George Soros, known as Popper's student and admirer, argues that *reflexivity* and *fallibility* are the most important features of the new age we live. Reflexivity is related to our decision making mechanisms (say, buying/selling in the stock market). The decision depends on both knowledge and expectation, and they influence reality [475].

> George Soros
>
> Could the recognition of our imperfect understanding serve to establish the open society as a desirable form of social organization? I believe it could, although there are formidable difficulties in the way. We must promote a belief in our own fallibility to the status that we

> normally confer on a belief in ultimate truth. But if ultimate truth is not attainable, how can we accept our fallibility as ultimate truth?
>
> This is an apparent paradox, but it can be resolved. The first proposition, that our understanding is imperfect, is consistent with a second proposition: that we must accept the first proposition as an article of faith. The need for articles of faith arises exactly because our understanding is imperfect. If we enjoyed perfect knowledge, there would be no need for beliefs. But to accept this line of reasoning requires a profound change in the role that we accord our beliefs....
>
> To derive a political and social agenda from a philosophical, epistemological argument seems like a hopeless undertaking. Yet it can be done. There is historical precedent. The Enlightenment was a celebration of the power of reason, and it provided the inspiration for the Declaration of Independence and the Bill of Rights. The belief in reason was carried to excess in the French Revolution, with unpleasant side effects; nevertheless, it was the beginning of modernity. We have now had 200 years of experience with the Age of Reason, and as reasonable people we ought to recognize that reason has its limitations. The time is ripe for developing a conceptual framework based on our fallibility. Where reason has failed, fallibility may yet succeed.
>
> From the "The Capitalist Threat" [474].

Interestingly, while Soros benefited literally very much from his deep understanding of the heterogeneous nature of the world (economy), as the Reader knows, Thomas Friedman, an influential journalist of the New York Times, wrote a best-seller about the globalization (I found strange that Soros was not mentioned) with the title "The World Is Flat" [185], stating that there is a tendency to reduce this heterogeneity, mostly due to the development of Internet technologies. This makes possible to include China and India in the complex supply chain.

Towards a New Synthesis? From the "two cultures" to the third

C. Snow pointed out the gap between the cultures of literary intellectuals and of scientists [470]. Actually he blamed mostly non-scientists.

10.3 Complexity Explained: In Defense of (Bounded) Rationality

> Snow's critique:
>
> "A good many times I have been present at gatherings of people who, by the standards of the traditional culture, are thought highly educated and who have with considerable gusto been expressing their incredulity of scientists. Once or twice I have been provoked and have asked the company how many of them could describe the Second Law of Thermodynamics. The response was cold: it was also negative. Yet I was asking something which is the scientific equivalent of: Have you read a work of Shakespeare's?
>
> I now believe that if I had asked an even simpler question – such as, What do you mean by mass, or acceleration, which is the scientific equivalent of saying, Can you read? – not more than one in ten of the highly educated would have felt that I was speaking the same language. So the great edifice of modern physics goes up, and the majority of the cleverest people in the western world have about as much insight into it as their neolithic ancestors would have had."

I used to annoy my social scientist friends by asking about the laws of thermodynamics, but abandoned this endeavor, since I realized that it is counterproductive, and does not help to narrow the gap between our way of thinking. Probably I am not right. Steven Pinker, Harvard psychologist, working on cognition and languages, and who writes also excellent popular books, just wrote (27 May 2007) in the New York Times Book Review:

> Steven Pinker about scientific illiteracy:
>
> People who would sneer at the vulgarian who has never read Virginia Woolf will insouciantly boast of their ignorance of basic physics. Most of our intellectual magazines discuss science only when it bears on their political concerns or when they can portray science as just another political arena.
>
> From [408].

Pinker is a member of the informal group coordinated by John Brockman. Thanks to Brockman, the works and thoughts of a set of scientific intellectuals seem to be integrated in a "third culture" [75, 76], see also the website http://www.edge.org.

The third culture movement is optimistic with the hope to understand and explain the real world of humans, machines, societies and the Universe. Science, technology and business is now interconnected and forms a global culture.

> The new humanists
>
> "...something radically new is in the air: new ways of understanding physical systems, new ways of thinking about thinking that call into question many of our basic assumptions. A realistic biology of the mind, advances in physics, information technology, genetics, neurobiology, engineering, the chemistry of materials -all are challenging basic assumptions of who and what we are, of what it means to be human. The arts and the sciences are again joining together as one culture, the third culture..."
>
> From John Brockman: Introduction: The New Humanists, in [76].

Instead of Summary

I think, the perspective what a pluralistic theory of complex systems offers, is the key to understand, explain and control the world which seemed to be derailed. We may have the chance to get a better insight into our own cognitive-emotional structures. Analysis based on the combination of biological and social approaches will help to explain and control our choices and decision makings at individual level, and the evolution of norms and values in societies. Increasing knowledge about our (and others) genes, brains and minds, our willingness to cooperate and compete both individually and in groups should help to understand the world and our role in it.

But of course, flat or not, that will be a different world.

References

1. ABBOTT, E. *Flatland: A Romance of Many Dimensions*, unabridged, 1992 ed. Dover Publications, 1884.
2. ABRAHAM, T. (Physio)logical circuits: the intellectual origins of the McCulloch-Pitts neural networks. *Journal Historical Behavioural Science 38* (2002), 3–25.
3. ADORJÁN, P., BARNA, G., ÉRDI, P., AND K, O. A statistical neuralfield approach to orientation selectivity. *Neurocomputing 26-27* (1999), 313–318.
4. ALBERT, R. Scale- free networks in cell biology. *Journal of Cell Science 118* (2005), 4947–4957.
5. ALBERT, R., AND BARABÁSI, A. Statistical mechanics of complex networks. *Reviews of Modern Physics 74* (2002), 47–97.
6. ALBEVERIO, S., JENTSCH, V., AND KANTZ, H. *Extreme Events in Nature and Society*. Springer, Berlin, 2006.
7. ALBEVERIO, S., AND PITTERBARG, V. Mathematical methods and concepts for the analysis of extreme events. In *Extreme Events in Nature and Society*, S. Albeverio, V. Jentsch, and H. Kantz, Eds. Springer, Berlin, 2006.
8. ALDANA-GONZALEZ, M., COPPERSMITH, S., AND KADANOFF, L. Boolean dynamics with random couplings. In *In Perspectives and Problems in Nonlinear Science. A celebratory volume in honor of Lawrence Sirovich*, E. Kaplan, J. Marsden, and S. KR, Eds. Springer, 2003.
9. ALDOUS, D. J. Stochastic models and descriptive statistics for phylogenetic trees, from yule to today. *Statist. Sci.* (2001), 23–34.
10. ALJUNDI, A., DEKEYSER, J.-L., KECHADI, M.-T., AND SCHERSON, I. A study of an evaluation methodology for unbuffered multistage interconnection networks. In *Proceedings of 17th International Parallel and Distributed Processing Symposium, IPDPS'03* (Nice, France, 2003).
11. ALMASSY, N., EDELMAN, G., AND O, S. Behavioral constraints in the development of neuronal properties: a cortical model embedded in a real-world device. *Cereb Cortex.* (1998).
12. ALONZO1, S., AND SINERVO, B. Mate choice games, context-dependent good genes, and genetic cycles in the side-blotched lizard, uta stansburiana. *Behavioral Ecology and Sociobiology 49*, 176–186.
13. ANDERSON, P. More is different. *Science 177* (1972), 393–396.

14. ANDERSSON, G. *Criticism and the History of Science. Kuhn's, Lakatos's and Feyerabend's Criticisms of Critical Rationalism*. Philosophy of History and Culture, 13. Brill, Leiden, 1994.
15. ANDERSSON, H., AND BRITTON, T. *Stochastic Epidemic Models and Their Statistical Analysis*. Springer, 2000. Lecture Notes in Statistics.
16. ANDRIEVSKII, B., AND FRADKOV, A. Control of chaos: Methods and applications. i. methods. *Autom. Remote Control 64*, 5 (2003), 673–713.
17. ARADI, I., AND ÉRDI, P. Computational neuropharmacology: dynamical approaches in drug discovery. *Trends in Pharmacological Sciences 27* (2006), 240–243.
18. ARBIB, M. *The Metaphorical Brain 2: Neural Networks and Beyond*. Wiley-Interscience, New York, 1989
19. ARBIB, M. Warren McCulloch's search for the logic of the nervous system. *Perspectives in Biology and Medicine 43* (2000), 193–216.
20. ARBIB, M. Neuroinformatics. In *The handbook of brain theory and neural networks, second edition*, M. Arbib, Ed. MIT press, 2003, pp. 741–745.
21. ARBIB, M. Schema theory. In *The Handbook of Brain Theory and Neural Networks*, M. Arbib, Ed., 2nd ed. MIT Press, 2003), pp. 993–998.
22. ARBIB, M., ÉRDI, P., AND SZENTÁGOTHAI, J. *Neural Organization: Structure, Function and Dynamics*. MIT Press, 1997.
23. ARNOLD, L., HORSTHEMKE, W., AND LEFEVER, R. White and coloured external noise and transition phenomena in nonlinear systems. *Z. Physik B* (1978).
24. ARNOLD, V. *Catastrophe Theory*, 3rd ed. Springer, Berlin, 1992.
25. ARTHUR, W. Positive feedbacks in the economy. *Scientific American* (February 1990), 92–99.
26. ARTHUR, W. Inductive reasoning and bounded rationality. *American Economic Review (A.E.A. Papers and Proc.) 84* (1994), 406–411.
27. ARTHUR, W. Cognition: the black box of economics. In *The Complexity Vision and the Teaching of Economics*, D. Colander, Ed. Edward Elgar Publishers, 2000.
28. ASHBY, W. *Design for a Brain*. John Wiley, New York, 1952.
29. ASHBY, W. *An Introduction to Cybernetics*. John Wiley, New York, 1956.
30. ASPRAY, W. *John von Neumann and the Origins of Modern Computing*. The MIT Press, 1990.
31. ATTARDI, D., AND SPERRY, R. Preferential selection of central pathways by regeneration optic fibers. *Experimental Neurology 7* (1963), 46–64.
32. AUMANN, R. J., AND MASCHLER, M. Game theoretic analysis of a bankruptcy problem from the Talmud. *Journal of Economic Theory 36* (1985), 95–213.
33. AUSTIN, A. Sharing a cake. *Mathematical Gazette 6*, 437 (1982), 212–215.
34. AXELROD, R. *The Evolution of Cooperation*. Basic Books, 1984.
35. BACHELIER, L. Theorie de la speculation (thesis). *Annales Scientifiques de l'École Normale Superieure III-17* (1900), 21–86.
36. BAILEY, C., GIUSTETTO, M., HUANG, Y., HAWKINS, R., AND KANDEL, E. Is heterosynaptic modulation essential for stabilizing Hebbian plasticity and memory? *Nat Rev Neurosci 1*, 1 (2000), 11–20.
37. BAJCSY, R. Active perception. *Proc. IEEE 6* (1988).
38. BAK, P. *How Nature Works*. Springer, Berlin, 1996.
39. BAK, P., TANG, C., AND WIESENFELD, K. Self-organized criticality: an explanation of 1/f noise. *Physics Review Letters 59* (1987).

40. BALL, P. *Critical Mass: How One Thing Leads to Another*. Farrar, Straus and Giroux, 2004.
41. BALLARD, D. Animate vision. *Artificial Intelligence Journal 48* (1991).
42. BAR-YAM, Y. *Dynamics of Complex Systems*. Perseus Books, Reading, Mass, 1997.
43. BARABÁSI, A. *Linked: The New Science of Networks*. Perseus Books, 2002.
44. BARABÁSI, A.-L., JEONG, H., RAVASZ, E., NÉDA, Z., SCHUBERTS, A., AND VICSEK, T. Evolution of the social network of scientific collaborations. *Physica A 311* (2002), 590–614.
45. BARAS, L., AND SHCHUR, L. Periodic orbits of the ensemble of sinai-arnold cat maps and pseudorandom number generation. *Physics Reviews E 73* (2006), 036701.
46. BARKOW, J., ; COSMIDES, L., AND TOOBY, J. *The Adapted Mind: Evolutionary Psychology and The Generation of Culture*. Oxford University Press, 1992.
47. BARNA, G., GROBLER, T., AND ERDI, P. Statistical model of the hippocampal CA3 region II. The population framework: model of rhythmic activity in the CA3 slice. *Biol Cybern 79*, 4 (1998), 309–21.
48. BATES, E., AND ELMAN, J. The ontogeny and phylogeny of language: A neural network perspective. In *Biology, brains, and behavior: The evolution of human development*, S. Parker, J. Langer, and M. McKinney, Eds. School of American Research Press, Sante Fe, 2000.
49. BAUER, F. The Kermack-McKendrick epidemic model revisited. *Math. Biosci 198* (2005).
50. BAUM, L., AND FRAMPTON, P. Turnaround in cyclic cosmology. *Phys. Rev. Lett. 98* (2007).
51. BAUMGARTNER, F., AND JONES, B. Positive and negative feedback in politics. In *Policy Dynamics*, F. Baumgartner and B. Jones, Eds. University of Chicago Press, 2002.
52. BAZSÓ, F., KEPECS, A., LENGYEL, M., PAYRITS, S., SZALISZNYÓ, K., ZALÁNYI, L., AND P, E. Single cell and population activities in cortical-like systems. *Reviews in the Neurosciences 10* (1999).
53. BEHRENS, D., CAULKINS, J., AND FEICHTINGER, G. A model of chaotic drug markets and their control. *Nonlinear Dynamics Psychol Life Sci 8* (2004).
54. BEINHOCKER, E. *Origin of Wealth: Evolution, Complexity, and the Radical Remaking of Economics*. Harvard Business School Press, 2006.
55. BENNETT, M., SCHATZ, M., ROCKWOOD, H., AND K, W. Huygens's clocks. *Proceedings of the Royal Society A: Mathematical, Physical and Engineering Sciences 458* (2002), 563–579.
56. BERNARD, C. *Lecons sur les phnomnes de la Vie*. Baillire, Paris, 1878.
57. BERNSTEIN, D. Feedback control: an invisible thread in the history of technology. *IEEE Control Systems Magazine 22* (2002), 53–68.
58. BEUTER, A., BARBO, E., RIGAL, R., AND BLANCHET, P. Characterization of subclinical tremor in Parkinson's disease. *Mov. Disord. 20* (2005), 945–950.
59. BEUTER, A., AND K, V. Tremor: Is Parkinson's disease a dynamical disease? *Chaos 5* (1995), 35–42.
60. BI, G., AND POO, M. Synaptic modifications in cultured hippocampal neurons: dependence on spike timing, synaptic strength, and postsynaptic cell type. *J Neurosci 18*, 24 (1998), 10464–72.
61. BICKLE, J. *Philosophy and Neuroscience. A Ruthlessly Reductive Account*. Kluwer, Dordrecht, 2003.

62. BILKE, S., AND SJUNNESSON, F. Stability of the Kauffman model. *Physical Review E 65* (2002).
63. BLACK, F., AND SCHOLES, M. The pricing of options and corporate liabilities. *Journal of Political Economy 81* (1973), 637–654.
64. BLISS, T., AND LOMO, T. Long-lasting potentiation of synaptic transmission in the dentate area of the anaesthetized rabbit following stimulation of the perforant path. *Journal of Physiology (London) 232* (1973), 331–356.
65. BOGUÑÁ, M., AND PASTOR-SATORRAS, R. Epidemic spreading in correlated complex networks. *Phys. Rev. E 66* (2002), 047104.
66. BONCHEV, D., AND BUCK, G. Quantitative measures of network complexity. In *Complexity in Chemistry, Biology and Topology*, D. Bonchev and D. Rouvray, Eds. Kluwer Academic, New York, 2006. in press.
67. BORGES, J. *Other Inquisitions 1937–52*. Hardback Souvenir Press, 1064.
68. BOWER, J., AND BOLOURI, H., Eds. *Computational Modeling of Genetic and Biochemical Networks*. MIT Press, 2001.
69. BRAMS, S. J., JONES, M. A., AND KLAMER, C. Better ways to cut a cake. *Notices of the AMS 53*, 11 (2006), 1314–1321.
70. BRAMS, S. J., AND TAYLOR, A. D. An envy-free cake division protocol. *The American Mathematical Monthly 102*, 1 (1995), 9–18.
71. BRAMS, S. J., AND TAYLOR, A. D. *Fair Division – From cake-cutting to dispute resolution*. Cambridge University Press, 1996.
72. BREAKSPEAR, M., TERRY, J., FRISTON, K., WILLIAMS, L., BROWN, K., BRENNAN, J., AND GORDON, E. A disturbance of nonlinear interdepen dence in scalp EEG of subjects with first episode schizophrenia. *Neuroimage 20* (2003), 466–478.
73. BRESSLER, S., AND KELSO, J. Cortical coordination dynamics and cognition. *Trends Cogn Sci 5*, 1 (2001), 26–36.
74. BRETHERTON, I. The origins of attachment theory: John Bowlby and Mary Ainsworth. *Developmental Psychology 28* (1992).
75. BROCKMAN, J., Ed. *The Third Culture: Beyond the Scientific Revolution*. Simon and Schuster, 1995.
76. BROCKMAN, J., Ed. *The New Humanists: Science at the Edge*. Barnes & Noble Book, New York, 2003.
77. BROOK, A., AND STAINTON, R. *Knowledge and Mind. A Philosophical Introduction*. Bradford Books. MIT Press, Cambridge, MA and London, 2000.
78. BROOKS, R. *Flesh and Machines: How Robots Will Change Us*. Pantheon Books, New York, 2002.
79. BRUSH, S. Irreversibility and indeterminism: Fourier to Heisenberg. *Journal of the History of Ideas 37* (1976), 603–30.
80. BRUSH, S. *The Kind of Motion we call Heat: A History of the Kinetic Theory of Gases in the 19th Century*. North-Holland, Amsterdam, 1976.
81. BRUSH, S. Nietzsche's recurrence revisited: the French connection. *J. History of Philosophy XIX* (1981), 235–238.
82. BULL, J., AND CHARNOV, E. On irreversible evolution. *Evolution 39* (1985), 1149–1155.
83. BUNGE, M. *Causality and Modern Science*, 3rd ed. Dover Publications, New York, 1979.
84. BUNGE, M. *The Mind–Body Problem*. Pergamon Press, 1980.
85. BUZSAKI, G. *Rhythms of the Brain*. Oxford Univ. Press, 2006.

86. C, G. Introduction to random boolean networks. In *Ninth International Conference on the Simulation and Synthesis of Living Systems (ALife IX)*, M. Bedau, P. Husbands, T. Hutton, S. Kumar, and H. Suzuki, Eds. 2004.
87. CABEZA, C., AND KINGSTONE, A. *Handbook on functional neuroimaging of cognition*. MIT Press., 2001.
88. CAJAL, R. *Histologie du systéme nerveux de l'Homme et des vertébrés*. Maloine, Paris, 1909.
89. CALLAWAY, D., NWEMAN, M., STROGATZ, S., AND WATTS, D. Network robustness and fragility: Percolation on random graphs. *Phys. Rev. Lett. 85* (2000), 5468–5471.
90. CANETTI, E. *Auto-da-Fé*. Farrar, Straus and Giroux, New York, eight edition, 1935/2000.
91. CANNON, W. Organization for physiological homeostasis. *Physiology Reviews 9* (1929), 399–431.
92. CASTETS, V., DULOS, E., BOISSONADE, J., AND DE KEPPER, P. Experimental evidence of a sustained standing turing-type nonequilibrium chemical pattern. *Physics Review Letters 64* (1990), 2953–2956.
93. CASTI, J. Complexity: An introduction.
94. CASTI, J. *Paradigm lost. Tackling the Unanswered Mysteries of Modern Sciences*. Avon Books, New York, 1989.
95. CHANG, W., AND SMYTH, D. The existence and persistence of cycles in a non-linear model: Kaldor's 1940 model re-examined. *The Review of Economic Studies* (1971).
96. CHANGEUX, J., COURREGE, O., AND DANCHIN, A. A theory of epigenesis of neuronal networks by selective stabilization of synapses. *Proc. Natl. Acad. Sci USA 70* (1973).
97. CHANGEUX, J., AND DANCHIN, A. Selective stabilization of developing synapses as a mechanism for the specification of neuronal networks. *Nature 264* (1976).
98. CHANGEUX, J.-P. *Neuronal Man*. Oxford University Press, New York Oxford, 1985.
99. CHURCHLAND, P. *Neurophilosophy: Toward a Unified Science of the Mind-Brain*. The MIT Press, Cambridge, MA, 1986.
100. CLINE, H. Sperry and Hebb: oil and vinegar? *TRENDS in Neurosciences 26* (2003), 655–661.
101. COBB, L. Stochastic catastrophe models and multimodal distributions. *Behavioral Science 23* (1978), 360–373.
102. COLLIER, J., MCINERNEY, D., SSH SS, S., MAINI, P., DJ, G., , HOUSTON, P., AND STERN, C. A cell cycle model for somitogenesis: mathematical formulation and numerical simulation. *Journal of theoretical Biology 207* (2000).
103. COLLINS, J. J., AND STEWART, I. Coupled nonlinear oscillators and the symmetries of animal gaits. *J Nonlinear Sci 3* (1993a), 349–392.
104. COLLINS, J. J., AND STEWART, I. Hexapodal gaits and coupled nonlinear oscillator models. *Biol. Cybernetics 68* (1993b), 287–298.
105. COLLINS, J. J., AND STEWART, I. A group-theoretic approach to rings of coupled biological oscillators. *Biol. Cybernetics 71* (1994), 95–103.
106. COMMISSION, F. T. To promote innovation: The proper balance of competition and patent law and policy, report, October 2003.
107. CONRAD, M. The brain-machine disanalogy. *BioSystems 22* (1989), 197–213.

108. CONRAD, M., AND PATTEE, H. Evolution experiments with an artificial ecosystem. *Journal of Theoretical Biology 28* (1970), 393–409.
109. CONRAD, M., AND RIZKI, M. The artificial worlds approach to emergent evolution. *BioSystems 23* (1989), 247–260.
110. CONWAY, F., AND SIEGELMAN, J. *Dark Hero of the Information Age: In Search of Norbert Wiener, the Father of Cybernetics*. Basic Books, 2004.
111. CORKIN, S. What's new with the amnesic patient H.M.? *Nature Reviews Neuroscience 3*, 2 (2002), 153–60.
112. CORNISH-BOWDEN, A., CÁRDENAS, M., LETELIER, J.-C., SOTO ANDRADE, J., AND GUINEZ ABARZÚA, F. Understanding the parts in terms of the whole. *Biol. Cell 96* (2004).
113. CRAIK, K. *The Nature of Explanation*. Cambridge Univ. Press, 1943.
114. CRICK, F. *Of Molecules and Men*. University of Washington Press, Seattle, 1966.
115. CRICK, F., AND KOCH, C. Towards a neurobiological theory of consciousness. *Seminars Neurosci. 2*, 263-275 (1990).
116. CROOK, S., AND COHEN, A. Central pattern generators. In *The Book of GENESIS: Exploring Realistic Neural Models with the GEneral NEural SImulation System*, J. Bower and D. Beeman, Eds. Springer-Verlag, New York, 1998, pp. 131–147.
117. CRUTCHFIELD, J. P., AND MITCHELL, M. The evolution of emergent computation. *Proceedings of the National Academy of Sciences, USA 92* (1995).
118. CSÁRDI, G., STRANDBURG, K., ZALÁNYI, L., TOBOCHNIK, J., AND ÉRDI, P. Modeling innovation by a kinetic description of the patent citation system. *Physica A 374* (2007), 783–793.
119. DA SILVA, A., AND DE MELO MENDES, V. Value-at-risk and extreme returns in asian stock markets. *Int. J. Business 8*, 1 (2003).
120. DAGOTTO, E. Complexity in strongly correlated electronic systems. *Science 309* (2005), 257–262.
121. DAMASIO, A. *Descartes' Error: Emotion, Reason, and the Human Brain*. Harper Perennial, 1995.
122. DARVAS, G., AND FARKAS, F. An artist's works through the eyes of a physicist: Graphic illustration of particle symmetries. *Leonardo 39* (2006), 51–57.
123. DARWIN, C. *On the Origin of Species by Means of Natural Selection, or the Preservation of Favoured Races in the Struggle for Life*, avenel 1979 ed. Random House Value Publishing, 1859.
124. DAS, A., AND GILBERT, C. Long-range horizontal connections and their role in cortical reorganization revealed by optical recording of cat primary visual cortex. *Nature 375* (1995), 780–784.
125. DAWKINS, R. Sociobiology: the debate continues. *New Scientist* (1985).
126. DAWKINS, R. *The Blind Watchmaker: Why the Evidence of Evolution Reveals a Universe Without Design*. W.W. Norton, 1986.
127. DE ROSNAY, J. *The Macroscope: A New World Scientific System*. Harper & Row, New York, 1979.
128. DE SAUSSURE, F. *Cours de linguistique générale*. Payot, 1916. in French.
129. DENNETT, D. The interpretation of texts, people and other artifacts. *Philosophy and Phenomenological Research. L, Suppl, Fall* (1990).
130. DENNETT, D. *Consciousness Explained*. Penguin Books Ltd., 1991.
131. DITTRICH, P ZIEGLER, J., AND BANZHAF, W. Artificial chemistries-a review. *Artificial Life 7* (2001), 225–275.

132. DITZINGER, T., AND HAKEN, H. Osillations in the erception of mbiguous atterns. *Biol. Cybernetics 61* (1989), 279–287.
133. DITZINGER, T., AND HAKEN, H. The impact of fluctuations on the recognition of ambiguous patterns. *Biol. Cybernetics 63* (1990), 453–456.
134. DOKOUMETZIDIS, A., ILIADIS, A., AND MACHERAS, P. Nonlinear dynamics and chaos theory: concepts and applications relevant to pharmacodynamics. *Pharm. Res. 18* (2001), 415–426.
135. DOYLE, F., AND STELLING, J. Systems interface biology. *J. R. Soc. Interface 3* (2006).
136. DROSSEL, B., AND MCKANE, A. Modelling food webs. In *Handbook of graphs and networks*, S. Bornholdt and H. Schuster, Eds. Wiley-VCH, Berlin, 2002.
137. DUPUY, J. *The Mechanization of the Mind*. Princeton University Press, 2000.
138. EBERSOLE, J. In search of seizure prediction: a critique. *Clin Neurophysiol. 116* (2005), 489–492.
139. ECCLES, J. *The Physiology of Synapses*. Springer Verlag, 1964.
140. ECCLES, J., ITO, M., AND SZENTÁGOTHAI, J. *The Cerebellum as a Neuronal Machine*. Springer Verlag, Berlin, Heildeberg, New York, 1967.
141. ECKHORN, R., BAUER, R., JORDAN, W., BROSCH, M., KRUSE, W., MUNK, M., AND REITBOEK, H. Coherent oscillations: a mechanism of feature linking in the visual cortex? *Biol. Cybern. 60* (1988), 121–130.
142. EDELMAN, G. Group selection and phasic reentrant signalling: A theory of higher brain function. In *The Mindful Brain: Cortical Organization and the Group-Selective Theory of Higher Brain Function*, G. Edelman and V. Mountcastle, Eds. The MIT Press, Cambridge, MA, 1978, pp. 55–100.
143. EDELSTEIN-KESHET, L. *Mathematical Models in Biology*. Random House, New York, 1988.
144. EDWARDS, A. The fundamental theorem of natural selection. *Biol. Rev. 69* (1994).
145. EGUILUZ, V., AND KLEMM, K. Epidemic threshold in structured scale-free networks. *Phys. Rev. Lett. 89* (2002), 108701.
146. EIGEN, M. Selforganisation of matter and the evolution of biological macromolecules. *Naturwiss. 58* (1971), 466–565.
147. EIGEN, M., AND SCHUSTER, P. *The Hypercycle: A principle of natural self-organization*. Springer Verlag, Heidelberg, 1979.
148. EINSTEIN, A. über die von der molekularkinetischen theorie der wärme geforderte bewegung von in ruhenden flüssigkeiten suspensierten teilchen. *Ann. Phys. (Leipzig) 17* (1905), 549–560.
149. ELIADE, M. *The Myth of the Eternal Return: Or, Cosmos and History*. Princeton University Press, 1954.
150. ELLNERE, S., AND GUCKENHEIMER, J. *Dynamics Models in Biology*. Princeton Univ. Press, Princeton and London, 2006.
151. ELMAN, J., BATES, E. ANDJOHNSON, M., KARMILOFF-SMITH, A., PARISI, D., AND PLUNKETT, K. *Rethinking innateness: A connectionist perspective on development*. MIT Press/Bradford Books., 1996.
152. EPSTEIN, I., AND POJMA, J. *An Introduction to Nonlinear Chemical Dynamics: Oscillations, Waves, Patterns, and Chaos*. Oxford University Press, 1988.
153. EPSTEIN, J. *Nonlinear Dynamics, Mathematical Biology, and Social Science*. Santa Fe Inst. Studies in Sciences of Complexity. Addison-Wesley, 1997.
154. EPSTEIN, J. *Generative Social Science: Studies in Agent-Based Computational Modeling*. Princeton University Press, Princeton, 2006.

155. EPSTEIN, J., AND AXTELL, R. *Growing Artificial Societies: Social Science From The Bottom Up*. Brookings Institute Press, 1996.
156. ERDŐS, P., AND RÉNYI, A. On the random graphs. *Publ. Institute of Mathematics University of Debrecen 6* (1959).
157. ERDŐS, P., AND RÉNYI, A. On the evolution of random graphs. *Institute of Mathematics, Hungarian Academy of Sciences 5* (1960).
158. ÉRDI, P. The brain as a hermeneutic device. *Biosystems 38* (1996), 179–189.
159. ÉRDI, P. Neural and mental development: selectionism, constructivism, hermeneutics. In *Neuronal bases and psychological aspects of consciousness*, C. Taddei-Ferrati and C. Musio, Eds. World Scientific, Singapore, 1999, pp. 507–518.
160. ÉRDI, P., AND BARNA, G. Self-organizing mechanism for the formation of ordered neural mappings. *Biological Cybernetics 51* (1984), 93–101.
161. ÉRDI, P., AND KISS, T. The complexity of the brain: Structural, functional and dynamic modules. In *Emergent Neural Computational Architectures based on Neuroscience*, S. Wermter, J. Austin, and D. Willshaw, Eds. Springer Verlag, Heidelberg, 2000.
162. ÉRDI, P., KISS, T., TÓTH, J., UJFALUSSY, B., AND ZALÁNYI, L. From systems biology to dynamical neuropharmacology: Proposal for a new methodology. *IEE Proceedings in Systems Biology 153*, 4 (2006).
163. ÉRDI, P., AND SOMOGYVÁRI, Z. Post-hebbian learning algorithms. In *The Handbook of Brain Theory and Neural Networks*, 2nd ed. MIT Press, 2002, pp. 898–901.
164. ÉRDI, P., AND TÓTH, J. *Mathematical Models of Chemical Reactions. Theory and applications of deterministic and stochastic models*. Manchester Univ. Press., Princeton Univ. Press., 1989.
165. ÉRDI, P., AND TSUDA, I. Hermeneutic approach to the brain: process versus device? *Theoria et Historia Scientiarum VI* (2002), 307–321.
166. ERMENTROUT, G. B. The behavior of rings of coupled oscillators. *J. Math. Biol. 23* (1985), 55–74.
167. ERMENTROUT, G. B., AND KOPELL, N. Frequency plateaus in a chain of weakly coupled oscillators. *SIAM J. Math Anal. 15* (1984), 215–237.
168. ETKIN, M., PITTENGER, C., POLAN, J., AND KANDEL, E. Toward a neurobiology of psychotherapy: Basic science and clinical applications. *J Neuropsychiatry Clin Neurosci 17* (2005).
169. FALOUTSOS, M., FALOUTSOS, P., AND FALOUTSOS, C. On power-law relationships of the internet topology. *Computer Communications Review 29* (1999), 251–262.
170. FARKAS, F., AND ÉRDI, P. 'Impossible' forms: Experimental graphics and theoretical associations. *Leonardo 18* (1985), 179–183.
171. FEINBERG, M. Mathematical aspects of mass action kinetics. In *Chemical Reactor Theory: A Review*, L. Lapidus and N. Amundson, Eds. Prentice-Hall, Englewood Cliffs, 1970.
172. FEINBERG, M. On chemical kinetics of a certain class. *Arch. Rational Mech. Anal. 46* (1972), 1–41.
173. FELDMAN, D., AND CRUTCHFIELD, J. Measures of statistical complexity: Why. *Physics Letters A 238* (1997), 244–252.
174. FINGELKURTS, A., AND FINGELKURTS, A. Making complexity simpler: multivariability and metastability in the brain. *Int J Neurosci 114*, 7 (2004), 843–62.

175. FLAUBERT, G. *Bouvard and Pecuchet.* Piguine books, 1976.
176. FLEISCHER, J., GALLY, J., EDELMAN, G., AND KRICHMAR, J. Retrospective and prospective responses arising in a modeled hippocampus during maze navigation by a brain-based device. *Proc Natl Acad Sci U S A 104* (2007).
177. FLOREANO, D., SUZUKI, M., AND MATTIUSSI, C. Active vision and receptive field development in evolutionary robots. *Evolutionary Computation 13* (2005).
178. FOGEL, D. *Evoloutionary Computation. Towards a new philosphy of machine intelligence.* IEEE Inc. New York, 1995.
179. FÖLDY, C., SOMOGYVÁRI, Z., AND ÉRDI, P. Hierarchically organized minority games. *Physica A 323* (2003), 735–742.
180. FONAGY, P., AND TARGET, M. ??!! *Journal of the American Psychoanalytic Association 55* (2007).
181. FOSTER, D., AND PEYTON, Y. Stochastic evolutionary game dynamics. *Theoret. Population Biol. 38* (1990).
182. FRANK, S., AND SLATKIN, M. Fishers fundamental theorem of natural selection. *Trends in Ecology and Evolution 7* (1992).
183. FREEMAN, W., KOZMA, R., AND WERBOS, P. Biocomplexity: Adaptive behavior in complex stochastic dynamical systems. *BioSystems 59* (2001), 109–123.
184. FREUND, T. Interneuron diversity series: Rhythm and mood in perisomatic inhibition. *Trends in Neurosciences 28*, 9 (September 2003), 489–495.
185. FRIEDMAN, T. *The World Is Flat: A Brief History of the Twenty-first Century.* Farrar, Straus, Reese, and Giroux, 2005, 2005.
186. FRISTON, K. Transients, metastability, and neuronal dynamics. *Neuroimage 5* (1997), 164–71.
187. FRISTON, K. The labile brain. II. transients, complexity and selection. *Philos Trans R Soc Lond B Biol Sci 355* (2000), 237–52.
188. FRISTON, K., HARRISON, L., AND PENNY, W. Dynamic causal modelling. *NeuroImage 19* (2003), 1273–1302.
189. FUCHS, T. Neurobiology and psychotherapy: an emerging dialogue. history and philosophy. *Current Opinion in Psychiatry 17* (2004).
190. FUKUYAMA, F. *The End of History and the Last Man.* Free Press, 1992.
191. GALBRAITH, J. *A Short History of Financial Euphoria.* Whittle, 1990.
192. GALISON, P. *Eintsein's clock, Poincaré's map. Empires of Time.* W. W. Norton Co., New York, London, 2003.
193. GARDNER, M. Critical degenerateness in linear systems. Tech. Rep. Tech. Rep. No. 5.8, Biological Computer Laboratory, University of Illinois, Urbana,Illinois 61801, 1968.
194. GARDNER, M., AND ASHBY, W. Connectance of large, dynamic cybernetic systems: Critical values for stability. *Nature* (1970).
195. GASPARD, P. Rössler systems. In *Encyclopedia of Nonlinear Science*, A. Scott, Ed. Routledge, New York, 2005, pp. 808–811.
196. GELDER, T. J. The dynamical hypothesis in cognitive science. *Behavioral and Brain Sciences 21* (1998)).
197. GELL-MANN, M. *The Quark and the Jaguar: Adventures in the Simple and the Complex.* Owl Books, 1995.
198. GERITZ, S., KISDI, E., MESZÉNA, G., AND METZ, J. Evolutionarily singular strategies and the adaptive growth and branching of the evolutionary tree. *Evol. Ecol. 12* (1998).
199. GERSTEIN, G., AND MANDELBROT, B. Random walk models for the spike activity of a single neuron. *Biophys J. 71* (1964), 41–68.

200. GIBSON, J. *The Senses Considered as Perceptual Systems.* Houghton Mifflin, Bosten, MA, 1966.
201. GILBERT, C. Circuitry, architecture, and functional dynamics of visual cortex. *Cereb Cortex 3*, 5 (1993), 373–86.
202. GILBERT, C., AND WIESEL, T. Clustered intrinsic connections in cat visual cortex. *J Neurosci 3*, 5 (1983), 1116–33.
203. GILBERT, C. D. Horizontal integration in the neocortex. *Trends Neurosci. 8* (1985), 160–165.
204. GILLESPIE, D. *Markov Processes. (An introduction to physical scientists.).* Academic Press, 1992.
205. GILLET, V., AND WILLETT, P. Chemoinformatics techniques for processing chemical structure databases. In *Computer Applications in Pharmaceutical Research and Development*, S. Ekins, Ed. John Wiley and Sons, 2006.
206. GLUZMAN, S., AND SORNETTE, D. Log-periodic route to fractal functions. *Phys. Rev. E 65* (2002), 036142.
207. GOLDENFELD, N., AND KADANOFF, L. Simple lessons from complexity. *Science 284* (1999), 87–89.
208. GOODHILL, G., AND CARREIRA-PERPINAN, M. Cortical columns. *Encyclopedia of Cognitive Science 1* (2002), 845–851.
209. GOODWIN, M. A growth cycle. In *Socialism, Capitalism and Economic Growth*, C. Feinstein, Ed. Cambridge University Press, 1967.
210. GOTO, S., OKUNO, Y., HATTORI, M., NISHIOKA, T., AND KANEHISA, M. LIGAND: database of chemical compounds and reactions in biological pathways. *Nucleic Acids Res 30* (2002), 402–404.
211. GOULD, S. Dollo on Dollo's law: Irreversibility and the status of evolutionary laws. *Journal of the History of Biology 3* (1970), 189–212.
212. GOULD, S. *Ontogeny and Phylogeny.* Harvard Univ. Press, Cambridge MA, 1977.
213. GRAFEN, A. Fisher the evolutionary biologist. *Journal of the Royal Statistical Society: Series D (The Statistician) 52* (2003).
214. GRANOVETTER, M. The strength of weak ties. *American Journal of Sociology 78* (1973), 1360–1380.
215. GRANOVETTER, M. The strength of weak ties: A network theory revisited. *Sociological Theory 1* (1983), 201–233.
216. GRASSBERGER, P. On the critical behavior of the general epidemic process and dynamical percolation. *Math. Biosci. 63* (1983), 157–172.
217. GRASSBERGER, P. Toward a quantitative theory of selfgenerated complexity. *Intl. J. Theo. Phys. 25*, 9 (1986), 907–938.
218. GRAY, B., AND ROBERTS, M. Analysis of chemical kinetic systems over the entire parameter space i. the Sal'nikov thermokinetic oscillator. *Proceedings of the Royal Society of London. Series A, Mathematical and Physical Sciences 416*, 1851 (1988), 391–402.
219. GRAY, C., AND SINGER, W. Stimulus-specific neuronal oscillations in orientation columns of cat visual cortex. *Proc. Nat. Acad. Sci. USA 86* (1989), 1698–1702.
220. GRAY, C., AND VIANA DI PRISCO, G. Properties of stimulus-dependent rhythmic activity of the visual cortical neurons in the alert cat. Tech. Rep. 16, Soc. Neurosci. Abs., 1993.
221. GRAY, J. A., AND MCNAUGHTON, N. *The neuropsychology of anxiety*, second ed. Oxford Psychology Series. Oxford University Press, 2000.

222. GRÖBLER, T., BARNA, G., AND ÉRDI, P. Statistical model of the hippocampal CA3 region. I. The single-cell module: bursting model of the pyramidal cell. *Biol Cybern 79*, 4 (1998), 301–8.
223. GROSSMAN, J. The evolution of the mathematical research collaboration graph. *Congressus Numerantium 158* (2002).
224. GUTH, A. A golden age of cosmology. In *The new humanists. Science at the edge.*, J. Brockman, Ed. Barnes and Noble Books, New York, 2003.
225. H-G, G. *Truth and Method*. Sheed and Ward, London, 1976.
226. HAJÓS, M., HOFFMANN, W. E., ORBÁN, G., KISS, T., AND ÉRDI, P. Modulation of septo-hippocampal θ activity by $GABA_A$ receptors: An experimental and computational approach. *Neuroscience 126*, 3 (2004), 599–610.
227. HAKEN, H. Synergetics: Formation of ordered structures out of chaos. *Leonardo 15* (1971), 66–67.
228. HAKEN, H. *Synergetics: An Introduction*, vol. 1 of *Series in Synergetics*. Springer, Berlin, 1977.
229. HAKEN, H., KELSO, J., AND BUNZ, H. A theoretical model of phase transitions in human hand movements. *Biol Cybern 51*, 5 (1985), 347–56.
230. HALL, B. Exploring the patent explosion. *Journal of Technology Transfer 30* (2005), 35–48.
231. HALL, B., JAFFE, A., AND TRAJTENBERG, M. The NBER patent citation data file: Lessons, insights and methodological tools. In *Patents, Citations, and Innovations: A Window on the Knowledge Economy*, A. Jaffe and M. Trajtenberg, Eds. MIT Press, 2003.
232. HARTWELL, L., HOPFIELD, J., LEIBLER, S., AND MURRAY, A. From molecular to modular cell biology. *Nature 402* (1999), C47–C52.
233. HEBB, D. *The organization of the behavior*. New York: Wiley, 1949.
234. HEGSELMANN, R., AND KRAUSE, U. Opinion dynamics and bounded confidence: models, analysis and simulation. *Journal of Artificial Societies and Social Simulation 5*, 30 (2002), paper 2.
235. HESSE, H. *The Glass Bead Game*. Holt, Rinehart and Winston, 1943/1969.
236. HEYLIGHEN, F., AND JOSLYN, C. Cybernetics ans second-order cyberentics. In *Encyclopedia of Physical Science & Technology*, R. Meyers, Ed., 3rd ed. Academic Press, New York, 2001.
237. HODGKIN, A., AND HUXLEY, A. A quantitative description of membrane current and its application to conduction and excitation in nerve. *J. Physiol. 117* (1952), 500–544.
238. HOFSTADTER, D. *Gödel, Escher, Bach: an Eternal Golden Braid*. Penguin Books, 1979.
239. HOLLAND, J. *Adaptation in Natural and Artificial Systems*. University of Michigan Press, Ann Arbor, 1975.
240. HOLYST, J., AND URBANOWICZ, K. Chaos control in economical model by time-delayed feedback method. *Physica A 287* (2000).
241. HORN, F. Necessary and sufficient conditions for complex balancing in chemical kinetics. *Arch. Rational Mech. Anal. 49* (1972), 172–186.
242. HRASKÓ, P. Quantum theory, holism, reductionism (in Hungarian). In *Theory Model Tradition (in Hungarian)*, P. Érdi and J. Tóth, Eds. MTA KFKI, Budapest, 1992.
243. HUBEL, D., AND WIESEL, T. Ferrier lecture. Functional architecture of macaque monkey visual cortex. *Proc R Soc Lond B Biol Sci 198*, 1130 (1977), 1–59.

244. HUBERMAN, B., AND ADAMIC, L. Evolutionary dynamics of the world wide web. Tech. rep., Xerox Palo Alto Research Center, 1999.
245. HUBERMAN, B., AND HOGG, T. Complexity and adaptation. *Physica D 22* (1986), 376–384.
246. HUBERMAN, B., AND LUMER, E. Dynamics of adaptive systems. *IEEE Transactions on Circuits and Systems 37* (1990), 547–550.
247. HUHN, Z., ORBÁN, G., LENGYEL, M., AND ÉRDI, P. Dendritic spiking accounts for rate and phase coding in a biophysical model of a hippocampal place cell. *Neurocomputing 15* (2005), 250–262.
248. IASEMIDIS, L., SHIAU, D., PARDALOS, P., CHAOVALITWONGSE, W., NARAYANAN, K., PRASAD, A., TSAKALIS, K., CARNEY, P., AND SACKELLARES, J. Long-term prospective on-line real-time seizure prediction. *Clin. Neurophysiol. 116* (2005), 32–44.
249. IBERALL, A. A field and circuit thermodynamics for integrate physiology i. information to the general notions. *Am. J. Physiol. 233* (1978), 171–180.
250. IKEGAMI, T., AND KANEKO, K. Evolution of host-parasitoid through homeochaotic dynamics. *Chaos 2* (1992), 397–407.
251. IZHIKEVICH, E. *The Geometry of Excitability and Bursting.* The MIT Press, Cambridge, Mass., 2007.
252. JACKSON, E. Controls of dynamic flows with attractors. *Phys. Rev. A 44* (1991), 4839–4853.
253. JACKSON, E. *Exploring Nature's Dynamics.* Wiley Series in Nonlinear Science. John Wiley and Sons, 2001.
254. JACKSON, E., AND KODOGEORGIU, A. Entrainment and migration controls of two-dimensional maps. *Physica D 54* (1991), 253–265.
255. JACOB, F., AND MONOD, J. Genetic regulatory mechanisms in the synthesis of proteins. *Journal of Molecular Biology 3* (1961), 318–356.
256. JAFFE, A., AND LERNER, J. *Innovation and Its Discontents: How Our Broken Patent System is Endangering Innovation and Progress, and What to Do About It.* Princeton University Press, 2004.
257. JEONG, H., MASON, S., BARABÁSI, A., AND OLTVAI, Z. Lethality and centrality in protein networks. *Nature 411* (2001), 41–42.
258. JEONG, H., NÉDA, Z., AND BARABÁSI, A.-L. Measuring preferential attachment in evolving networks. *Europhys. Lett. 61* (2003), 567–572.
259. JEONG, J. EEG dynamics in patients with Alzheimer's disease. *Clin. Neurophysiol. 115* (2004), 1490–1505.
260. JIRSA, V. Connectivity and dynamics of neural information processing. *Neuroinformatics 2* (2004).
261. JIRSA, V., AND DING, M. Will a large complex system with time delays be stable? *Phys Rev Lett 93*, 7 (2004), 070602.
262. JIRSA, V., FUCHS, A., AND KELSO, J. Connecting cortical and behavioral dynamics: bimanual coordination. *Neural Comput 10*, 8 (1998), 2019–45.
263. KAMPIS, G. *Self-Modifying Systems in Biology and Cognitive Science. A new framewrok to Dynamics, Information and Complexity.* Pergamon Press, Oxford, 1991.
264. KANDEL, E. Biology and the future of psychoanalysis: A new intellectual framework for psychiatry revisited. *Am. J. Psychiatry 156* (1999).
265. KANEKO, K. Clustering, coding, switching, hierarchical ordering, and control in network of chaotic elements. *Physica D 41* (1990), 137–172.

266. KANEKO, K. Dominance of milnor attractors and noise-induced selection in a multiattractor system. *Phys. Rev. Lett. 78* (1997), 2736–2739.
267. KANEKO, K., AND TSUDA, I. *Complex systems: chaos and beyond a constructive approach with applications in life sciences.* Springer, New York, 2001.
268. KANEKO, K., AND TSUDA, I. Chaotic itinerancy. *Chaos 13* (2003), 926–936.
269. KAUFFMAN, S. Metabolic stability and epigenesis in randomly connected nets. *Journal of Theoretical Biology 22* (1969).
270. KAUFFMAN, S. *Origins of Order: Self-Organization and Selection in Evolution.* Oxford University Press, 1993.
271. KAUFFMAN, S. Complexity and genetic networks, 2003. Exystence Project News.
272. KELLY, G. *The psychology of personal constructs.* Norton, New York, 1955.
273. KELSO, J. *Dynamic patterns. The self-organization of brain and behavior.* A Bradford Book, The MIT Press, 1995.
274. KELSO, J., FUCHS, A., LANCASTER, R., HOLROYD, T., CHEYNE, D., AND WEINBERG, H. Dynamic cortical activity in the human brain reveals motor equivalence. *Nature 392*, 6678 (1998), 814–8.
275. KELSO, J., AND ZANONE, P. Coordination dynamics of learning and transfer across different effector systems. *J Exp Psychol Hum Percept Perform 28*, 4 (2002), 776–97.
276. KESTEVEN, M. On the mathematical theory of clock escapements. *Am. J. Physics 46* (1978), 125–129.
277. KHEOWAN, P., KANTRASIRI, S., UTHAISAR, C., GÁSPÁR, V., AND MÜLLER, S. Spiral wave dynamics controlled by a square-shaped sensory domain. *Chem. Phys. Lett. 389* (2004).
278. KINNEY, G., PATINO, P., MERMET-BOUVIER, Y., STARRETT, JR, J., AND GRIBKOFF, V. Cognition-enhancing drugs increase stimulated hippocampal theta rhythm amplitude in the urethane-anesthetized rat. *J Pharmacol Exp Ther 291*, 1 (1999), 99–106.
279. KIRK, C., AND RAVEN, J. *The Presocratic Philosophers.* Cambridge University Press, 1957.
280. KISS, T., AND ÉRDI, P. From electric patterns to drugs: perspectives of computational neuroscience in drug design. *BioSystems 86* (2006), 46–52.
281. KISS, T., ORBÁN, G., AND ÉRDI, P. Modelling hippocampal theta oscillation: applications in neuropharmacology and robot navigation. *International Journal of Intelligent Systems 21*, 9 (2006), 903–917.
282. KISS, T., ORBÁN, G., AND P, E. Modelling hippocampal theta oscillation: applications in neuropharmacology and robot navigation. *International Journal of Intelligent Systems 21* (2006), 903–917.
283. KITANO, H. Systems biology: A brief overview. *Science 295* (2002).
284. KLIMESCH, W. EEG alpha and theta oscillations reflect cognitive and memory performance: a review and analysis. *Brain Res Brain Res Rev 29*, 2-3 (1999), 169–95.
285. KOESTLER, A. *The sleepwalkers: a history of man's changing vision of the universe.* Hutchinson, London, 1959. with an introd. by Herbert Butterfield.
286. KOHLSDORF, T., AND WAGNER, G. Evidence for the reversibility of digit loss: a phylogenetic study of limb evolution in the genus bachia (gymnophthalamidae: squamata). *Evolution 60*, 9 (2006), 1896–1912.

287. KOPELL, N. Toward a theory of modeling central pattern generators. In *Neural control of rhythmic movements in vertebrates*, A. Cohen, S. Rossignol, and S. Grillner, Eds. Wiley: New York, 1988.
288. KOPELL, N., AND ERMENTROUT, G. Phase transition and other phenomena in chains of coupled oscillators. *SIAM J.Appl. Math. 50* (1990), 1014–1052.
289. KOTSIS, G. Interconnection topologies for parallel processing systems. *PARS Mitteilungen 11* (1993), 1–6.
290. KOTTER, R. Online retrieval, processing, and visualization of primate connectivity data from the cocomac database. *Neuroinformatics 2* (2004).
291. KOUTSOYIANNIS, D. On the appropriateness of the gumbel distribution in modelling extreme rainfall. In *Hydrological Risk: recent advances in peak river flow modelling, prediction and real-time forecasting. Assessment of the impacts of land-use and climate changes*, A. Brath, A. Montanari, and E. Toth, Eds. Editoriale Bios, Castrolibero, Italy, 2004, pp. 303–319.
292. KOZA, J. *Genetic Programming: On the Programming of Computers by Means of Natural Selection.* MIT Press, 1992.
293. KOZA, J. *Genetic Programming II: Automatic Discovery of Reusable Programs.* MIT Press, 1994.
294. KOZA, J., BENNETT, F. I., ANDRE, D., AND KEANE, M. *Genetic Programming III: Darwinian Invention and Problem Solving.* Morgan Kaufmann, 1999.
295. KOZA, J., KEANE, M., STREETER, M., MYDLOWEC, W., YU, J., AND LANZA, G. *Genetic Programming IV: Routine Human-Competitive Machine Intelligence.* Kluwer Academic Publishers, 2003.
296. KOZMA, R., PULJIC, M., BALISTER, P., BOLLOBÁS, B., AND FREEMAN, W. Phase transitions in the neuropercolation model of neural populations with mixed local and non-local interactions. *Biological Cybernetics 92* (2005), 367–379.
297. KRAKAUER, D. Robustness in biological systems – a provisional taxonomy. Santa Fe Institute Working Paper 03-02-008, 2003. Also in: Complex Systems Science in Biomedicine. Kluwer, 2005.
298. KRAKAUER, D., AND PLOTKIN, J. Principles and parameters of molecular robustness. In *Robust design: a repertoire for Biology, Ecology and Engineering*, E. Jen, Ed. Oxford University Press, 2005.
299. KRAWIEC, A., AND SZYDLOWSKI, M. The Kaldor-Kalecki model of business cycle as a two-dimensional dynamical system. *Journal of Nonlinear Mathematical Physics* (2001).
300. KREITER, A., AND SINGER, W. Oscillatory neuronal responses in the visual cortex of the awake macaque monkey. *Eur. J. Neurosci. 4* (1992), 369–375.
301. KRUSE, P., CARMESIN, H.-O., PAHLKE, L., STRÜBER, L., AND STADLER, M. Continuous phase transitions in the perception of multistable visual patterns. *Biological Cybernetics 75* (1996), 321–330.
302. KURAMOTO, Y. *Chemical oscillations, waves, and turbulence.* Springer, Berlin, 1984.
303. LAKATOS, K., TOTH, I., NEMODA, Z., NEY, K., SASVARI-SZEKELY, M., AND GERVAI, J. Dopamine D4 receptor (DRD4) gene polymorphism is associated with attachment disorganization in infants. *Molecular Psychiatry* (2000).
304. LANDES, D. *Revolution in Time: Clocks and the Making of the Modern World.* Belknap Press of Harvard Univ. Press, Cambridge, MA, 1983.
305. LAUGHLIN, R., AND PINES, D. The theory of everything and its critique. *PNAS 97* (2000), 28–31.

306. LeBaron, B. Agent-based computational finance. In *Handbook of Computational Economics, Vol. 2: Agent-Based Computational Economics*, L. Tesfatsion and K. Judd, Eds. North-Holland, 2006.
307. LeBaron, L., Arthur, B., and Palmer, R. Time series properties of an artificial stock market model. *Journal of Economic Dynamics and Control 23* (1999).
308. Lengyel, I., and Epstein, I. Modeling of Turing structures in the chlorite-iodide-malonic acid-starch reaction system. *Science 251* (1991), 650–652.
309. Lengyel, M., Szatmáry, Z., and Érdi, P. Dynamically detuned oscillations account for the coupled rate and temporal code of place cell firing. *Hippocampus 13* (2003), 700–714.
310. Lenski, R., Ofria, C., Pennock, R., and Adami, C. The evolutionary origin of complex features. *Nature 423* (2003).
311. Lepschy, A., Mian, G., and Viaro, U. Feedback control in ancient water and mechanical clocks. *IEEE Trans. Education 35* (1992), 3–10.
312. Letelier, J.-C., Soto-Andrade, J., Guinez Abarzúa, F., and Cornish-Bowden, A Cárdenas, M. Organizational invariance and metabolic closure: analysis in terms of (M,R) systems. *J. theor. Biol. 238* (2006).
313. Lévi-Strauss, C. *Anthropologie structurale (Structural Anthropology, trans.1968*. Plon, Paris, transl: Allen Lane, London, 1958.
314. Li, X., Ouyang, G., Yao, X., and Guan, X. Dynamical characteristics of pre-epileptic seizures in rats with recurrence quantification analysis. *Physics Letters A 333* (2004).
315. Lieberman, E., Hauert, C., and Nowak, M. Evolutionary dynamics on graphs. *Nature 2005* (2005).
316. Liljenström, H. Neural stability and flexibility: a computational approach. *Neuropsychopharmacology 28* (2003), 564–573.
317. Liljeros, F., Edling, C., Amaral, L., He, S., and Åberg, Y. The web of human sexual contacts. *Nature 411* (2001), 907–908.
318. Linde, A. Inflationary theory versus ekpyrotic/cyclic scenario. Presented at Stephen Hawking's 60th birthday conference, 2002.
319. Lisman, J. Bursts as a unit of neural information: making unreliable synapses reliable. *Trends in Neuroscience 20* (1997), 38–43.
320. Llinas, R. The intrinsic electrophysiological properties of mammalian neurons: insights into central nervous system function. *Science 242* (1988), 1654–1664.
321. Lopez da Silva, F., Blanes, W., Kalitzin, S., Parra, J., Suffczynski, P., and Velis, D. Epilepsies as dynamical diseases of brain systems: basic models of the transition between normal and epileptic activity. *Epilepsia 44 (Suppl. 12)* (2003), 72–83.
322. Loskutov, A. Dynamics control of chaotic systems by parametric destochastization. *J. Phys. A: Math. Gen. 26* (1993), 4581–4594.
323. Luria, A. *The mind of a mnemonist: A little book about a vast memory. Lynn Solotaroff.* Basic Books, New York, 1968.
324. Lyons-Ruth, K., and members of the Change Process Study Group. Implicit relational knowing: its role in development and psychoanalytic treatment. *Infant Mental Health Journal 19* (1998).
325. Ma, N. Five rules for the evolution of cooperation. *Science 314* (2006).
326. Mallery, J., Hurwitz, R., and Duffy, G. Hermeneutics: From textual explication to computer understanding? In *The Encyclopedia of Artificial Intelligence*, S. Shapiro, Ed. Wiley & Sons, New York, 1987.

327. MANTEGNA, R., AND STANLEY, H. Scaling behaviour in the dynamics of an economic index. *Nature 376* (1995), 46–49.
328. MARCUS, M. Book review. dark hero of information age: In search of norbert wiener, the father of cybernetics. *Notices of the American Mathematical Society 53* (2006), 574–579.
329. MARKRAM, H., LUBKE, J., FROTSCHER, M., ROTH, A., AND SAKMANN, B. Physiology and anatomy of synaptic connections between thick tufted pyramidal neurones in the developing rat neocortex. *J Physiol 500 (Pt 2)* (1997), 409–40.
330. MARX, K. *Das Kapital. Kritik der politischen Oekonomie. Band I*. Meissner, Hamburg, 1867.
331. MATURANA, H., AND VARELA, F. *Autopoiesis and Cognition*. Reidel, Dordrecht, 1980.
332. MAY, R. Will a large complex system be stable? *Nature* (1972).
333. MAY, R. *Stability and complexity in model ecosystems*. Princeton University Press, Princeton, NJ, 1973.
334. MAY, R. Simple mathematics models with very complicated dynamics. *Nature 61* (1976), 459–67.
335. MAYR, O. *The Origins of Feedback Control*. MIT Press, Cambridge, 1970.
336. MAYR, O. Maxwell and the origins of cybernetics. *Isis 62* (1971), 425–444.
337. MCCANN, K. The diversity-stability debate. *Nature 405* (2000), 228–233.
338. MCCANN, K., HASTINGS, A., AND HUXEL, G. Weak trophic interactions and the balance of nature. *Nature 395* (1998), 794–798.
339. MCCULLOCH, W. Agatha tyche: Of nervous nets: The lucky reckoners. In *Mechanization of Thought Processes*. Stationery Office, London, 1959, pp. 611–25.
340. MCCULLOCH, W. What is a number that a man may know it, and a man that he may know a number? *Gen. Sem. Bul. 26–27* (1961), 2–18. also in: McCulloch WS (1965); Embodiments of Mind. MIT Press.
341. MCCULLOCH, W. *Embodiments of Mind*. MIT Press, 1965.
342. MCCULLOCH, W., AND PITTS, W. A logical calculus of the ideas immanent in nervous system. *Bulletin of Mathematical Biophysics 5* (1943), 115–133.
343. MCCULLOCH, W. S., AND PITTS, W. H. A logical calculus of the ideas immanent in nervous activity. *Bull. Math. Biophys.* (1943).
344. MCNAUGHTON, N., AND GRAY, J. A. Anxiolytic action on the behavioural inhibition system implies multiple types of arousal contribute to anxiety. *Journal of Affective Disorders 61* (2000), 161–176.
345. MEDAWAR, P. The evolution of a proof, Nov 1973.
346. MEINHARDT, H. *Models of biological pattern formation*. Academic Press, London, 1982.
347. MEINHARDT, H. A boundary model for pattern formation in vertebrate limbs. *J Embryol Exp Morphol. 76* (1983), 115–137.
348. MEINHARDT, H. Models of segmentation. In *Somites in developing embryos*, H. Bellairs, A. Edie, and J. Lash, Eds. Plenum Press, New York, 1986.
349. MEINHARDT, H., AND GIERER, A. Applications of a theory of biological pattern formation based on lateral inhibition. *J Cell Sci. 15* (1974), 321–46.
350. MERRILL, S., LEVIN, R., AND MB, M., Eds. *A Patent System for the 21st Century, National Research Council of the National Academies*. National Academies Press, 2004.

351. MERTON, R. *The Sociology of Science.* University of Chicago Press, Chicago, 1973.
352. MILNOR, J. On the concept of attractor. *Comm.Math. Phys. 99* (1985). Correction and remarks ,ibid 102 (1985) 517-519.
353. MINSKY, M., AND PAPERT, S. *Perceptrons: An Introduction to Computational Geometry*, expanded ed. MIT Press, Cambridge, MA, 1988.
354. MITCHELL, M. A complex-systems perspective on the "computation vs. dynamics" debate in cognitive science. In *Proceedings of the 20th Annual Conference of the Cognitive Science Society;Cogsci98* (1998), G. MA and S. Derry, Eds.
355. MITCHELL, M., HRABER, P., AND CRUTCHFIELD, J. Revisiting the edge of chaos: Evolving cellular automata to perform computations. *Complexity 7* (1993).
356. MITZENMACHER, M. A brief history of generative models for power law and lognormal distributions. *Internet Math 1* (2003), 226–251.
357. MONOD, J., AND JACOB, F. Telenomic mechanisms in cellular metabolism, growth and differentiation. *Cold Springer Harbor Symp. Quant. Biol. 26* (1961), 389–401.
358. MONTROLL, E., AND SHLESINGER, M. On 1/f noise and other distributions with long tails. *Proc. Natl. Acad. Sci. USA 79* (1982).
359. MOON, F., AND STIEFEL, P. Coexisting chaotic and periodic dynamics in clock escapements. *Philosophical Transactions of the Royal Society A 364* (2006), 2539 – 2563.
360. MORANGE, M. *A History of Molecular Biology.* Harvard University Press, Cambridge, MA, 1998. Trans. By Matthew Cobb.
361. MORANGE, M. What history tell us. *J. Biosci. 30* (2005), 313–316.
362. MORMANN, F., KREUZ, T., RIEKE, C., ANDRZEJAK, R., AND KRASKOV, A. On the predictability of epileptic seizures. *Clin. Neurophysiol. 116* (2005), 569–587.
363. MORMANN, F., KREUZ, T., RIEKE, C., ANDRZEJAK, R., KRASKOV, A., DAVID, P., ELGER, C., AND LEHNERTZ, K. On the predictability of epileptic seizures. *Clin. Neurophysiol. 116* (2005), 569–587.
364. MOTOIKE, I., AND YOSHIKAWA, K. Information operations with an excitable field. *Phys.Rev E 59* (1999).
365. MOUNTCASTLE, V. Modality and topographic properties of single neurons of cat's somatic sensory cortex. *J Neurophysiol 20*, 4 (1957), 408–34.
366. MULLER, V., LUTZENBERGER, W., PULVERMULLER, F., MOHR, B., AND BIRBAUMER, N. Investigation of brain dynamics in Parkinson's disease by methods derived from nonlinear dynamics. *Exp. Brain Res. 137* (2001), 103–110.
367. MUMFORD, L. *Technics and Civilization.* Harcourt, New York, 1934.
368. MURRAY, J. *Mathematical Biology.* Springer Verlag, 1989.
369. MURTHY, V., AND FETZ, E. Coherent 25-35Hz oscillations in the sensorimotor cortex of the awake behaving monkey. *Proc. Natl. Acad. Sci. 89*, 5670-5674 (1992).
370. NANDRINO, J., DODIN, V., MARTIN, P., AND HENNIAUX, M. Emotional information processing in first and recurrent major depressive episodes. *J. Psychiatr. Res. 38* (2004), 475–484.
371. NEEDHAM, J. The missing link in horological history: A Chinese contribution. *Proceedings of the Royal Society of London. Series A 250* (1959), 147–179.
372. NEEDHAM, J., WANG, L., AND DE SOLLA PRICE, D. *Heavenly Clockwork: The Great Astronomical Clocks of Medieval China.* Cambridge Univ. Press, 1960.

373. NÉGYESSY, L., NEPUSZ, T., KOCSIS, L., AND BAZSÓ, F. Prediction of the main cortical areas and connections involved in the tactile function of the visual cortex by network analysis. *European Journal of Neuroscience 23*, 2 (2006), 1919–1930.
374. NEISSER, U. *Cognition and Reality*. W.H. Freeman, San Francisco, 1976.
375. NEWELL, A., AND SIMON, H. Computer science as empirical enquiry: Symbols and search. *Communications of the ACM 19*, 3 (1976), 113–126.
376. NEWMAN, M. Clustering and preferential attachment in growing networks *Phys. Rev. E 64* (2001), 025102.
377. NEWMAN, M. Scientific collaboration networks: I. network construction and fundamental results. *Phys. Rev. E 64* (2001), 016131.
378. NEWMAN, M. Scientific collaboration networks: II. shortest paths, weighted networks, and centrality. *Phys. Rev. E 64* (2001), 016132.
379. NEWMAN, M. The structure of scientific collaboration networks. *Proc. Natl. Acad. Sci. USA 98* (2001), 404–409.
380. NEWMAN, M. The structure and function of complex networks. *SIAM Review 45* (2003), 167–256.
381. NEWMAN, M. Power laws, Pareto distributions and Zipf's law. *Contemporary Physics 46* (2005), 323–351.
382. NEWMAN, M., BARABÁSI, A., AND DJ, W., Eds. *The Structure and Dynamics of Complex Networks*. Princeton University Press, Princeton, 2006.
383. NI, W., AND TANG, M. Turing patterns in the Lengyel-Epstein system for the cima reaction. *Trans Am. Math. Soc. 357* (2005), 3953–3969.
384. NOSZTICZIUS, Z., HORSTHEMKE, W., MCCORMICK, W., AND SWINNEY, H. Sustained chemical waves in an annular gel reactor:a chemical pinwheel. *Nature 32* (1987), 619–620.
385. NOWAK, M. *Evolutionary Dynamics. Exploring the Equations of Life*. The Belknap Press of Harvard University Press, 2006.
386. NOWAK, M., AND K SIGMUND, K. Evolution of indirect reciprocity. *Nature 437* (2005).
387. NOWAK, M., AND SIGMUND, K. Evolutionary dynamics of biological games. *Science 303* (2004).
388. NOWAK MA, KM PAGE, K. S. Fairness versus reason in the ultimatum game. *Science 298* (2000), 1773–1775.
389. O'KEEFE, J., AND DOSTROVSKY, J. The hippocampus as a spatial map. Preliminary evidence from unit activity in the freely moving rat. *Brain Res 34* (1971), 171–175.
390. O'KEEFE, J., AND RECCE, M. L. Phase relationship between hippocampal place units and the eeg theta rhythm. *Hippocampus 3*, 3 (1993), 317–330.
391. ÖRKÉNY, I. *One Minute Stories*. Brandl and Schlesinger, Sydney, 1994. translated by Sollosy, J.
392. OTT, E., GREBOGI, C., AND YORKE, J. Controlling chaos. *Phys. Rev. Lett. 64* (1990), 1196–1199.
393. OUYANG, Q., AND SWINNEY, H. Transition from a uniform state to hexagonal and striped turing patterns. *Nature 352* (1991), 610–612.
394. PAGE, K., AND NOWAK, M. Unifying evolutionary dynamics. *Journal of Theoretical Biology 219* (2002).
395. PAGELS, H. *The Dreams of Reason: The Computer and the Rise of the Sciences of Complexity*. Simon & Schuster, New York, 1988.

396. PALMER, R., ARTHUR, W., HOLLAND, J., LEBARON, B., AND TAYLER, P. Artificial economic life: A simple model of a stock market. *Physica D 75* (1994).
397. PAN, W., AND MCNAUGHTON, N. The medial supramammillary nucleus, spatial learning and the frequency of hippocampal theta activity. *Brain Res 764*, 1-2 (1997), 101–8.
398. PANCS, R., AND VRIEND, R. Schelling's spatial proximity model of segregation revisited. University of London, Department of Economics, Working Paper No. 487, 2003.
399. PANCS, R., AND VRIEND, R. Schelling's spatial proximity model of segregation revisited. *Journal of Public Economics* (2007).
400. PASTOR-SATORRAS, R., AND VESPIGNANI, A. Epidemic speading in scale-free networks. *Phys. Rev. Lett. 86* (2001), 3200–3203.
401. PATTEN, B. *Foundations of Embryology*. McGraw-Hill Book Company, New York, London, Toronto, 1958.
402. PEITGEN, H.-O., AND RICHTER, D. *The Beauty of Fractals: Images of Complex Dynamical Systems*. Springer-Verlag, New York, 1986.
403. PETŐ, K. Modern attachment theory and psychoanalysis. a possible mechanism of action of psychoanalysis (in Hungarian). In *Lifecycles (in Hungarian)*, K. Pető, Ed. Animula, Budapest, 2005, pp. 95–102.
404. PETTERS, D. *Designing Agents to Understand Infants*. PhD thesis, Department of Computer Science, University of Birmingham, 2006. Ph.D. thesis in Cognitive Science.
405. PEZARD, L., NANDRINO, J., RENAULT, B., ELMASSIOUI, F., ALLILAIRE, J., MÜLLER, J., VARELA, F., AND MARTINERIE, J. Depression as a dynamical disease. *Biol. Psychiatry 39* (1996), 991–999.
406. PIAGET, J. *L'Equilibration des Structures Cognitives. Problème Central du Développement*. Presses Universitaires de France. Paris, 1975.
407. PICCININI, G. The first computational theory of mind and brain: A close look at McCulloch and Pitts's 'logical calculus of ideas immanent in nervous activity'. *Synthese 141* (2004), 175–215.
408. PINKER, S. The known world. *New York Times Book Review CLVI* (2007).
409. POINCARÉ, H. Sur le problème des trois corps et les équations de la dynamique. *Acta Mathematica 13* (1890), 1–270.
410. POLANYI, M. *The Tacit Dimension*. Doubleday and Co, 1966.
411. POPPER, K. *The Logic of Scientific Discovery*. Routledge, 1959.
412. POPPER, K., AND ECCLES, J. *The self and its brain*. Springer Verlag, Berlin, 1977.
413. PRICE, G. Fishers fundamental theorem made clear. *Ann. Hum. Genet. 36* (1972).
414. PRIMAS, H. Realism and quantum mechanics, 1994.
415. PRIMMETT, D., NORRIS, W., CARLSON, G., KEYNESM, R., AND STERN, C. Periodic segmental anomalies induced by heat shock in the chick embryo are associated with the cell cycle. *Development 105* (1989), 119–30.
416. PYRAGAS, K. Continuous control of chaos by self-controlling feedback. *Phys. Lett. A 170* (1992), 421–428.
417. PYRAGAS, K., AND TAMASEVICIUS, A. Experimental control of chaos by delayed self-controlling feedback. *Phys. Lett. A 180* (1993), 99–102.
418. QUARTZ, S., AND SEJNOWSKI, T. The neural basis of cognitive development: A constructivist manifesto. *Behav. Brain Sci. 20(4)* (1997).

419. REDNER, S. Citations statistics from 110 years of Physical Review. *Physics Today 58* (2005).
420. REED, W. The pareto, zipf and other power laws. *Economics Letters 74* (2001).
421. REED, W. A stochastic model for the spread of a sexually transmitted disease which results in a scale-free network. *Math Biosci 201* (2006).
422. REED, W., AND HUGHES, B. From gene families and genera to incomes and internet file sizes: why power-laws are so common in nature. *Phys. Rev. E 66* (2002), 067103.
423. REED, W., AND HUGHES, B. On the size distribution of live genera. *Journal of Theoretical Biology 217* (2002), 125–135.
424. REED, W., AND HUGHES, B. On the distribution of family names. *Physica A 319* (2003), 579–590.
425. REINER, M. The Deborah number. *Physics Today 17* (1964).
426. RÉNYI, A., AND BRÓDY, A. On the regulation of prices (in Hungarian). *Publ. Math. Inst. Hung. Acad. Sci. 1* (1956), 325âĂŞ335.
427. RÉNYI, A., AND SZENTÁGOTHAI, J. The probability of synaptic transmission in simple models of interneuronal synapses with convergent coupling; (in Hungarian). *Publ. Math. Inst. Hung. Acad. Sci. 1* (1956), 83–91.
428. RESCORLA, R., AND WAGNER, A. A theory of Pavlovian conditioning: Variations in the effectiveness of reinforcement and nonreinforcement. In *Classical Conditioning II*, B. A. and P. W, Eds. Appleton-Century-Crofts., 1972.
429. RICCIARDI, L. Diffusion models of single neurones activity. In *Neural Modeling and Neural Networks*, F. Ventriglia, Ed. Pergamon Press, Oxford, 1994.
430. RICKART, C. *Structuralism and structures: a mathematical perspective*. World Scientific, Singapore, 1995.
431. RIEGLER, A. When a cognitive system embodied? *Cognitive Systems Research 3* (2002).
432. ROBBINS, P., AND ZACKS, J. Attachment theory and cognitive science. *Journal of the American Psychoanalytic Association 55* (2007).
433. ROBERTS, P. Computational consequences of temporally asymmetric learning rules: I. Differential hebbian learning. *J Comput Neurosci 7*, 3 (1999), 235–46.
434. ROBERTSON, J., AND WEBB, W. *Cake-Cutting Algorithms: Be Fair If You Can*. AK Peters Ltd, 1998.
435. ROCHESTER N, HOLLAND J, H. L., AND W, D. Tests on a cell assembly theory of the action of the brain, using a large scale digital computer. *IRE Transactions of Information Theory IT-2* (1956), 80–93.
436. ROEHNER, B. *Patterns of speculation. A study in observational econophysics*. Cambridge University Press.Cambridge, 2002.
437. ROSE, S., KAMIN, L., AND LEWONTIN, R. *Not in Our Genes: Biology, Ideology and Human Nature*. Pantheon Books, 1985.
438. ROSEN, R. *Anticipatory Systems: Philosophical, Mathematical and Methodological Foundations*. Pergamon Press, 1985.
439. ROSEN, R. On information and complexity. In *Complexity, Language and Life: Mathematical Approaches*. Springer, Berlin, 1985, pp. 174–195.
440. ROSEN, R. *Life Itself: A Comprehensive Inquiry into the Nature, Origin, and Fabrication of Life*. Columbia Univ. Press, 1991.
441. ROSEN, R. Drawing the boundary between subject and object: comments on the mind-brain problem. *Theoret. Medicine 14* (1993).
442. ROSENBLATT, F. *Principles of Neurodynamics*. Spartan, Washington, DC, 1962.

443. ROSENBLUETH, A., N, W., AND BIGELOW, J. Behaviour, purpose and teleology. *Phil. Sci. 10* (1943), 18–24.
444. ROSENZWEIG, M., AND MACARTHUR, R. Graphical representation and stability condiions of predator-prey interaction. *American Naturalist 97* (1963).
445. RÖSSLER, O. Chaotic behavior in simple reaction systems. *Z. Naturforsch. A 31* (1976), 259–264.
446. RÖSSLER, O. An equation for continuous chaos. *Phys. Lett. A 57* (1976), 397–398.
447. RÖSSLER, O. Chaos and starnge attractors in chemical kinetics. In *Synergetics. Far from Equlibrium*, A. Pacault and C. Vidal, Eds. Springer, 1979, pp. 107–113.
448. RÖSSLER, O. The chaotic hierarchy. *Z. Naturforsch 38a* (1983), 788–801.
449. ROUP, A., BERNSTEIN, D., NERSESOV, S., HADDAD, W., AND CHELLABOINA, W. Limit cycle analysis of the verge and foliot clock escapement using impulsive differential equations and poincaré maps. *Int J. Control 76* (2003), 1685–1698.
450. ROZENFELD, A., COHEN, R., BEN AVRAHAM, D., AND HAVLIN, S. Scale-free networks on lattices. *Phys. Rev. Lett. 89* (2002), 218701.
451. SAICHEV, A., AND D. SORNETTE, D. Power law distribution of seismic rates: theory and data analysis. *Eur. Phys. J. B* (2006).
452. SAKAGUCHI, H. Oscillatory and excitable behaviors in a population of model neurons. *Prog. Theoret. Phys. 79* (1988), 1061–1068.
453. SAL'NIKOV, I. Y. Contribution to the theory of the periodic homogenous chemical reactions: Ii a thermokinetic self-excited oscillating model. *Zhurnal Fizicheskoi Khimii 23* (1949), 258–272. in Russian.
454. SAMMIS, C., AND SORNETTE, D. Positive feedback, memory, and the predictability of earthquakes. *Proc. Natl. Acad, USA 99, Suppl.1* (2002).
455. SATTLER, R. *Biophilosophy: Analytic and Holistic Perspectives*. Springer, Berlin, 1986.
456. SCHAFFHAUSEN, J. The day his world stood still. http://www.brainconnection.com/topics/?main=fa/hm-memory4.
457. SCHELLING, T. Dynamic models of segregation. *J. Math Sociol. 1* (1971), 119–132.
458. SCHELLING, T. *Micromotives and Macrobehavior*. Norton, 1978.
459. SCHELTER, B., WINTERHALDER, M., MAIWALD, T., BRANDT, A., SCHAD, A., SCHULZE-BONHAGE, A., AND TIMMER, J. Testing statistical significance of multivariate time series analysis techniques for epileptic seizure prediction. *Chaos 16* (2006), 013108.
460. SCHRÖDINGER, E. *What Is Life? The Physical Aspect of the Living Cell*. Cambridge University Press, Cambridge, 1944.
461. SCHUSTER, P., AND SIGMUND, K. Replicator dynamics. *J. Theor. Biol. 100* (1983), 533–538.
462. SELVERSTON, A., AND MOULINS, M. *The crustacean stomatogastric system*. Springer-Verlag, Berlin, 1987.
463. SHANNON, C. The mathematical theory of communication. *Bell System Tech. J. 27* (1948), 379–423 and 623–656.
464. SHIM, E. A note on epidemic models with infective immigrants and vaccination. *Math. Biosci and Engineering* ((in press)).
465. SIEKMEIER, P., AND HOFFMAN, R. Enhanced semantic priming in schizophrenia: a computer model based on excessive pruning of local connections in association cortex. *Br. J. Psychiatry 180* (2002), 345–350.

466. SIMON, H. On a class of skew distribution functions. *Biometrika 42* (1955).
467. SIMON, H. The architecture of complexity. *Proc Am Phil Soc 106* (1962), 467–482.
468. SKINNER, B. *Beyond Freedom and Dignity*. Knopf, New York, 1971.
469. SMART, J. Physicalism and emergence. *Neuroscience 6* (1981), 109–113.
470. SNOW, C. *The Two Cultures*. Cambridge Univ. Press, 1959.
471. SOMOGYVÁRI, Z., BARNA, B., SZÁSZ, A., SZENTE, M., AND ÉRDI, P. Slow dynamics of epileptic seizure: analysis and model. *Neurocomputing 38-40* (2001), 921–926.
472. SORNETTE, D. Predictability of catastrophic events: Material rupture, earthquakes, turbulence, financial crashes, and human birth. *Proc. Natl. Acad. Sci. USA 99* (2002).
473. SORNETTE, D. *Why Stock Markets Crash: Critical Events in Complex Financial Systems*. Princeton Univ. Press, Princeton NJ, 2003.
474. SOROS, G. The capitalist threat. *Atlantic Monthly 279*, 2 (1997).
475. SOROS, G. *The Age of Fallibility*. Public Affairs, New York, 2006.
476. SPERRY, R. A modified concept of consiousness. *Psychol. Rev. 76* (1969), 532–536.
477. SPERRY, R. Mind–brain interaction: mentalism yes, dualism no. *Neuroscience 5* (1980), 195–206.
478. SPORNS, O., AND TONONI, G., Eds. *Selectionism and the brain*. Int.Rev. Neurobiol. Vol. 37. Acad. Press., San Diego, 1994.
479. SPRAGUE, N., BALLARD, D., AND ROBINSON, A. Modeling embodied visual behaviors. *ACM Transactions on Applied Perception xx* (2007).
480. SPROTT, J. Dynamical models of love. *Nonlinear Dynamics, Psychology, and Life Sciences 8* (2004).
481. STAM, C., MONTEZ, T., JONES, B., ROMBOUTS, S., VAN DER MADE, Y., PIJNENBURG, Y., AND SCHELTENS, P. Disturbed fluctuations of resting state EEG synchronization in alzheimer 's disease. *Clin. Neurophysiol. 116* (2005), 708–715.
482. STANTON, P., AND SEJNOWSKI, T. Associative long-term depression in the hippocampus induced by hebbian covariance. *Nature 339* (1989), 215–218.
483. STAUFFER, D., AND SAHIMI, M. Discrete simulation of the dynamics of spread of extreme opinions in a society. *Physica D 364* (2006).
484. STEINHARDT, P., AND TUROK, N. A cyclic model of the universe. *Science 296* (2002), 1436–1439.
485. STEINHARDT, P., AND TUROK, N. The cyclic model simplified, 2004.
486. STELLING, J., SAUER, U., SZALLASI, Z., DOYLE, F. R., AND J, D. Robustness of cellular functions. *Cell 118* (2004), 675–685.
487. STEPHAN, K., BALDEWEG, T., AND FRISTON, K. Synaptic plasticity and dysconnection in schizophrenia. *Biol Psychiatry 59*, 10 (2006), 929–39.
488. STEPHAN, K., HARRISON, L., KIEBEL, S., DAVID, O., PENNY, W., AND FRISTON, K. Dynamic causal models of neural system dynamics: current state and future extensions. *Journal of Biosciences 32* (2007), 129–144.
489. STEPHAN, K., KAMPER, L., BOZKURT, A., BURNS, G., YOUNG, M., AND KÖTTER, R. Advanced database methodology for the Collation of Connectivity data on the Macaque brain (CoCoMac). *Philos Trans R Soc Lond B Biol Sci. 356*, 1412 (2001), 1159–86.

490. STERN, C., FRASER, S., KEYNES, R., AND PRIMMETT, D. A cell lineage analysis of segmentation in the chick embryo. *Development 104* (1988), Suppl:231–244.
491. STROGATZ, S. *Nonlinear Dynamics and Chaos: With Applications to Physics, Biology, Chemistry and Engineering*. Addison Wesley, Reading, 1994.
492. STROMQUIST, W. How to cut a cake fairly. *American Mathematical Monthly 87* (1980), 640–644.
493. STROMQUIST, W. How to cut a cake fairly. Addendum. *American Mathematical Monthly 88* (1981), 613–614.
494. SUTTON, R., AND BARTO, A. Toward a modern theory of adaptive networks: expectation and prediction. *Psychol Rev 88*, 2 (1981), 135–70.
495. SZALISZNYÓ, K., AND ÉRDI, P. Hippocampal rhythm generation. In *The Handbook of Brain Theory and Neural Networks*, M. Arbib, Ed., 2nd ed. MIT Press, 2003), pp. 898–901.
496. SZÉKÁCS, I. *Psychoanalysis and Natural Science (in Hungarian)*. Párbeszéd Kiadó, 1991.
497. SZENTÁGOTHAI, J. Specificity versus (quasi-)randomness in cortical connectivity. In *Architectonics of the Cerebral Cortex Connectivity*, M. Brazier and H. Petsche, Eds. Raven Press, New York, 1978, pp. 77–97.
498. SZENTÁGOTHAI, J. The modular architectonic principle of neural centers. *Rev. Physiol. Biochem. Pharmacol. 98* (1983), 11–61.
499. SZENTÁGOTHAI, J. Downward causation? *Ann. Rev. Neurosci. 7* (1984), 1–11.
500. SZENTÁGOTHAI, J. Specificity versus (quasi-)randomness revisited. *Acta Morphologica Hungarica 38* (1990), 159–167.
501. SZENTÁGOTHAI, J., AND ÉRDI, P. Self-organization in the nervous system. *Journal of Social and Biological Structures 12* (1989), 367–384.
502. SZILI, L., AND TÓTH, J. Necessary condition of the Turing instability. *Phys. Rev. E 48* (1993), 183–186.
503. SZILI, L., AND TÓTH, J. On the origin of Turing instability. *J. Math. Chem. 22* (1997), 39–53.
504. TATUM, B. *Why Are All The Black Kids Sitting Together in the Cafeteria? A Psychologist Explains the Development of Racial Identity*. Basic Books, 1999.
505. TAYLOR, P., AND JONKER, L. Evolutionarily stable strategies and game dynamics. *Mathematical Biosciences 40* (1978).
506. TAYLOR, R. Order in Pollock's chaos. *Scientific American* (2002), 117–121.
507. TAYLOR, R. Fractal expressionism – where art meets science. In *Art And Complexity*, J. Casti and A. Karlqvist, Eds. Elsevier Science, Amsterdam, 2003, pp. 117–144.
508. TAYLOR, R., MICOLICH, A., AND JONAS, D. Fractal expressionism. Can science be used to further our understanding of art? *Physics World 12*, 25 (1999).
509. TEMIN, H., AND MIZUTANI, S. Rna-dependant DNA polymerase in virions of Rous sarcoma virus. *Nature 226* (1970).
510. TEMIN, P., AND VOTH, H. Riding the South Sea bubble. *American Economic Review 94* (2004).
511. TESFATSION, L. Agent-based computational economics: Modelling economies as complex adaptive systems. *Information Science 149* (2003), 263–269.
512. THAGARD, P. *Mind. Introduction to Cognitive Science. Second Edition*. The MIT Press. A Bradfor Book, 2005.
513. THELEN, E., AND BATES, E. Connectionism and dynamic systems: are they really different? *Developmental Science 6:4* (2003).

514. THELEN, E., AND SMITH, L. *A Dynamic Systems Approach to the Development of Cognition and Action*. MIT Press, Cambridge, Mass, 1994.
515. TONONI, G., EDELMAN, G., AND SPORNS, O. Complexity and coherency: integrating information in the brain. *Trends in Cognitive Science 2* (1998), 474–484.
516. TÓTH, J., AND HÁRS, V. Orthogonal transforms of the Lorenz- and Rössler-equations. *Physica D 19* (1986), 135–144.
517. TRAUB, R., AND MILES, R. *Neuronal Networks of the Hippocampus*. Cambridge Univ. Press, 1991.
518. TRIVERS, R. The evolution of reciprocal altruism. *Quarterly Review of Biology 46* (1971).
519. TRUEMAN, W., PFEIL, B., KELCHNER, S., AND YEATES, D. Did stick insects really regain their wings? *Systematic Entomology 29*, 2 (2004), 138.
520. TRUESDELL, C. *The Tragicomical History of Thermodynamics*. Springer, New York, Heidelberg, 1980.
521. TSUDA, I. A hermeneutic process of the brain. *Prog. Theor. Phys. Suppl. 79* (1984).
522. TSUDA, I. Dynamic link of memory - chaotic map in nonequilibrium neural networks. *Neural Networks 5* (1992).
523. TSUDA, I. Toward an interpretation of dynamic neural activity in terms of chaotic dynamical systems. *Behavioral and Brain Sciences 24* (2001), 793–810.
524. TSUDA, I., FUJI, H., TADOKORO, S., YASUOKA, Y., AND YAMAGUTI, Y. Chaotic itinerancy as a mechanism of irregular changes between synchronization and desynchronization in a neural network. *J. of Integrative Neuroscience 3* (2004), 159–182.
525. TSUDA, I., IWANAGA, H., , AND TAKARA, T. Chaotic pulsation in human capillary vessels and its dependence on mental and physical conditions. *Int. J. Bifurcation and Chaos* (1992), 313–326.
526. TUCKWELL, H., AND WAN, F. Time to first spike in Hodgkin-Huxley stochastic systems. *Physica A - Stat. Mech. Applic. 351* (2005).
527. UJFALUSSY, B., AND KISS, T. How do glutamatergic and GABAergic cells contribute to synchronization in the medial septum? *Journal of Computational Neuroscience 21* (2006), 343–357.
528. UJFALUSSY, B., KISS, T., ORBÁN, G., HOFFMANN, W., ÉRDI, P., AND HAJÓS, M. Pharmacological and computational analysis of alpha-subunit preferential GABA-A positive allosteric modulators on the rat septo-hippocampal activity. *Neuropharmacology 52* (2007), 733–74.
529. VALLACHER, R., AND NOWAK, A. *Dynamical Systems in Social Psychology*. Academic Press, San Diego, London, 1994.
530. VAN DONGEN, S. A stochastic uncoupling process for graphs. Tech. Rep. INS-R0011, National Research Institute for Mathematics and Computer Science in the Netherlands, 2000.
531. VAN OOYEN, A. Competition in the development of nerve connections: a review of models. *Network 12*, 1 (2001), R1–47.
532. VAN REGENMORTEL, M. Biological complexity emerges from the ashes of genetic reductionism. *J. Mol. Recognit. 17* (2004), 145–148.
533. VAN REGENMORTEL, M. Reductionsim and complexity in molecular biology. *EMBO reports 5* (2004), 1016–1020.

534. VARELA, F., THOMPSON, E., AND ROSCH, E. *The Embodied Mind: Cognitive Science and Human Experience.* The MIT Press, 1991.
535. VENTRIGLIA, F. Kinetic approach to neural systems: I. *Bull Math Biol 36*, 5-6 (1974), 535–44.
536. VENTRIGLIA, F. Computational simulation of activity of cortical-like neural systems. *Bull Math Biol 50*, 2 (1988), 143–85.
537. VENTRIGLIA, F. *Neural Modeling and Neural Networks.* Pergamon Press, 1994, ch. Towards a kinetic theory of cortical-like neural fields.
538. VICSEK, T. Complexity: The bigger picture. *Nature 418* (2002).
539. VIZI, S. Role of high-affinity receptors and membrane transporters in nonsynaptic communication and drug action in the central nervous system. *Pharmacol. Rev. 52* (2000), 63–90.
540. VON DER MALSBURG, C. The correlation theory of brain functions. Tech. Rep. 81-2, Dept. Neurobiol. Max Planck Inst. Biophys. Chem., Göttingen, 1981.
541. VON NEUMANN, J. *Mathematical Foundatins of Quantum Mechanics.* Princeton University Press, Princeton, NJ, 1932.
542. VON NEUMANN, J. Probabilistic logics and the synthesis of reliable organisms from unreliable components. In *Automata Studies*, C. Shannon and J. McCarthy, Eds. Princeton University Press, Princeton, NJ, 1956, pp. 43–98.
543. VON NEUMANN, J. *The Computer and the Brain.* Yale Univ. Press, New Haven, 1958.
544. VON NEUMANN, J., AND MORGENSTERN, O. *Theory of Games and Economic Behavior.* Princeton Univ Press, 1944.
545. VON SMOLUCHOWSKI, M. Zur kinetischen theorie der brownschen molekularbewegung und der suspensionen. *Ann. Phys. (Leipzig) 21* (1906), 756–780.
546. WAGENMAKERS, E., MOLENAAR, P., GRASMAN, R., HARTELMAN, P., AND VAN DER MAAS, H. Transformation invariant stochastic catastrophe theory. *Physica D 211* (2005).
547. WAGNER, G. The logical structure of irreversible systems-transformations: A theorem concerning Dollo's law and chaotic movement. *J. theor. Biol. 96* (1982), 337–346.
548. WARD, L. *Dynamical Cognitive Science.* MIT Press, 2002.
549. WARREN, C., LM, S., AND SOKOLOV, I. Geography in a scale-free network model. *Phys. Rev. E 66* (2002), 056105.
550. WATTS, D. *Small Worlds: The Dynamics of Networks Between Order and Randomness.* Princeton Studies in Complexity. Princeton University Press, Princeton, N.J., 1999.
551. WATTS, D., AND STROGATZ, S. Collective dynamics of 'small-world' network. *Nature 393* (1998).
552. WEIDLICH, W. *Sociodynamics. A Systematic Approach to Mathematical Modelling in the Social Sciences.* Harwood Acad. Publ., 2000.
553. WEINBERG, S. Sokal's hoax. *The New York Review of Books XLIII*, 13 (1996).
554. WEISS, T. A model for firing patterns of auditory nerve fibers. Tech. Rep. 418, Cambridge, Massachusetts, Research Laboratory of Electronics, MIT, 1964.
555. WENG, J., MCCLELLAND, J., PENTLAND, A., SPORNS, O., STOCKMAN, I., SUR, M., AND THELEN, E. Autonomous mental development by robots and animals. *Science* (2000).
556. WHITING, M., BRADLER, S., AND MAXWELL, T. Loss and recovery of wings in stick insects. *Nature 421* (2003), 264–267.

557. WHITING, M., AND WHITING, A. Is wing recurrence really impossible?: A reply to Trueman et al. *Systematic Entomology 29* (2004), 140–141.
558. WIENER, N. *Cybernetics or Control and Communication in the Animal and the Machine*. Hermann, MIT Press, Wiley and Sons, 1948.
559. WIENER, N., AND ROSENBLUTH, A. The mathematical formulation of the problem of conduction of impulses in a network of connected excitable elements, specifically in cardiac muscle. *Arch Inst Cardiol Mex 16* (1946).
560. WIGNER, E. On the distributions of the roots of certain symmetric matrices. *Ann. Math. 67* (1958).
561. WILBUR, A., AND RINZEL, J. A theoretical basis for large coefficient of variation and bimodality in neuronal interspike distribution. *Journal of Theoretical Biology 105* (1983), 345–368.
562. WILLSHAW, D., AND PRICE, D. Models for topographic map formation. In *Modelling Neural Development*, A. van Ooyen, Ed. MIT Press, 2003, pp. 213–245.
563. WILSON, E. *Sociobiology: The New Synthesis*. Belknap Press, 1975.
564. WINFREE, A. Spiral waves of chemical activity. *Science 175* (1972).
565. WINOGRAD, S., AND COWAN, J. *Reliable Computation in the Presence of Noise*. MIT Press, Cambridge, MA, 1963.
566. WOLFENSTEIN, L. Lessons from Kepler and the theory of everything. *PNAS 100* (2003), 5001–5003.
567. WOLKENHAUER, O. Systems biology: the reincarnation of systems theory applied in biology? *Briefings in Bioinformatics 2* (2001), 258–270.
568. WOODALL, D. Dividing a cake fairly. *J. Math. Anal. Appl. 78* (1980), 233–247.
569. WOODCOCK, A., AND DAVIS, M. *Catastrophe Theory*. E. P. Dutton, New York, 1978.
570. WORRALL, J. Imre Lakatos (1922-1974) : Philosopher of mathematics and philosopher of science. *Z. Allgemeine Wissenschaftstheorie 5* (1974).
571. YAGIL, G. On the structural complexity of simple biosystems. *J. Theor. Biol. 112* (1982), 1–29.
572. YAGIL, G. Complexity and order in chemical and biological systems. In *Unifying Themes in Complex Systems*, Y. Bar-Yam and T. Toffoli, Eds. Perseus Books Group, 2000, pp. 645–654.
573. YAMAGUCHI, A., AOKI, K., AND MAMITSUKA, H. Graph complexity of chemical compounds in biological pathways. *Genome Informatics 14* (2003), 376–377.
574. YAMAMOTO, J. Relationship between hippocampal theta-wave frequency and emotional behaviors in rabbits produced with stresses or psychotropic drugs. *Japanese Journal of Pharmacology 76* (1998), 125–127.
575. YATES, F. Physical causality and brain theories. *Am. J. Physiol. 238* (1980), R277–R290.
576. YILDIRIM, N., AND MACKEY, M. Feedback regulation in the lactose operon: A mathematical modeling study and comparison with experimental data. *Biophysical Journal 84* (2003), 2841–2851.
577. ZAHLER, R., AND SUSSMAN, H. Claims and accomplishments of applied catastrophe theory. *Nature 269*, 10 (1977), 759–763.
578. ZAIKIN, N., AND ZHABOTINSKY, A. Concentration wave propagation in two-dimensional liquid phase self oscillating system. *Nature 225* (1970).

579. ZANETTE, D., AND MANRUBIA, S. Vertical transmission of culture and the distribution of family names. *Physica A: Statistical Mechanics and its Applications 295* (2001), 1–8.
580. ZEEMAN, E. *Catastrophe Theory: Selected Papers 1972-1977*. Addison-Wesley, Reading, 1977.
581. ZIEMKE, T. Cybernetics and embodies cognition: on the construction of realities in organisms and robots. *Kybernetes 34* (2005), 118–128.

Index

action-perception loop, 295
active vision, 296
adaptive landscape, 103
age of fallibility, 361
agent-based modeling, 305–317
Alzheimer's disease, 275
Anasazi Native Americans, 314
Arnold's cat map, 90
arrow of time, 59
artificial chemistry, 307
artificial intelligence, 175–178
artificial life, 306
artificial societies, 311–315
atomism, 25, 27
attachment theory, 300, 301
attractor, *60*, 91, 92, 98, 100, 137, 237, 247, 249, 272, 295, 358
    fixed point, 48, 49, 51, *60*, 77, 80, 81, 92, 98, 111, 150, 267
    Milnor, 99
    periodic, 17, 55, *60*, 82, *83*, 85, 139, 237
    strange, 17, *60*, 97, 98, 212, 351
autocatalysis, *9–11*, 84, 86, 198

Belousov-Zhabotinsky reaction, 84
bifurcation, 17, 51, 85, *85*, 93, 133, 246, 265, 266, 271
    Hopf, *85*, 160
bistability, 10, 50, 267
Black–Scholes model, 196
brain theory
    bottom up, 237, *242*
    top down, 237, *242*

brain-mind problem, 32, *289*
Brownian motion, 41, 115, *187*, 252
    geometric, 196
Brusselator model, 47, *84*, 85, 86
business cycles, 83, *159*, 356
butterfly effect, 15

cake division, 318
catastrophe theory, 10, 21, 50, *51*, 121, 191
causality, *109*, 110, 112, 113
    circular, 7, 8, 41, 112, 291, 357
    linear, 112
causation
    downward, 32
    upward, 32
cellular automata, 45, 138, 139, 154, 174, 202, 306
central pattern generator, 261, 265
chaos, 9, 15–17, 21, 33, 80, *87*, 88, 91–94, 101, 118, 137–139, 163, 212, 271, 298, 355, 358
    conservative, 90
    controlling of, 96–98, 161, 162, 271
    dissipative, 90
chaotic itinerancy, *98*, 100, 268, 298
chemical kinetics, 21, 92, *113*, 115, 120, 121, 144, 147, 217, 222, 251, 355, 358
    deterministic models, 55, 114, 116, 217, 355
    stochastic models, 55, 116, 134, 218, 355
chemical waves, 130

classical conditioning, 284
clockwork universe, 66, 75
cognitive neuroscience, 238, 302
cognitive science, 20, 290, 291, 294, 300–303
collaboration graph, 225–227
collective phenomena, 332, 349, 354, 358
combat models, 148–152
complexity
  algorithmic, 3, 201
  architecture of, 177
  biological, 31, 132
  cognitive, 3
  cyclomatic, 209
  dynamical, 3
  of graphs, 208
  of patterns, 202
  structural, 1, 3, 201, 206, 216, 237, 298, 358
computational economics
  agent based, 316
computational neuropharmacology, 275
computational theory of mind, 290
constant of motion, 68, 70
constructivism, 293, 294
coordination
  bimanual, 266
coordination dynamics, 266, 267, 294
cortical columns, 261
cortical rhythms
  normal, 269
  pathological, 269
coupled oscillators, 94, 261, 262, 265
critical state, 329, 343
criticality
  intermittent, 329, 343, *345*, 348, 359
  self-organized, 328, 329, 343, 345, 347, 359
cybernetics, 8, 20, 22, 34, *37*, 39–45, 77, 129, 130, 133, 140, 169, 170, 175, 176, 289, 290, 294, 300, 307, 347
  second order, 43
cyclic universe, 57, 59, 105, 106
cyclomatic number, 209

Deborah number, 7
description
  causal, 110–112
  teleological, 110–112
differential laws, 67, 68
digital evolution, 307–308
disconnection syndrome, 272, 274
disorder
  neurological, 237, 273, 356
  psychiatric, 237, 273
dissipative structures, 20, 35, 45, 47, 121
distribution
  bimodal, *191*
  Fréchet, 339, 340
  Gaussian, *185*, 189, 191, 356
  Gumbel, 339, *339*, 340
  log-normal, 195, 340
  long tail, *191–199*
  Pareto, 191, 194, 196
  power law, 192–199, 221, 358
  Weibull, 340
Dollo's law, 100–102
drug market, 161–163, 357
dynamical causal modeling, 272
dynamical disease, 270, 273, 274
dynamical systems, 58, 60, 79, 80, 85, 86, 91, 98, 109, *109*, 110, 112, 113, 133, 135, 141, 223, 241, 249, 255, 271, 294, 295, 305, 312, 316, 354

edge of chaos, 137, 139
embodied cognition, 291–292, 295, 303
emergence, 7, 19, 22, 28, 121, 122, 246, 294, 300, 317, 322, 358
epidemics, 144, 145, 223
  social, 146
epilepsy, 240, *270*, 331, 349
epileptic seizures, 270, 328, 331
Erdős number, 225–227
escapement
  deadbeat, 65
  verge-and-foliot, 61–64
eternal recurrence, 57–59
evolution, 19, 22, 33, 34, 120, 147
  direction of, 100–104
  of cooperation, 322
evolutionary dynamics, 147–148, 322, 326
evolutionary psychology, 325
extreme value analysis, 338

fair division, 319
fallibility, 361
feedback, 7, 8, 12, 32, 33, 41, 61, 64, 77, 86, 96, 97, 112, 161, 265, 316, 355, 357–359
   negative, 8, 11, 41, 163, 347, 348
   positive, 8–12, 152, 162, 163, 179, 198, 301, 346–349, 355
feedback control, 40, 61, 65, 77
financial crisis, 329, 347, 348
finite time singularities, 345–347
first passage time problem, 253
Fokker-Planck equation, 188, 218
fractal, 17, *212*, 351, 358
fractal art, 213
fractal expressionism, 214
fundamental theorem of natural selection, 103

Gödel's theorem, 170
gait transition, 265
Galton–Watson branching process, 198
game theory, 171–173
   classical, 318–322
   evolutionary, 322–327
gamma rhythm, 270
genetic algorithm, 308, 323
genetic code, 30
genetic programming, 309
goal-directed behavior, 40
graph
   distances, 204, 208, 221
   random, 220–221
   theory, 204
group selection, 327
Gutenberg–Richter law, 331
Gutenberg-Richter law, 330

Haken–Kelso–Bunz model, 266
heat conduction, 76
hermeneutic device, 289, 296, 298
hermeneutics, 289, *297*
hierarchy, 1, 14, 34, 101, 177, 181, 198, 239, 241, 255, 290, 309, 356, 358
hippocampus, 240, 243, 246, 270, 280–282, 285, 288, 290
holism, 4, 25
homeochaos, 33
homeostasis, 33, 41

impossible forms, 215–216
integrability, 71, 90
integral laws, 66, 67
interactionist dualism, 238, 289
invariance principles, 72, 73
irreversibility, 46, 55, 58, 59, 75, 77, 78, *78*, 79, 101, 102, 112

Julia set, 213

KAM theory, 87, 90
Kermack-McKendrick model, 145
kin selection, 325

learning rule, 43, 180, 242, 257
   Hebbian, 245, 283–288
limit cycle, 47, 60, *83*, 85, 86, 160, 261, 263, 264
linear stability analysis, 51, 81, 85, 140
Liouville process, 188
log-periodic corrections, 349
logical paradoxes, 7, *12*, 170
long term potentiation, 285
long-term depression, 285
Lotka–Volterra model, 147
   generalized, *83*, 150
Lotka-Volterra model, *81–83*, 159
love models, 151–154
Lyapunov exponent, *88*

Macy conferences, 40, 43, 169
Mandelbrot set, 213
master equation, 116
materialistic monism, 289
May-Wigner theorem, 140
McCulloch–Pitts model, 271
McCulloch-Pitts model, *38*, 39, 173
mechanical clock, 61, 66, 94
metastability, 268
minority game, 180–182
mixing, 90, 144
modular architectonic principle, 260
modularity, *33*
modus ponens, 166
modus tollens, 166
molecular biology, 29, 30, 32, 34, 42
mood regulation, 276
multirhythmicity, 249
multiscale modeling, 278
multistable perception, 50

Nash equilibrium, 172, 318
networks
  biological, 221
  epidemic on, 223
  genetic, 135–140
  in cell biology, 221
  metabolic, 221
  patent citation, 227
  random Boolean, 135
  scale-free, 221
  small world, 220
  social, 227
neural code, 279
  double, 279
neural rhythms, 269
neurodynamics, 275, 355
neuroinformatics, 241
neuron doctrine, 247
neurostatics, 258
noise-induced
  ordering, 217
noise-induced transition, 247
nonlinear science, 45, 113

oil price, 53
opinion dynamics, 157–159
organizational principles, 28, 130, 243–247
Ornstein–Uhlenbeck process, 190
oscillation, 51, 55, 64, 65, 68, 74, 75, *81–87*, 105, 127, 154, 249, 262, 269, 271, 275–277, 281, 282
  conservative, 71
  limit cycle, 47
  of single cells, 249

Parkinson disease, 274
patent citation, 227
pattern formation, 120, 122, 247
  biological, 120, 123
  spatiotemporal, 120
pendulum, 44, 60, 83, 111
pendulum clock, 64, 75
Perceptron, 43
phase code, 268, 269
phase synchrony, 268, 269
phase transition, 45, 47, 48, 51, 268, 359
place cell, 246, 280, 283
Poincaré map, 93

predictability, 16, 19, 23, 66, *328*
preferential attachment, 221
price peaks, 344
Principia Mathematica, 167, 176
principle of least action, 111
prisoner dilemma, 321
  iterated, 323
psychoanalysis, 299
psychotherapy, 237

quantum physics, 25
quasiperiodic motion, 86, 89

randomness, 78, 87, 138, 139, 201, 214, 220, 268, 309
rationality, 360, 361
  bounded, 177, 178, 180, 316, 361
  resigning, 361
reaction-diffusion model, 120, *124*, 129, 275
reasoning
  deductive, 165
  inductive, 166
reciprocity
  direct, 325
  indirect, 326
  network, 326
recurrence
  theorem, 58, 78
  time, 78, 91
reductionism, 25
  genetic, 29
reflexivity, 361

Santa Fe Artificial Stock Market, 316
schizophrenia, 274
science war, 360
segregation, 154
  dynamics, 154–156
selective stabilization hypothesis, 293
self-organization, 19, 44, 45, 216, 217, 246, 356
self-referential, 13, 43, 131, 298
self-similarity, 15, 90, 213
single cell modeling, 250–255
slaving principle, 47
sociobiology, 325
somitogenesis, 123

stability, 11, 32, 75, 79–82, 85, 87, 119, 121, 140–142, 148, 153, 355
  ecological, 141
state space, *60*, 79, 86, 87, 92, 109, 114, 115, 137, 157, 187, 196
steady state, 10, 61, 86
steam engine, 77
stochastic resonance, 254
stock market crash, 328, *331*
strange loops, 7, *14*
structural stability, 82, 86
structuralism, 2
Sugarscape, 311, 315
symmetries, 71, 73
synchronization, 264, 270, 275, 349
synergetics, 45, *47*, 55
systems
  closed, 5
  open, 5
systems biology, 34, *130*, 134
systems theory, 5, 20, 22, 34, *35*, 130

thermodynamics, 46, 59, 76, 78, 102
  first law of, 77
  second law of, 77
theta rhythm, 269, 277, 283
third culture, 363
three body problem, 70, 80
time concepts
  cyclic, 57
  linear, 59
topographic order, 244
Turing structures, 121

Ultimatum Game, 324
unpredictability, 19, 345

water clocks, 61
Wiener index, 204
Wiener process, 188, 189

Yule process, 197

Zipf's law, 192

Printing: Krips bv, Meppel, The Netherlands
Binding: Stürtz, Würzburg, Germany